REGULATION OF G PROTEIN–COUPLED RECEPTOR FUNCTION AND EXPRESSION

RECEPTOR BIOCHEMISTRY AND METHODOLOGY

SERIES EDITORS

David R. Sibley
Molecular Neuropharmacology Section
Experimental Therapeutics Branch
NINDS
National Institutes of Health
Bethesda, Maryland

Catherine D. Strader
Department of CNS and
  Cardiovascular Research
Schering-Plough Research
  Institute
Kenilworth, New Jersey

New Volumes in Series

Receptor Localization: Laboratory Methods and Procedures
Marjorie A. Ariano, *Volume Editor*

Identification and Expression of G Protein–Coupled Receptors
Kevin R. Lynch, *Volume Editor*

Structure–Function Analysis of G Protein–Coupled Receptors
Jürgen Wess, *Volume Editor*

Regulation of G Protein–Coupled Receptor Function and Expression
Jeffrey L. Benovic, *Volume Editor*

Founding Series Editors
J. Craig Venter   Len C. Harrison

REGULATION OF
G PROTEIN–COUPLED
RECEPTOR FUNCTION
AND EXPRESSION

Edited by

JEFFREY L. BENOVIC
Kimmel Cancer Center
Thomas Jefferson University
Philadelphia, PA

A JOHN WILEY & SONS, INC., PUBLICATION

New York • Chichester • Weinheim • Brisbane • Singapore • Toronto

Copyright © 2000 by Wiley-Liss, Inc. All rights reserved.

Published simultaneously in Canada.

For ordering and customer service, call 1-800-CALL-WILEY.

Library of Congress Cataloging-in-Publication Data:

Regulation of G protein–coupled receptor function and expression /
    edited by Jeffrey L. Benovic.
        p.   cm. — (Receptor biochemistry and methodology)
    Includes index.
    ISBN 0-471-25277-8 (cloth). — ISBN 0-471-25269-7 (alk. paper)
    1. G proteins—Receptors. 2. G proteins—Receptors—Research—
Methodology.   I. Benovic, Jeffrey L., 1953–   . II. Series :
Receptor biochemistry and methodology (Unnumbered)
QP552.G16R44   2000
572′.6—dc21
                                        99-30367
                                        CIP

Printed in the United States of America.

10 9 8 7 6 5 4 3 2 1

CONTENTS

SERIES PREFACE

The activation of cell surface receptors serves as the initial step in many important physiological pathways, providing a mechanism for circulating hormones or neurotransmitters to stimulate intracellular signaling pathways. Over the past 10–15 years, we have witnessed a new era in receptor research, arising from the application of molecular biology to the field of receptor pharmacology. Receptors can be classified into families on the basis of similar structural and functional characteristics, with significant sequence homology shared among members of a given receptor family. By recognizing parallels within a receptor family, our understanding of receptor-mediated signaling pathways is moving forward with increasing speed. The application of molecular biological tools to receptor pharmacology now allows us to consider the receptor–ligand interaction from the perspective of the receptor as a compliment to the classic approach of probing the binding pocket from the perspective of the ligand.

Against this background, the newly launched Receptor Biochemistry and Methodology series will focus on advances in molecular pharmacology and biochemistry in the receptor field and their application to the elucidation of the mechanism of receptor-mediated cellular processes. The previous version of this series, published in the mid-1980s, focused on the methods used to study membrane-bound receptors at that time. Given the rapid advances in the field over the past decade, the new series will focus broadly on molecular and structural approaches to receptor biology. In this series, we interpret the term *receptor* broadly, covering a large array of signaling molecules including membrane-bound receptors, transporters, and ion channels, as well as intracellular steroid receptors. Each volume will focus on one aspect of receptor biochemistry and will contain chapters covering the basic biochemical and pharmacological properties of the various receptors, as well as short reviews covering the theoretical background and strategies underlying the methodology. We hope that the series will provide a valuable overview of the status of the receptor field in the late 1990s, while also providing information that is of practical utility for scientists working at the laboratory bench. Ultimately, it is our hope that this series, by pulling together molecular and biochemical information from a diverse array of receptor fields, will facilitate the integration of structural and functional insights across receptor families and lead to a broader understanding of these physiologically and clinically important proteins.

DAVID R. SIBLEY
CATHERINE D. STRADER

PREFACE

Current research into receptor regulation does not appear to suffer from a lack of interest or effort. Recent advances in molecular and cell biology that enable the cloning, expression, and mutagenesis of signal transduction proteins have prompted an explosion of knowledge in the field of receptor regulation, facilitating the discovery of new classes of regulatory proteins and providing a basis and means for manipulating receptor function through multiple intracellular targets. In this volume, we describe many of the current techniques that are used to define the molecular mechanisms involved in receptor regulation. The information presented features G protein–coupled receptor (GPCR) signaling pathways, reflecting the prominence of these systems in the study of receptor regulation.

The book is divided into three major sections. The first covers various techniques that have proven useful for characterizing GPCR function. Individual chapters in this section focus on techniques such as epitope tagging (Chapter 1), measurement and analysis of receptor phosphorylation (Chapters 2–4), analysis of palmitoylation (Chapter 5), and assessment of receptor–G protein coupling (Chapter 6). The second section reviews the role of regulatory proteins in modulating receptor function and includes chapters on GPCR kinases (Chapter 7), arrestins (Chapter 8), and GPCR phosphatases (Chapter 9). The final section provides methods for studying receptor trafficking and expression. Included are chapters on the use of radioligand binding (Chapter 10), immunofluorescence (Chapter 11), and green fluorescent protein tagging (Chapter 12) to study GPCR trafficking as well as methods to study post-transcriptional regulation (Chapter 13).

I thank all the authors for their insightful and timely contributions to this volume; the senior editors of this series, Drs. David Sibley and Catherine Strader, for their guidance in putting this book together; and Ms. Colette Bean, my editor at Wiley, for her help in keeping things moving along. I would also like to thank the various members of my lab for their extensive input into the chapters on G protein–coupled receptor kinases and arrestins. I hope this book serves as a helpful guide to those interested in characterizing GPCR function and regulation.

JEFFREY L. BENOVIC
Philadelphia, PA

CONTRIBUTORS

JEFFREY L. BENOVIC, Kimmel Cancer Center, Thomas Jefferson University, Philadelphia, PA, USA

MICHEL BOUVIER, Department of Biochemistry, Universite de Montreal, Montreal, Quebec, Canada

RICHARD A. CERIONE, Department of Pharmacology, Cornell University, Ithaca, NY, USA

SUSAN R. GEORGE, Departments of Medicine and Pharmacology, University of Toronto, Toronto, Ontario, Canada

VSEVOLOD V. GUREVICH, Sun Health Research Institute, Sun City, AZ, USA

M. MARLENE HOSEY, Department of Molecular Pharmacology and Biological Chemistry, Northwestern University Medical School, Chicago, IL, USA

HUI JIN, Department of Pharmacology, University of Toronto, Toronto, Quebec, Canada

JAMES H. KEEN, Kimmel Cancer Center, Thomas Jefferson University, Philadelphia, PA, USA

UWE KLEIN, Departments of Psychiatry and Cellular & Molecular Pharmacology, University of California–San Francisco, San Francisco, CA, USA

MONIQUE LAGACÉ, Department of Biochemistry, Universite de Montreal, Montreal, Quebec, Canada

ROBERT J. LEFKOWITZ, Howard Hughes Medical Institute, Departments of Medicine and Biochemistry, Duke University Medical Center, Durham, NC, USA

QUIBO LI, Department of Pharmacology, Cornell University, Ithaca, NY, USA

ROBERT P. LOUDON, Fox Chase Cancer Center, Philadelphia, PA, USA

CRAIG C. MALBON, Department of Pharmacology, SUNY at Stony Brook Health Science Center, Stony Brook, NY, USA

BRIAN O'DOWD, Department of Pharmacology, University of Toronto, Toronto, Quebec, Canada

HIROSHI OHGURO, Department of Ophthalmology, Sapporo Medical University School of Medicine, Sapporo, Japan

MICHAEL J. ORSINI, Kimmel Cancer Center, Thomas Jefferson University, Philadelphia, PA, USA

KRZYSZTOF PALCZEWSKI, Departments of Ophthalmology and Pharmacology, University of Washington Medical School, Seattle, WA, USA

JULIE A. PITCHER, Department of Biochemistry, Duke University Medical Center, Durham, NC, USA

ALEXEY N. PRONIN, Kimmel Cancer Center, Thomas Jefferson University, Philadelphia, PA, USA

JUDITH A. PTASIENSKI, Department of Molecular Pharmacology and Biological Chemistry, Northwestern University Medical School, Chicago, IL, USA

FRANCESCA SANTINI, Kimmel Cancer Center, Thomas Jefferson University, Philadelphia, PA, USA

ROLAND D. STAUBER, Institute for Virology, University of Erlangen, Erlangen, Germany

STÉPHANIE ST-ONGE, Department of Biochemistry, Universite de Montreal, Montreal, Quebec, Canada

NADYA I. TARASOVA, Molecular Aspects of Drug Design Section, ABL-Basic Research Program, National Cancer Institute, Frederick Cancer Research and Development Center, Frederick, MD, USA

BABY G. THOLANIKUNNEL, Department of Medicine, Medical University of South Carolina, Charleston, SC, USA

MYRON L. TOEWS, Department of Pharmacology, University of Nebraska Medical Center, Omaha, NE, USA

J. PRESTON VAN HOOSER, Department of Ophthalmology, University of Washington Medical School, Seattle, WA, USA

MARK VON ZASTROW, Departments of Psychiatry and Cellular & Molecular Pharmacology, University of California–San Francisco, San Francisco, CA, USA

STEPHEN A. WANK, Digestive Disease Branch, National Institute of Diabetes and Digestive and Kidney Diseases, National Institutes of Health, Bethesda, MD, USA

REGULATION OF G PROTEIN–COUPLED RECEPTOR FUNCTION AND EXPRESSION

PART I

TECHNIQUES FOR CHARACTERIZING RECEPTOR FUNCTION

CHAPTER I

EPITOPE TAGGING AND DETECTION METHODS FOR RECEPTOR IDENTIFICATION

UWE KLEIN and MARK VON ZASTROW

Regulation of G Protein–Coupled Receptor Function and Expression,
Edited by Jeffrey L. Benovic.
ISBN 0-471-25277-8   Copyright © 2000 Wiley-Liss, Inc.

1. INTRODUCTION

Epitope tagging is a useful technique that has been applied in recent years to the study of G protein–coupled receptors and associated signaling proteins. Epitopes are local structures present in proteins that are recognized by specific antibodies. Antigenic epitopes can be formed by either contiguous or noncontiguous amino acid sequences (Harlow and Lane, 1988). Some epitopes formed by a contiguous sequence of amino acids are relatively small and are recognized by their cognate antibodies even when inserted into the structure of heterologous proteins, allowing them to be used to add a specific antigenic "tag" to a recombinant protein molecule that is recognized by a well-characterized, readily available antibody. Many antigenic epitopes have been used in this manner. The present chapter focuses on the practical use of several useful epitope tags, with an emphasis on our experience in their use for biochemical and cell biological studies of G protein–coupled receptors.

1.A. History

Antibodies raised against full-length proteins and protein fragments have been used for many years in biochemistry and cell biology. A major limitation of this approach is that a new antibody must be raised and tested for each individual protein studied. Not all proteins are good immunogens, and antibodies generated often do not work well for specific applications. For example, it is a common experience that an antibody that works well for one application (e.g., immunoblotting) works poorly or not at all for other applications (e.g., immunoprecipitation or immunocytochemistry). The generation of useful antibodies against multiple proteins is therefore time consuming and cost intensive.

    The realization that an epitope formed by a contiguous amino acid sequence can still be recognized by its antibody when fused to a different, unrelated protein started the use of epitope tagging as a general experimental tool in cell biology (e.g., Munro and Pelham, 1984, 1987; Field et al., 1988; Geli et al., 1988). The isolation of cDNAs encoding G protein–coupled receptors made it possible to apply epitope tagging to this class of proteins, as demonstrated initially in studies of agonist-induced internalization (von Zastrow and Kobilka, 1992) and palmitoylation (Mouillac et al., 1992) of epitope-tagged β_2-adrenergic receptors.

1.B. Utility of Epitope Tagging

Immunochemical methods are useful for cell biological and physiological studies of receptor proteins. Although antibodies recognizing natively expressed receptors are highly valuable, it is not always practical or feasible to generate these reagents for each receptor protein studied. Epitope tagging can allow one to use a well-characterized antibody that is readily available and has favorable functional properties for a particular application. Epitope tagging can also facilitate the specific purification or analysis of mutated or modified proteins present in complex systems or extracts. For example, receptors containing specific point mutations can be co-expressed in the same cells with their wild-type counterpart, and the relative subcellular distribution and biochemical properties of these closely similar proteins can be specifically studied by virtue of different epitope tags attached to individual mutant receptors. Conversely, receptor proteins with substantially different biochemical or pharmacological properties can be tagged with the same epitope, thereby allowing one to use a standardized method for the detection, localization, or purification of a wide variety of receptor proteins.

I.C. Major Applications of Epitope Tagging in Receptor Biology

A wide variety of peptide epitope tags have been described, and well-characterized monoclonal and polyclonal antibodies recognizing specific epitope tag sequences are available from commercial suppliers. In principle, there is a nearly limitless number of potential applications for epitope tagging techniques in receptor biology. Several of the most commonly used applications are mentioned briefly below.

I.C.a. Use as Tags for Detection.
Epitope tags were used initially in G protein–coupled receptor research to study the subcellular distribution of receptors and to examine their membrane trafficking properties with immunofluorescence microscopy (Keefer and Limbird, 1993; von Zastrow and Kobilka, 1992). Further development of this basic approach has been used to comparatively study the membrane trafficking properties of closely related receptor subtypes with a combination of antibodies recognizing epitope tags and native receptor proteins (von Zastrow et al., 1993) or different epitope tags (Chu et al., 1997).

The use of extracellular epitopes allows quantitation of receptor proteins on the cell surface with analytical techniques like flow cytometry (Mostafapour et al., 1996) and enzyme-linked immunosorbent assay (ELISA) (Ishii et al., 1993) (methods supplementing conventionally used membrane-binding assays) thereby providing quantitative measurements of their internalization. Epitope-tagged G protein–coupled receptor proteins can be detected on Western blots after cellular extracts and protein samples are resolved with gel electrophoresis (Cvejic and Devi, 1997). Western blotting is a useful tool to confirm the presence of receptor protein in isolated subcellular organelles, immunoprecipitates, transfected cells, and fractions from purification.

I.C.b. Use of Epitope Tags For Biochemical Purification.
The availability of a broad spectrum of monoclonal antibodies has facilitated the design of highly efficient purification methods that can be carried out under gentle conditions to maximize the yield of functional receptor proteins. It is generally possible to use the same antibody to accomplish purifications ranging from preparative scale to analytical scale. Affinity chromatography with immobilized

antibodies has proved very useful for preparative scale purifications, while analytical scale purifications generally utilize immunoprecipitation procedures.

Depending on the protein abundance and fold purification desired, immunoaffinity chromatography may be the only purification step needed in a preparative scale purification. Alternatively, with immunoaffinity chromatography combined with other purification steps, it has been possible to purify functional G protein–coupled receptors nearly to homogeneity from a crude cell lysate expressing receptors at relatively low abundance (Kobilka, 1995).

In analytical scale purifications, immunoprecipitation can be used to rapidly obtain highly purified receptors from crude cell extracts. Purifications of this type can be used to investigate biochemical modifications or physical properties of the tagged protein (Emrich et al., 1993; Mouillac et al., 1992; Ng et al., 1993) or to identify protein interactions or dimerization in multiprotein complexes by co-immunoprecipitation (Hebert et al., 1996; Klein et al., 1997; Romano et al., 1996).

2. COMMONLY USED EPITOPES

2.A. Epitopes Recognized by Antibodies

An epitope is any continuous peptide or protein sequence that is recognized by an antibody. In principle, any epitope can therefore be used for tagging. The first example of epitope tagging that we are aware of used part of the neuropeptide substance P (Munro and Pelham, 1984), for which well-characterized antibodies were available. Later examples used peptide epitopes derived from contiguous stretches of protein sequence derived from a variety of natural proteins. Still other epitopes in current use are not derived from native proteins but were specifically "designed" *de novo* for favorable immunochemical or biochemical properties. In general, the most useful epitope tags are recognized by monoclonal antibodies, which are readily available and can be produced in relatively large quantities from the appropriate hybridoma cell lines grown in culture or ascites. The following discussion summarizes four of the most commonly used epitope tags for immunopurification, immunoblotting, and immunocytochemical staining.

> *Flag*
>
> Recognition sequence: DYKDDDDK
>
> *Origin: De novo* designed for maximum hydrophilicity with target sequence for enterokinase (Hopp et al., 1988)
>
> *Comments:* The Flag epitope is one of the most popular epitopes used. Because it is *de novo* designed and is not derived from a protein sequence, antibodies recognizing the Flag epitope show very little cross-reactivity with endogenous proteins and are highly specific. The Flag epitope contains the target sequence for enterokinase (DDDDK↓X) and can therefore be removed via enzymatic cleavage. To obtain the free N-terminal aspartate residue (D), which is necessary for recognition by the M1 antibody, the tag needs to be fused C-terminal of a suitable signal sequence, which is removed by the appropriate signal

peptidase during or after synthesis of the protein, thereby "unmasking" the epitope in the cell. Vectors for expression in *Escherichia coli,* yeast, and mammalian cells with the appropriate signal peptides N-terminal of the Flag sequence are commercially available (Eastman Kodak, Scientific Imaging Systems, New Haven, CT).

Antibodies

M1: IgG_{2b}; requires the epitope at the very N terminus of the protein with free N-terminal aspartate residue (epitope must be fused C-terminal of a cleavable signal sequence); binding of the M1 antibody to Flag is Ca^{2+} dependent, which allows mild disruption of the immunocomplex with EDTA (Prickett et al., 1989) (Kodak; BAbCO)

M2: IgG_1; most universal in all applications; recognizes Flag independent of its location in the tagged protein (Kodak; BAbCO)

M5: IgG_1; highest affinity for Flag positioned at the very N terminus of the fusion protein with a preceding methionine residue. Met-Flag fusion proteins are designed by placing the ATG start codon right before the Flag coding sequence (Kodak; BAbCO)

Polyclonal rabbit IgG anti-Flag raised against Met-Flag peptide (Zymed)

HA

Recognition sequence: YPYDVPDYA

Origin: Antibodies raised against 36-mer peptide sequence derived from influenza hemagglutinin, HA1(75–110) (Niman et al., 1983); the complete antigenic determinant is contained in the short nine amino acid sequence YPYDVPDYA (Wilson et al., 1984)

Comments: Many cloning vectors are now commercially available for tagging proteins with the HA epitope, together with well-characterized monoclonal and polyclonal antibodies

Antibodies

12CA5: Mouse monoclonal IgG_{2b} (Boehringer Mannheim)

16B12 (HA.11): Mouse monoclonal IgG_1. Second-generation antibody from BAbCO

HA.11: Rabbit polyclonal (BAbCO)

Polyclonal rabbit IgG, raised against nine amino acid peptide conjugate (Zymed)

EE

Recognition sequence: EEEEYMPME

Origin: Antibody raised against an internal region of polyoma virus medium T antigen (Grussenmeyer et al., 1985)

Comments: Historically, the EE epitope has been particularly popular in two geographic areas, Freiburg in Germany and the San Francisco Bay area in California due to local availability of the antibody, which is now licensed by a commercial supplier (BAbCO)

Antibodies

Glu-Glu, mouse monoclonal IgG_1 (BAbCO).

c-myc

Recognition sequence: EQKLISEEDL

Origin: Epitope corresponds to amino acids 410–419 of the human c-myc protein

Comments: c-myc is a popular epitope that has been used for many years for protein tagging (Munro and Pelham, 1987). The hybridoma cell line for the monoclonal antibody 9E10 is available through American Type Culture Collection (ATCC CRL1729, MYC 1-9E10.2). Anecdotal reports suggest that somewhat variable results can be obtained in the efficiency of immunoprecipitation by this antibody, and this antibody has potential background problems in cell staining experiments due to cross-reactivity with endogenous c-myc gene product

Antibodies

9E10: mouse monoclonal IgG_1 (Evan et al., 1985) (BAbCO; Boehringer Mannheim; Zymed)

Polyclonal rabbit IgG (BAbCO)

2.B. Other Epitopes

In addition to epitopes recognized by antibodies, there are a variety of peptide tags added to proteins that are recognized specifically by other proteins or reagents. Probably the most popular is the His^6 tag (HHHHHH), which allows purification of tagged proteins with metal-chelating chromatography (Lilius et al., 1991). The method takes advantage of the strong affinity of an engineered poly-histidine peptide to Ni^{2+} cations, which can be bound to a metal chelate–affinity resin. His tags have been used successfully in many protein purifications, including G protein–coupled receptors (Janssen et al., 1995; Kobilka, 1995). A variety of His tagging vectors are commercially available from different suppliers, and recently a monoclonal antibody recognizing the His^6 tag was described (Lindner et al., 1997).

Another interesting peptide tag that has not yet found broad application is the Strep tag. The Strep tag is a nine amino acid peptide (AWRHPQFGG) with intrinsic streptavidin-binding activity. This sequence was identified by screening a random peptide library for peptides with high affinity for streptavidin (Schmidt and Skerra, 1993). Recombinant Strep-tagged proteins can be directly purified by affinity chromatography from cell extracts using immobilized streptavidin and can be eluted under mild conditions using biotin or diaminobiotin (Schmidt and Skerra, 1994).

2.C. General Considerations About Selection of a Tag

Selection of a specific epitope tag should be governed by several considerations, depending on the application in mind. When epitope tagging is used for immunoaffinity purification, for instance, it is generally desirable to select an epitope–antibody combination that can be dissociated under relatively mild conditions to facilitate isolation of functional protein in the eluate. A particularly good system in this regard is the combination of the Flag epitope with the

M1 antibody, which binds with high affinity in the presence of Ca^{2+} and can be eluted relatively readily at neutral pH and physiological salt concentrations by chelating Ca^{2+} with EDTA or EGTA. Conversely, the 12CA5 antibody directed against the HA tag binds to its epitope with such high affinity that it can be eluted only with harsh conditions, making the HA/12CA5 combination a good tool for stable immobilization of a protein on a column or other matrix (Wilson and Cox, 1990).

When two different epitopes are used in one experiment (e.g., for co-localization or co-immunoprecipitation of two recombinant proteins), the availability of two different classes of antibodies (raised in different species or different IgG isotypes) recognizing the epitopes may be useful to allow simultaneous detection of individual epitopes and to avoid cross-reactivity. A favorable combination is a mouse monoclonal antibody for detection of one epitope and a rabbit polyclonal antibody for detection of the other epitope because of the availability of species-specific secondary antibodies and because IgGs derived from these species tend to bind to protein A with markedly different affinities. Alternatively, IgG subclass-specific antibodies can be used, or antibodies can be chemically derivatized with accessory moieties to facilitate their specific purification or detection (e.g., digoxigenin, biotin, fluorophores).

3. DESIGN OF EPITOPE-TAGGED PROTEINS

The most fundamental requirement of epitope tagging techniques is the availability of cloned cDNA encoding the protein of interest or at least precise knowledge of the polypeptide sequence to allow design and construction of an appropriate tagged expression construct with synthetic DNA. Epitope tagging of a protein is generally accomplished by fusing a DNA sequence encoding the desired epitope to the cDNA encoding the protein of interest. Accordingly, this experimental approach requires a basic familiarity with the use of recombinant DNA techniques (Sambrook et al., 1992). An excellent overview about epitope tagging of proteins is provided in a manual from Boehringer Mannheim (1996).

3.B. Position of the Tag in the Target Protein

In general, epitopes can be added to the C terminus, to the N terminus, or within the coding sequence of the target protein. However, addition of the epitope in an unfavorable position might lead to loss of function of the target protein, or the epitope might be buried in the three-dimensional structure of the resulting fusion protein and might become unaccessible for the antibody. Adding the epitope to either end of the target protein will most likely lead to less interference with its function and maximal accessibility to the epitope-specific antibody, but the position of the epitope needs to be considered carefully for each individual case, depending on the nature of the target protein and epitope of choice. For proteins with solved three-dimensional structure or defined functional domains, it may be quite straightforward to design functional fusion proteins in which the epitope is accessible. In other cases, it may be useful to look for protein domains that are highly divergent, because divergent sequences are often exposed on the surface of proteins and may be more likely to tolerate structural modification caused by the addition of an epitope tag. If nothing points to a favorable position of the epitope within the target protein sequence, the best position needs to be found empirically by trial and error. In G protein–coupled receptors, previous studies suggest that epitope tagging may be accomplished by insertion of polypeptide sequences into the N-terminal extracellular domain, the third cytoplasmic loop, and the C-terminal cytoplasmic tail. Although there is ample reason for caution, it is sometimes surprising how tolerant proteins can be of added peptide epitopes. Some proteins tolerate even rather large polypeptides or entire proteins without loss of function (e.g., Gs or GFP fusion proteins with the β_2-adrenergic receptor) (Barak et al., 1997; Bertin et al., 1994).

Identification of the best position for the epitope within the target protein sequence is difficult if the function of the protein in question is unknown (e.g., for newly identified proteins). In the specific case of G protein–coupled receptors, functionality of the epitope-tagged protein is typically assessed by ligand-binding and signaling assays conducted in transfected cells. However, the potential danger of interference of the epitope with the function of a protein should always be kept in mind. If possible, results obtained with epitope-tagged proteins should be verified with wild-type (nontagged) proteins as control. Alternatively, if this is not possible, it may be useful to compare constructs in which the epitope tag is placed in different regions of the protein to confirm that the results obtained are not dependent on the insertion of the epitope tag sequence in a specific protein domain.

In our experience, a single copy of the epitope tag fused to the protein of interest is sufficient for its detection in most cases. However, in some cases, recognition of the epitope is quite poor, perhaps due to steric hindrance in the three-dimensional structure of the target protein, or the protein of interest may be expressed in extremely low abundance. In these cases, the addition of multiple copies of the epitope tag or inclusion of polypeptide spacers between epitope and target protein may improve the efficiency of antibody binding (Borjigin and Nathans, 1994; Kast et al., 1996).

3.B. Considerations About the Application in Mind

In addition to steric and functional aspects, the position chosen for the epitope will depend on the specific application that is under consideration. For example, if the purpose of the epitope is to detect a G protein–coupled receptor on the surface of nonpermeabilized cells or to label receptor proteins at the cell surface with antibody *in vivo* for flow cytometric experiments, the epitope should be attached to an extracellular domain such as the N terminus. Attaching two different epitopes to both the C- and N-terminal ends of the target receptor protein allows the investigation of its orientation in a cell membrane or vesicle (Emrich et al., 1994). Addition of a C-terminal tag to a receptor protein allows one to check for complete translation of the tagged protein and to identify frameshift mutations or premature stop codons in the sequence, but also requires fixation and permeabilization of the cell before labeling with the antibody. In an elegant approach to study the activation and cleavage of the thrombin receptor, two epitopes flanking the thrombin cleavage site were engineered into the human thrombin receptor to allow the uncleaved and protease-activated forms of the receptor protein to be distinguished immunochemically (Ishii et al., 1993).

3.C. Increasing Expression by Adding a Signal Sequence to the N Terminus

Expression of a G protein–coupled receptor in cells is often hampered by poor transport of the mature protein to the plasma membrane. The addition of an N-terminal signal sequence together with the epitope of choice was shown to enhance translocation of the receptor into the endoplasmic reticulum membrane, thereby increasing expression of functional receptor in transfected cells (Guan et al., 1992). Fusion of the Flag epitope C-terminal of the influenza hemagglutinin signal sequence at the N terminus of the β_2-adrenergic receptor furthermore allows detection of only properly processed receptor (i.e., with cleaved signal sequence) when using the M1 antibody, because M1 requires the free N-terminal aspartate residue for recognition (Guan et al., 1992).

4. CONSTRUCTION AND EXPRESSION OF EPITOPE-TAGGED G PROTEIN–COUPLED RECEPTORS

To fuse an epitope to a G protein–coupled receptor, the cDNA sequence encoding the receptor protein needs to be modified by fusion of the respective DNA sequence encoding the epitope to the receptor coding sequence. There are multiple ways to add an epitope tag to a protein of interest. Among the easiest to use are

- Addition of an epitope to a target protein by polymerase chain reaction
- Insertion of an epitope-encoding synthetic adaptor
- Subcloning of the target cDNA into a tagging vector

4.A. Addition of an Epitope to a Target Protein by Polymerase Chain Reaction

Perhaps the easiest and fastest way to add an epitope to a target protein is by oligonucleotide-directed mutagenesis with polymerase chain reaction (PCR). One of the two PCR primers used should contain the DNA sequence encoding the epitope, in addition to the sequence complementary to the target gene. In the case of an N-terminal epitope, the 5′ forward primer needs to encode a unique restriction site, the ATG start codon in the context of a Kozak consensus translation initiation sequence (Kozak, 1987), the epitope, and sufficient 5′ sequence of the target gene for binding of the primer to the target gene, in this order. In the case of a C-terminal epitope, the 3′ (reverse) primer should contain sufficient 3′ sequence of the target gene for priming, the epitope, a stop codon, and a unique restriction site for cloning.

The PCR product is then inserted into the eukaryotic expression vector of choice with the unique restriction sites engineered to the 5′ and 3′ ends of the gene. This approach was used to add the HA epitope tag to the β_2-adrenergic

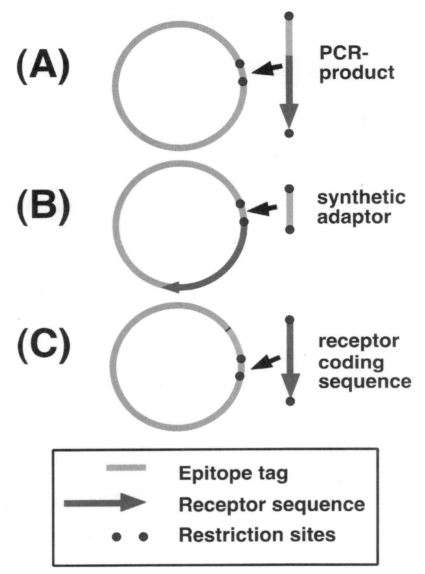

(A)

PCR-product

(B)

synthetic adaptor

(C)

receptor coding sequence

▬▬▬	**Epitope tag**
➡	**Receptor sequence**
● ●	**Restriction sites**

Figure 1.1. Approaches to the addition of an epitope tag to a target protein. There are three main approaches to the tagging of a protein with an epitope. Shown schematically is the addition of an epitope to the N terminus (5′ end) of a target protein sequence. **A:** Addition of an epitope to the target protein by PCR. The 5′ forward primer for amplification of the target protein cDNA sequence encodes the epitope after a start codon in the context of a Kozak consensus sequence. The PCR product is inserted into a mammalian expression vector via unique restriction sites. **B:** Insertion of an epitope-encoding synthetic adaptor. The epitope is inserted 5′ of the target protein cDNA sequence as a synthetic adaptor oligonucleotide duplex. The oligonucleotide needs to include a start codon in the context of a Kozak consensus sequence and needs to be in frame with the downstream target gene. **C:** Subcloning of the target cDNA into a tagging vector. The tagging vector already contains the sequence of the epitope wanted and suitable sites for insertion of the target gene (receptor coding sequence).

receptor (von Zastrow and Kobilka, 1992) and is shown schematically in Figure 1.1A.

4.B. Insertion of an Epitope-Encoding Synthetic Adaptor

If the cDNA of the target protein of interest is already inserted into a eukaryotic expression vector, the epitope can be fused to the protein by insertion of a synthetic adaptor oligonucleotide duplex encoding the epitope into suitable sites of the vector. If an N-terminal epitope tag with a start codon (ATG) is to be added, it should contain a Kozak consensus translation initiation sequence (e.g., 5'-CCA CC **ATG** G-3') to increase translation efficiency of the fusion protein (Kozak, 1987). In the case of a C-terminal epitope, the synthetic adaptor needs to encode a stop codon for termination of translation (TAA, TAG, or TGA). A synthetic adaptor was used to insert the HA epitope into the N-terminal end and third intracellular loop of the α_{2C}-adrenergic receptor (von Zastrow et al., 1993). The approach is shown schematically in Figure 1.1B.

4.C. Subcloning of the Target cDNA into a Tagging Vector

A very convenient and rapid way to epitope tag a protein of interest is through the use of a tagging vector. A tagging vector already contains the sequence of the epitope wanted and suitable restriction sites for insertion of the target protein to be tagged (Fig. 1.1C). A variety of tagging vectors with convenient multicloning sites for insertion of target proteins are commercially available, and customized tagging vectors can be constructed easily by modification of standard plasmid vectors with site-directed mutagenesis and insertion of a synthetic adaptor oligonucleotide duplex encoding the epitope (see Section 4.B.)

4.D. Expression of Epitope-Tagged Proteins in Eukaryotic Cells

One of the basic requirements of epitope tagging is that the experimental system studied be amenable to the addition or inclusion of the epitope-tagged protein. In most cases, this is accomplished by DNA transfection techniques to introduce an expression construct encoding the epitope-tagged protein into the cell type of interest. Recombinant DNA can be introduced into a number of cultured cell types quite readily. The introduction of recombinant DNA encoding epitope-tagged proteins into less readily transfected cell types or tissues can be more difficult and may require the use of specialized expression systems or transgenic technologies.

A variety of methods have been developed and are frequently used to introduce a recombinant cDNA vector into cultured mammalian cells. Some of the most popular methods include calcium phosphate–mediated or DEAE-dextran–mediated transfection, electroporation, liposome-mediated transfection, and direct microinjection (Sambrook et al., 1992). Liposome-mediated transfection protocols with complex lipid mixtures have become quite popular within the last few years, and a variety of easy-to-use kits are now commercially available. In our hands, the introduction of cDNA into adherent cultured cell lines via calcium phosphate co-precipitation is a reliable and inexpensive approach, typically yielding transfection efficiencies ranging from 20% to 60% with HEK 293 cells and the mammalian expression vector pcDNA3 (Graham and van der Eb, 1973; Sambrook et al., 1992).

If the purpose of the transfection experiment is to raise a cell line stably expressing the epitope-tagged protein, the expression vector used should include the gene for a selection marker like geneticin (G418) or hygromycin, allowing growth in the respective antibiotic only of transfected and expressing cells. For raising cell lines stably expressing two proteins after co-transfection, it is advantageous to use expression plasmids containing two different selection markers.

Protocol: Calcium-Phosphate–Mediated Transfection of Cultured, Adherent HEK 293 Cells

Materials Needed
- HBS (Hepes-buffered saline)

    25 mM Hepes

    140 mM NaCl

    5 mM KCl

    0.75 mM Na_2HPO_4

    6 mM dextrose

    Adjust to pH 7.05 with 0.5M NaOH and sterilize by passage through 0.22-μm membrane-filter
- 2M $CaCl_2$ solution, sterilized by passage through 0.22-μm membrane filter

Procedure
A. Transient transfection of cells
1. The cells to transfect are grown in a monolayer on a 60-mm dish to ~50% confluency. If the monolayer is grown too densely, transfection efficiency is greatly reduced.
2. 240 μl HBS is combined with 10 μl of DNA solution (1–50 μg plasmid cDNA) and mixed well. In our hands, purified cDNA works best (e.g., Qiagen plasmid purification), but reasonable transfection efficiencies can be obtained even with miniprepped cDNA after chloroform-phenol extraction and ethanol precipitation.
3. 15.5 μl 2 M $CaCl_2$ is slowly added in 1-μl steps by winding down a piston-displacement pipet (pipetman); the mix is gently agitated between single additions of $CaCl_2$.
4. The precipitation mix is incubated 20–30 minutes at room temperature to allow formation of a fine and uniform precipitate.
5. The calcium-phosphate/DNA precipitate is gently resuspended and added slowly to the cell monolayer with gentle agitation over 30 seconds.
6. Cells are incubated for 6–24 hours at 37°C in a humidified incubator in an atmosphere of 7% CO_2. The formed precipitate should now

be visible under the microscope as uniformly sized fine grains covering the dish.

7. The medium is replaced. Cells are ready for further analysis or re-seeding on 10-cm dishes and/or glass coverslips 1 day after the medium change (2 days after transfection).

B. Isolation of stable transformants

8. Cells after transfection are re-seeded sparsely on 10-cm dishes at two or three different densities and are placed on medium with the appropriate antibiotic added. We generally use the expression vector pcDNA3 (Invitrogen) and select for resistance to Geneticin.

9. Cells are grown under selection until well-separated single colonies are formed. Colonies are picked from the dish with a disposable pipet tip and are resuspended into the wells of a 24-well plate in selective media for further growth.

10. Single clones are screened for expression of the protein of interest with immunofluorescence microscopy, immunoblotting, or pharmacological methods. The uniformity of the cell clones can be assessed qualitatively by immunofluorescence microscopy or quantitatively by fluorescence flow cytometry (Mostafapour et al., 1996).

5. DETECTION AND PURIFICATION OF EPITOPE-TAGGED RECEPTORS

5.A. Immunopurification

One major advantage of epitope-tagged proteins is the easy use of immunoaffinity protocols for their isolation. Monoclonal antibodies directed against the epitope can be used to isolate epitope-tagged proteins out of crude cell extracts in one step. Immunoaffinity chromatography on a solid support with immobilized antibody is used to purify quantitative amounts of epitope-tagged protein that can be used in biochemical or biophysical applications. Immunoprecipitation is used to isolate small amounts of epitope-tagged protein for analytical purposes to determine their biochemical properties, metabolic processing, modifications, or interactions with other proteins on a qualitative or quantitative basis.

5.A.a. Immunoaffinity Chromatography. Epitope-tagged proteins can be quickly isolated from crude cellular extracts by immunoaffinity chromatography. The specific antibody recognizing the epitope tag is immobilized on a solid support, and an extract containing the epitope-tagged protein is passed over the column. Binding of the antibody to the peptide epitope of the fusion protein will retain the tagged protein on the column, while unbound proteins can be removed by simple washing. The protein of interest is then eluted from the column via a reagent disrupting the interaction between antibody and epitope or by addition of an excess of a displacing reagent such as a synthetic peptide corresponding to the epitope tag sequence.

A critical consideration for a successful purification of a G protein–coupled receptor in an active state is the choice of detergent for solubilization of the cell membranes. Although nonionic or zwitterionic detergents are generally preferred over ionic detergents, there is, unfortunately, no rational approach to determining the optimal detergent for a particular receptor purification. This determination must be made empirically in each individual case.

The use of epitope tags has facilitated the purification of G protein–coupled receptors considerably by allowing simple, one-step affinity procedures with good yield and specificity, thereby providing the enrichment needed (David et al., 1997; Gat et al., 1994; Grisshammer and Tucker, 1997; Janssen et al., 1995; Kobilka, 1995; Kwatra et al., 1995; Robeva et al., 1996). An example of a simple purification protocol is outlined below for the one-step purification of Flag-tagged β_2-adrenergic receptors from a digitonin extract of transfected mammalian cells with immobilized M1 anti-Flag antibody.

Protocol: Simple Purification of Flag-Tagged Proteins on Immobilized M1 Antibody

Materials Needed
- Lysis buffer
    - 10 mM Hepes, pH 7.4
    - Protease inhibitors
- Solubilization buffer
    - 20 mM Hepes, pH 7.4
    - 150 mM NaCl
    - 0.2% digitonin
    - 1 mM $CaCl_2$
    - Protease inhibitors
- Wash buffer
    - Solubilization buffer with 0.1% digitonin
- Anti-Flag M1 agarose affinity gel (Kodak IBI)

Binding Flag-Tagged Proteins to the Column
1. Cells expressing a Flag epitope tagged protein are lifted, collected by centrifugation, and resuspended in lysis buffer in a glass homogenizer; the lysate is centrifuged at 20,000*g* to prepare a crude membrane fraction.
2. The membrane pellet is solubilized in ~20 volumes of solubilization buffer in a glass homogenizer.
3. The membrane extract is clarified by centrifugation at 14,000*g* for 30 minutes.
4. The supernatant is loaded by gravity flow onto an antibody affinity column prepared by packing commercially available M1 agarose beads into a BioRad econo-column. Multiple passes of the sample over the column

---

or batch processing can be used to optimize the binding efficiency of receptor to the column.

Washing

The column is washed at least three times with wash buffer to remove unbound contaminants.

Elution

1. The antibody affinity column can be eluted with a variety of methods:

   With 2 mM EDTA to chelate Ca^{2+} ions

   At low pH with 0.1 M glycine, pH 3.5, or high pH with 0.1 M carbonate, pH 10.5

   By competition with Flag peptide (5 × column equivalents)

   With chaotropic agents such as KSCN

2. Method a is, by far, the preferred method for receptor elution from M1 affinity columns because it is gentle, efficient, and inexpensive. For other antibodies (such as M2 anti-Flag), elution by peptide competition (method c) is the preferred method for obtaining functional protein under mild conditions. The other methods, while generally efficient, should be used with caution if it is desired that the purified protein is eluted from the column in a functional condition.

In some cases, additional purification steps are required to purify G protein–coupled receptors to homogeneity. One example is the elegant use of two different affinity tags, an N-terminal Flag epitope and a C-terminal histidine tail, in the purification of the β_2-adrenergic receptor (Kobilka, 1995). The purification involves expression of the recombinant receptor in the baculovirus/Sf9 insect cell expression system and a two-step purification.

After cell lysis in hypotonic buffer, the crude cell membrane pellets are solubilized in n-dodecylmaltoside and subjected to affinity chromatography on immobilized Ni^{2+}, making use of the strong binding of the hexa-histidine tail to the Ni^{2+} cation. After elution with imidazole, the Ni resin–purified receptor is subjected to immunoaffinity chromatography on immobilized anti-Flag M1 antibody. Because the two epitopes used for purification are located at opposite ends of the receptor protein (N and C terminal, respectively), the two-step protocol yields only the full-length protein, removing proteolytic or improperly processed fragments. Even though the two-step protocol yields nearly pure receptor, a considerable amount of purified receptor protein is nonfunctional and does not bind its ligand. A final ligand affinity chromatography on immobilized alprenolol is then used to separate functional (binding) from nonfunctional receptor. This procedure has an overall yield of 28% of functional receptor with a 6,400-fold purification (Kobilka, 1995).

5.A.b. Immunoprecipitation. Immunoprecipitation is an extremely useful analytical method that can also be used preparatively in some cases. Because immunoprecipitation requires the availability of specific high-affinity antibodies, this method is well suited to the use of epitope-tagged proteins. Because immunoprecipitation is generally used as an end-stage analytical procedure rather than a purification method for functional receptor protein, the choice of detergent for immunoprecipitation of receptors is less critical than for immunoaffinity purification. In general, nonionic detergents such as Triton X-100 are well suited for immunoprecipitation experiments, and other detergents can be used alone or in combination to reduce nonspecific binding and to improve the efficiency of the washing steps. Most antibodies are denatured in the presence of high concentrations of ionic detergents, however, so these reagents should be used with caution in immunoprecipitation experiments.

Key steps in any immunoprecipitation protocol are (1) the formation of the antigen–antibody complex in solution and (2) the recovery of the complex by precipitation with a resin binding the antibody used. To reduce nonspecific adsorption of irrelevant proteins, it may be necessary to preclear the protein solution or cell extract containing the protein of interest with pre-immune serum or an unrelated antibody without binding activity for the target protein. The resin used for the precipitation should have high binding specificity for the antibody used.

Depending on the species the antibody was raised in and the particular IgG subclass, either protein A or protein G should be used (Harlow and Lane, 1988). The monoclonal EE antibody raised in mouse (IgG$_1$ subclass), for instance, shows only weak binding to protein A but binds strongly to protein G, while most rabbit IgGs bind quite efficiently to protein A. Recently, a chimeric fusion protein of the Fc-binding domains of protein A and protein G was introduced (Protein A/G Gel, Pierce), which has a more extended binding specificity than either protein A or protein G alone, reportedly binding virtually all IgG subclasses of mouse immunoglobulins with high affinity (Eliasson et al., 1988). An inexpensive alternative to the use of immobilized protein A are aldehyde-fixed *Staphylococcus aureus* cells, which can be obtained from several commercial sources.

Protocol: Immunoprecipitation

Materials Needed
- Triton X-100 extraction buffer
    0.5% (v/v) Triton X-100
    10 mM Tris-HCl, pH 7.5
    120 mM NaCl
    25 mM KCl
    2 mM EDTA, pH 8.0
    0.1 mM DTT
    Protease inhibitors
- High-stringency buffer (HSB)
    0.1% SDS
    0.5% Triton X-100

20 mM Tris-HCl, pH 7

120 mM NaCl

5 mM EDTA, pH 8.0

- High-salt wash buffer

  1M NaCl in HSB

- Low-salt wash buffer

  2 mM EDTA, pH 8.0

  10 mM Tris-HCl, pH 7.5

- 1 M Sucrose in HSB

- Antibody directed against the epitope used

- Protein A/G-resin (Pierce) or heat-killed and formalin-fixed cells of *S. aureus* (e.g., Pansorbin cells, Calbiochem)

Procedure

A. Preparation of cell extract

1. Cells expressing the epitope-tagged protein are grown on 10-cm tissue culture dishes and treated with the conditions of interest.

2. The dishes are chilled to 4°C in the cold room, cells are resuspended in 1 ml Triton X-100 extraction buffer by scraping and pipetting up and down. The suspension is transferred to 1.5-ml screwcap tubes and centrifuged at 14,000g in the cold room.

3. The supernatant is aliquoted and used for immunoprecipitation or frozen at −80°C.

B. Preclearing (optional) and complex formation

4. Add 30 µl of heat-killed, formalin-fixed *S. aureus* cells and 5 µl of preimmune or nonimmune serum to the samples; rotate 60 minutes at 4°C.

5. Spin samples in microfuge, and transfer supernatants to new tubes containing the appropriate amount of antiepitope antibody. The amount needed for the immunoprecipitation will vary with the antibody and epitope used. Rotate 60–120 minutes at 4°C.

C. Precipitation and washing of immune complexes

6. Add 50µl of the appropriate resin (protein A or protein G agarose, protein A/G resin, or *S. aureus*) to the mix; rotate 60 minutes at 4°C.

7. Spin samples in microfuge, aspirate supernatant, and add 800 µl of HSB buffer; mix and underlay sample with 200 µl 1 M sucrose in HSB.

8. Spin 2 minutes in microfuge, remove supernatant, and add 1 ml of high-salt wash buffer; rotate 10 minutes at 4°C.

9. Spin in microfuge, remove supernatant, and add 1 ml of low-salt wash buffer; rotate 10 minutes at 4°C.

10. Spin in microfuge, remove supernatant, and elute resin with 2 × SDS-PAGE sample buffer for 15 minutes at 37°C.

5.A.c. Co-Immunoprecipitation to Examine Protein–Protein Interactions.

As mentioned above, epitope tagging can be used to facilitate the study of protein–protein interactions by co-immunoprecipitation. In a typical experiment, both proteins in question are tagged with different epitopes and co-expressed in the same cells or mixed in a cell-free extract. Immunoprecipitation is carried out with an antibody recognizing the epitope tag present in protein No. 1, and the purified immunoprecipitate is assayed (usually by immunoblotting) for the presence of protein No. 2 in the immune complex with an antibody recognizing the epitope tag present in protein No. 2. Such a co-immunoprecipitation strategy has been used extensively to examine the interaction between a wide variety of protein partners.

Because of the extremely high sensitivity of these immunochemical methods, it is often possible to detect rather weak or inefficient protein interactions. For the same reason, however, it is essential to be alert to the possibility that false-positive results can be generated by nonspecific interactions with the immune complex, agarose beads, plastic tubes, or residual carry over volumes in the immunoprecipitation. Epitope tagging procedures are generally advantageous for these studies because they facilitate rigorous controls for immunochemical specificity. These controls may include demonstrating that protein No. 2 is not detected in precipitations performed in the absence of antibody recognizing protein No. 1 and that immunoprecipitations performed in the presence of the antibody recognizing protein No. 1 are negative if the epitope present in protein No. 1 is mutated, deleted, or substituted with another sequence.

One example of this approach, with controls for immunochemical specificity, is the detection of the interaction between the β_2-adrenergic receptor and eIF-2Bα in transfected HEK 293 cells (Klein et al., 1997) (Fig. 1.2). The detection of a specific co-immunoprecipitation of two proteins in this manner suggests, but does not prove, that these proteins interact *in vivo*. It can be much more difficult to determine whether the protein interaction detected by co-immunoprecipitation of epitope-tagged proteins expressed in transfected cells actually occurs with native proteins expressed in cells at normal levels or whether the protein interaction observed is "forced" by protein overexpression in a heterologous cell type. Addressing this question generally requires additional approaches, such as localization studies, to determine whether the proteins are co-localized in intact cells and functional studies to determine if manipulating one protein influences the functional properties of its putative protein partner.

5.B. Western Blotting

Western blotting is used to identify epitope-tagged proteins resolved by SDS-PAGE. After electrophoresis and transfer onto nitrocellulose membrane filters, the blot is incubated with an antibody directed against an epitope present in the protein of interest, followed by a secondary antispecies antibody conjugated to an enzyme such as alkaline phosphatase or horseradish peroxidase (HRP). The

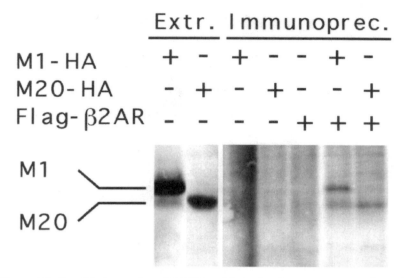

Figure 1.2. Specific interaction of eIF-2Bα with the β$_2$-adrenergic receptor shown by co-immunoprecipitation. HEX 293 cells or a 293 cell line stably expressing the Flag-tagged β$_2$-adrenergic receptor were transfected with C-terminally HA-epitope tagged versions of either full-length (M1-HA) or N-terminally truncated (M20-HA) forms of eIF-2Bα. Cells were lysed, and receptor complexes were immunoprecipitated with the M1 monoclonal antibody recognizing the Flag epitope and protein A Sepharose. Whole-cell extracts (lanes 1 and 2) or immunoprecipitates (lanes 3–7) were subjected to SDS-PAGE, blotted, and probed with the anti-HA monoclonal antibody HA.11. Epitope-tagged eIF-2Bα was detected with horseradish peroxidase–conjugated goat anti-mouse secondary antibody and enzyme-linked chemiluminescence (ECL, Amersham). Cells expressing only eIF-2Bα (as an M1 or M20 isoform, lanes 3 and 4, respectively) or only Flag-β$_2$-adrenergic receptor (lane 5) were used to control for specificity of the immunoprecipitation and detection.

most rapid and sensitive detection system currently used is based on an enhanced chemiluminescent reaction (ECL Western blotting system).

Critical for high sensitivity and low background are the quenching of endogenous peroxidase activity in the cell extracts (Navarre et al., 1996) and the efficient blocking of unoccupied sites on the filter before probing. After one round of detection, membranes can be re-probed several times with the same or other antibodies. To avoid cross reaction with already bound antibodies, membranes may be stripped of bound antibodies (Amersham ECL protocol), or bound HRP activity may be blocked by forming an insoluble reaction product on the blot (diaminobenzidine [DAB] reaction) (Krajewski et al., 1996).

Figure 1.3 shows an example of detection of the β$_2$-adrenergic receptor N-terminally tagged with the Flag epitope and stably expressed in HEK 293 cells by Western blotting with the M1 antibody.

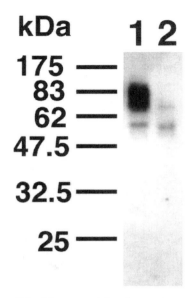

Figure 1.3. Detection of the Flag-tagged β_2-adrenergic receptor stably expressed in HEK 293 cells by Western blotting. HEK 293 cells stably expressing the Flag-tagged β_2-adrenergic receptor (lane 1) or untransfected HEK 293 cells as control (lane 2) were grown on 100-mm tissue culture dishes to confluency and lysed in 1 ml lysis buffer (20 mM Hepes, pH 7.4, 0.5% Triton X-100, 1 mM Pefabloc SC, 2 μg/ml aprotinin, 1 μg/ml leupeptin, 1 μg/ml pepstatin-A) for 15 minutes at 4°C. The lysates were centrifuged for 10 minutes at 20,000g; 40 μl of the supernatants were mixed with 20 μl 3 × SDS-PAGE sample buffer and incubated for 15 minutes at 37°C. Then 5 μl of the samples was resolved on a 12% SDS-PAGE and transferred to nitrocellulose. Flag-tagged β_2-adrenergic receptor was detected by probing the blot with the monoclonal M1 antibody (3 μg/ml) and HRP-conjugated goat antimouse secondary antibody (0.2 μg/ml), followed by enzyme-linked chemiluminescence (ECL, Amersham).

Protocol: Western Blotting

Materials Needed

- SDS-PAGE and blotting equipment
- Ponceau-S solution (staining of blot for total protein)
- 3% Hydrogen peroxide aqueous solution (block of endogenous peroxidases)
- TBS: Tris-buffered saline, pH 7.5
- TTBS: 0.5% Tween-20 in Tris-buffered saline, pH 7.5
- Blocking solution: 5% nonfat dry milk in TTBS
- Primary antibodies directed against epitope
- HRP-conjugated antispecies secondary antibody directed against primary antibody used

- ECL Western blotting system (Amersham Life Science)
- Film for exposure of ECL-developed blot (e.g., Kodak X-Omat AR)

Procedure

1. Separate protein sample by standard SDS-PAGE electrophoresis and transfer onto the nitrocellulose membrane.
2. Rinse blot in H_2O.
3. Ponceau-S stain the blot for total protein to check for complete transfer (optional).
4. Wash in TBS.
5. Incubate 15 minutes in 3% hydrogen peroxide solution to inactivate endogenous peroxidases present in the cell or tissue extracts (Navarre et al., 1996).
6. Wash in TTBS.
7. Block in 5% dry milk/TTBS at least 1 hour or overnight at room temperature.
8. Incubate with primary antibody in block solution at appropriate dilution for 1 hour at room temperature. The optimal dilution needs to be determined empirically or should be used according to the manufacturer's recommendations.
9. Wash extensively multiple times with TTBS over 1 hour at room temperature.
10. Incubate with secondary antibody in block solution at appropriate dilution (determine empirically or use manufacturer's recommended dilution) for 1 hour at room temperature.
11. Wash multiple times with TTBS.
12. Perform ECL reaction (Amersham) according to manufacturer's protocol.

5.C. Immunofluorescence Microscopy

One of the most commonly used applications of epitope tagging in the cell biology of G protein–coupled receptors is the detection and localization of epitope-tagged receptor proteins in transfected cells. After subjecting the cells to the respective treatment or condition of interest, the cells are fixed and the epitope-tagged receptor protein is detected with a combination of primary monoclonal antibody directed against the epitope and a secondary species-specific or subtype-specific antibody conjugated to a fluorophore that can be visualized by fluorescence microscopy. We routinely use two basic types of staining procedure to investigate the subcellular localization and membrane trafficking of G protein–coupled receptors:

1. "Staining" experiment: In a staining experiment, cells are first treated with agonist or subjected to the respective treatment and then fixed, per-

meabilized, and incubated with the primary antibody directed against the epitope, followed by a fluorophore-conjugated secondary antibody. This procedure will detect all receptor proteins present in the cell after treatment, irrespective of their location before treatment. Although receptors with N-terminal epitopes present in the plasma membrane can be detected without permeabilization of the cells, the detection of epitopes located at the cytoplasmic side of the receptor (C terminus) or of cytoplasmic proteins requires permeabilization before incubation with antibody (von Zastrow and Kobilka, 1992).

2. "Feeding" experiment: In a feeding experiment, cells expressing a G protein–coupled receptor with an extracellular epitope are decorated with a monoclonal antibody directed against the epitope in culture before fixation and permeabilization. This allows the subsequent detection only of the subset of receptors that was initially localized to the plasma membrane. After pre-binding the antibody, cells are treated with agonist or subjected to other treatments. Cells are then fixed, permeabilized, and incubated with the secondary antibody for detection. An example of a feeding experiment to visualize agonist-induced receptor internalization is shown in Figure 1.4 for a Flag-tagged vasopressin V2 receptor expressed in transfected HEK 293 cells.

Certain experiments require the co-localization of receptor proteins with other proteins or other receptors within the same specimen (Chu et al., 1997; Klein et al., 1997; von Zastrow et al., 1993). To facilitate co-localization experiments with dual-label fluorescence microscopy, the two proteins of interest are tagged with different epitopes, which are recognized by respective antibodies followed by different colored fluorophore-conjugated secondary antibodies. To avoid cross-reactivity of the secondary antibodies used, it is advantageous to use primary antibodies raised in different species (e.g., mouse and rabbit) or representing different IgG subclasses (e.g., IgG_1 and IgG_{2b}), thereby allowing antibody-specific detection with commercially available secondary antibodies. Alternatively, primary antibodies can be directly conjugated with fluorophores or chemical moieties (e.g., biotin or digoxigenin) to facilitate specific detection directly by fluorescence or by appropriate secondary reagents.

Specimens can be viewed by conventional epifluorescence microscopy with dichroic filter sets to selectively visualize the fluorescence emission from individual fluorophores. Dichroic filters designed for the visualization of specific fluorophores are available from several commercial sources (such as Chroma and Omega). Alternatively, specimens can be viewed by confocal fluorescence microscopy. Confocal microscopy, which is typically performed with a scanning laser image aquisition system, is capable of achieving increased spatial resolution relative to conventional microscopy, particularly in the z-axis. Thus confocal microscopy is often preferred for co-localization experiments or for the examination of relatively thick specimens.

Figure 1.5 shows co-localization of the Flag-tagged β_2-adrenergic receptor and HA-tagged eIF-2Bα in HEK 293 fibroblast cells (Klein et al., 1997). Co-localization of two proteins can be visualized in color-overlay pictures created from the individual fluorographs, as shown (Fig. 1.5C,D).

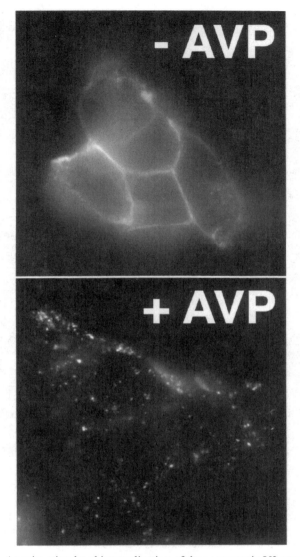

Figure 1.4. Agonist-stimulated internalization of the vasopressin V2 receptor demonstrated by immunofluorescence microscopy in a feeding experiment. The coding sequence of the human vasopressin V2 receptor (obtained from Dr. Mariel Birnbaumer, University of California, Los Angeles) was amplified by PCR and inserted into a tagging vector containing the influenza hemagglutinin signal sequence and the Flag epitope (Guan et al., 1992). The epitope-tagged receptor sequence was subcloned into the mammalian expression vector pcDNA3 and transfected into HEK 293 cells with calcium phosphate co-precipitation. Cells were treated in a feeding experiment as outlined in the protocol by incubating with the M1 antibody for 30 minutes at 37°C, without (–AVP) and with (+AVP) subsequent incubation with the agonist Arg[8] vasopressin (10 μM, 20 minutes). Cells were fixed and permeabilized, and bound M1 antibody was detected as described in the protocol for a feeding experiment. Flag-tagged vasopressin V2 receptor was visualized by epifluorescence microscopy.

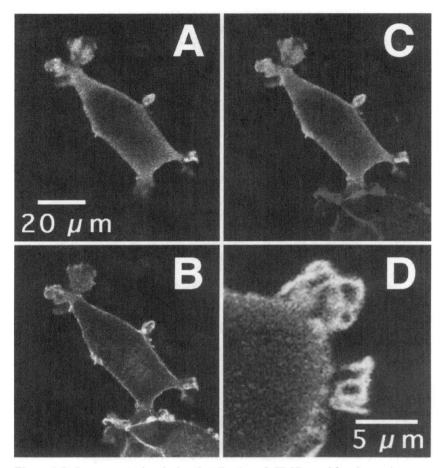

Figure 1.5. Immunocytochemical co-localization of eIF-2Bα and β$_2$-adrenergic receptor in the plasma membrane (Klein et al., 1997). HA epitope–tagged eIF-2Bα was transiently transfected into an HEK 293 cell line stably expressing the N-terminally Flag epitope–tagged β$_2$-adrenergic receptor. Cells were grown on glass coverslips, fixed, and permeabilized as described in the protocol for a staining experiment. HA epitope–tagged eIF-2Bα was detected with the monoclonal mouse antibody HA.11 (BAbCO, Richmond, CA), and β$_2$-adrenergic receptor was detected with receptor-specific rabbit antiserum (von Zastrow and Kobilka, 1992). After incubation with the secondary antibodies (FITC-conjugated donkey antimouse and Texas Red–conjugated donkey antirabbit, Jackson ImmunoResearch Laboratories, West Grove, PA), eIF-2Bα and β$_2$-adrenergic receptor were localized by dual-color confocal fluorescence microscopy with a BioRad MRC1000 and a Zeiss 100 NA1.3 objective. **A:** EIF-2Bα (green channel) localizes both in the cytoplasm and is observed in association with the plasma membrane, where it is concentrated in specialized regions. **B:** β$_2$-Adrenergic receptor (red channel) co-localizes with eIF-2Bα in these regions as shown in yellow in the two-color merged image (**C**). **D:** Another example of this co-localization is shown at higher magnification. (See color insert.)

Protocol: Immunostaining of Cells on Glass Coverslips for Fluorescence Microscopy

Materials Required
- Glass coverslips
- Phosphate-buffered saline, pH 7.5
- Tris-buffered saline, pH 7.5
- "Blotto" block solution
  - 3% non-fat dry milk,
  - 0.1% Triton X-100,
  - 50 mM Tris-HCl, pH 7.5
- Primary antiepitope antibody
- Secondary antibody conjugated to fluorophore
- Anti-fade mounting medium (e.g., Vectashield, Vector Laboratories Inc., Burlingame, CA)
- Nailpolish or plastic sealant

A. Procedure for staining experiment
  1. Transfected cells are grown on glass coverslips in six-well tissue culture plates in 2 ml of the appropriate medium.
  2. Cells are treated with ligand or other conditions of interest.
  3. The medium is quickly aspirated, and fixing solution is added (4% formaldehyde in PBS; this can be prepared by dissolving paraformaldehyde in heated PBS in a fume hood. We have also found that commercial formalin solution (37% formaldehyde) works satisfactorily for optical microscopy when diluted 1:10 with PBS). Cells are fixed for 10–30 minutes at room temperature or 4°C.
  4. The coverslips are washed three times with TBS, 5 minutes each.
  5. TBS is aspirated thoroughly from the coverslips, and any residual liquid is carefully removed from the edges of the coverslip.
  6. Blotto block solution is added, and the coverslip is incubated for 20 minutes at room temperature to permeabilize the cells and to block nonspecific sites.
  7. After aspirating the blotto, primary antibody directed against the epitope used is added to the coverslip and incubated for 60 minutes at room temperature. The optimal dilution for each antibody needs to be determined empirically or should be used according to the manufacturer's recommendations. We typically use the following concentrations: M1 anti-Flag, 3–5 μg/ml; HA.11 anti-HA, 5–7 μg/ml; 12CA5 anti-HA, 5 μg/ml; Glu-Glu anti-EE, 2 μg/ml.
  8. The antibody solution is removed, and the coverslip is washed thoroughly multiple times with TBS.
  9. Blotto block solution is added, and the coverslip is incubated for 5 minutes at room temperature.

10. After the blotto is aspirated, secondary antibody conjugated to the appropriate fluorophore is added to the coverslip and incubated for 20 minutes in the dark at room temperature. The optimal dilution for the secondary antibody needs to be determined empirically or should be used according to the manufacturer's protocol.

11. The coverslip is washed multiple times in TBS and mounted onto microscope glass slides with anti-fade mounting medium. The perimeter of the coverslip is sealed with nailpolish or a plastic sealant to prevent movement of the slip on the slide and to seal the specimen for microscopy and prolonged storage.

12. Coverslips should be stored in the dark at 4°C to prevent photobleaching and diffusion of the fluorophore-conjugated antibody.

B. Procedure for feeding experiment

1. Cells after transient transfection or cells stably expressing a G protein–coupled receptor with an extracellular accessible epitope are grown on glass coverslips in six-well tissue culture plates in 2 ml of the appropriate medium.

2. Primary antibody is added at the appropriate dilution and incubated for 30 minutes at 37°C in the incubator (decoration of receptors localized to the outer plasma membrane).

3. Cells are treated with ligand or other conditions of interest.

4. The medium is quickly aspirated, and fixing solution is added (4% formaldehyde in PBS; see comment in staining experiment protocol for preparation).

5. The coverslips are washed three times with TBS, 5 minutes each.

6. TBS is aspirated thoroughly from the coverslips, and any residual liquid is carefully removed from the edges of the coverslip.

7. Blotto block solution is added, and the coverslip is incubated for 20 minutes at room temperature to permeabilize the cells and to block nonspecific sites

8. Proceed from step 10 in the staining protocol (A).

6. CONCLUDING REMARKS

Over the past 10 years, the method of epitope tagging has become increasingly popular in receptor biology. The use of epitope-tagged G protein–coupled receptors has facilitated studies of receptor biochemistry, receptor localization in cells, identification and characterization of interacting proteins, and efficient purification of quantitative amounts of receptor protein. Future studies will develop additional applications for these techniques and will further help to unravel new questions about the structure and function of G protein–coupled receptors.

REFERENCES

Barak LS, Ferguson SS, Zhang J, Martenson C, Meyer T, Caron MG (1997): Internal trafficking and surface mobility of a functionally intact β_2-adrenergic receptor–green fluorescent protein conjugate. Mol Pharmacol 51:177–184.

Bertin B, Freissmuth M, Jockers R, Strosberg AD, Marullo S (1994): Cellular signaling by an agonist-activated receptor/$G_{s\alpha}$ fusion protein. Proc Natl Acad Sci USA 91:8827–8831.

Boehringer Mannheim (1996): Epitope Tagging: Basic Laboratory Methods. Mannheim, Germany: Boehringer Mannheim.

Borjigin J, Nathans J (1994): Insertional mutagenesis as a probe of rhodopsin's topography, stability, and activity. J Biol Chem 269:14715–14722.

Chu P, Murray S, Lissin D, von Zastrow M (1997): Delta and kappa opioid receptors are differentially regulated by dynamin-dependent endocytosis when activated by the same alkaloid agonist. J Biol Chem 272:27124–27130.

Cvejic S, Devi LA (1997): Dimerization of the delta opioid receptor: Implication for a role in receptor internalization. J Biol Chem 272:26959–26964.

David NE, Gee M, Andersen B, Naider F, Thorner J, Stevens RC (1997): Expression and purification of the *Saccharomyces cerevisiae* alpha-factor receptor (Ste2p), a 7-transmembrane-segment G protein–coupled receptor. J Biol Chem 272:15553–15561.

Eliasson M, Olsson A, Palmcrantz E, Wiberg K, Inganas M, Guss B, Lindberg M, Uhlen M (1988): Chimeric IgG-binding receptors engineered from staphylococcal protein A and streptococcal protein G. J Biol Chem 263:4323–4327.

Emrich T, Forster R, Lipp M (1993): Topological characterization of the lymphoid-specific seven transmembrane receptor BLR1 by epitope-tagging and high level expression. Biochem Biophys Res Commun 197:214–220.

Emrich T, Forster R, Lipp M (1994): Transmembrane topology of the lymphocyte-specific G-protein–coupled receptor BLR1: Analysis by flow cytometry and immunocytochemistry. Cell Mol Biol (Noisy le grand) 40:413–419.

Evan GI, Lewis GK, Ramsay G, Bishop JM (1985): Isolation of monoclonal antibodies specific for human c-myc proto-oncogene product. Mol Cell Biol 5:3610–3616.

Field J, Nikawa J, Broek D, MacDonald B, Rodgers L, Wilson IA, Lerner RA, Wigler M (1988): Purification of a RAS-responsive adenylyl cyclase complex from *Saccharomyces cerevisiae* by use of an epitope addition method. Mol Cell Biol 8:2159–2165.

Gat U, Nekrasova E, Lancet D, Natochin M (1994): Olfactory receptor proteins. Expression, characterization and partial purification. Eur J Biochem 225:1157–1168.

Geli V, Baty D, Lazdunski C (1988): Use of a foreign epitope as a "tag" for the localization of minor proteins within a cell: The case of the immunity protein to colicin A. Proc Natl Acad Sci USA 85:689–693.

Graham FL, van der Eb AJ (1973): Transformation of rat cells by DNA of human adenovirus 5. Virology 54:536–539.

Grisshammer R, Tucker J (1997): Quantitative evaluation of neurotensin receptor purification by immobilized metal affinity chromatography. Protein Expr Purif 11:53–60.

Grussenmeyer T, Scheidtmann KH, Hutchinson MA, Eckhart W, Walter G (1985): Complexes of polyoma virus medium T antigen and cellular proteins. Proc Natl Acad Sci USA 82:7952–7954.

Guan XM, Kobilka TS, Kobilka BK (1992): Enhancement of membrane insertion and function in a type IIIb membrane protein following introduction of a cleavable signal peptide. J Biol Chem 267:21995–21998.

Harlow E, Lane D (1988): Antibodies. A Laboratory Manual. Cold Spring Harbor, NY: Cold Spring Harbor Laboratory.

Hebert TE, Moffett S, Morello JP, Loisel TP, Bichet DG, Barret C, Bouvier M (1996): A peptide derived from a β_2-adrenergic receptor transmembrane domain inhibits both receptor dimerization and activation. J Biol Chem 271:16384–16392.

Hopp TP, Prickett KS, Price VL, Libby RT, March CJ, Cerretti DP, Urdal DL, Conlon PJ (1988): A short polypeptide marker sequence useful for recombinant protein identification and purification. Bio/Technology 6:1205–1210.

Ishii K, Hein L, Kobilka B, Coughlin SR (1993): Kinetics of thrombin receptor cleavage on intact cells. Relation to signaling. J Biol Chem 268:9780–9786.

Janssen JJ, Bovee GP, Merkx M, DeGrip WJ (1995): Histidine tagging both allows convenient single-step purification of bovine rhodopsin and exerts ionic strength–dependent effects on its photochemistry. J Biol Chem 270:11222–11229.

Kast C, Canfield V, Levenson R, Gros P (1996): Transmembrane organization of mouse P-glycoprotein determined by epitope insertion and immunofluorescence. J Biol Chem 271:9240–9248.

Keefer JR, Limbird LE (1993): The α_{2A}-adrenergic receptor is targeted directly to the basolateral membrane domain of Madin-Darby canine kidney cells independent of coupling to pertussis toxin-sensitive GTP-binding proteins. J Biol Chem 268:11340–11347.

Klein U, Ramirez MT, Kobilka BK, von Zastrow M (1997): A novel interaction between adrenergic receptors and the (-subunit of eukaryotic initiation factor 2B. J Biol Chem 272:19099–19102.

Kobilka BK (1995): Amino and carboxyl terminal modifications to facilitate the production and purification of a G protein–coupled receptor. Anal Biochem 231:269–271.

Kozak M (1987): At least six nucleotides preceding the AUG initiator codon enhance translation in mammalian cells. J Mol Biol 196:947–950.

Krajewski S, Zapata JM, Reed JC (1996): Detection of multiple antigens on Western blots. Anal Biochem 236:221–228.

Kwatra MM, Schreurs J, Schwinn DA, Innis MA, Caron MG, Lefkowitz RJ (1995): Immunoaffinity purification of epitope-tagged human β_2-adrenergic receptor to homogeneity. Protein Expr Purif 6:717–721.

Lilius G, Persson M, Bulow L, Mosbach K (1991): Metal affinity precipitation of proteins carrying genetically attached polyhistidine affinity tails. Eur J Biochem 198:499–504.

Lindner P, Bauer K, Krebber A, Nieba L, Kremmer E, Krebber C, Honegger A, Klinger B, Mocikat R, Pluckthun A (1997): Specific detection of His-tagged proteins with recombinant anti-His tag scFv-phosphatase or scFv-phage fusions. Biotechniques 22:140–149.

Mostafapour S, Kobilka BK, von Zastrow M (1996): Pharmacological sequestration of a chimeric β_3/β_2 adrenergic receptor occurs without a corresponding amount of receptor internalization. Recept Signal Transduct 6:151–163.

Mouillac B, Caron M, Bonin H, Dennis M, Bouvier M (1992): Agonist-modulated palmitoylation of β_2-adrenergic receptor in Sf9 cells. J Biol Chem 267:21733–21737.

Munro S, Pelham HR (1984): Use of peptide tagging to detect proteins expressed from cloned genes: Deletion mapping functional domains of *Drosophila* hsp 70. EMBO J 3:3087–3093.

Munro S, Pelham HR (1987): A C-terminal signal prevents secretion of luminal ER proteins. Cell 48:899–907.

Navarre J, Bradford AJ, Calhoun BC, Goldenring JR (1996): Quenching of endogenous peroxidase in western blot. Biotechniques 21:990–992.

Ng GY, George SR, Zastawny RL, Caron M, Bouvier M, Dennis M, O'Dowd BF (1993): Human serotonin 1B receptor expression in Sf9 cells: Phosphorylation, palmitoylation, and adenylyl cyclase inhibition. Biochemistry 32:11727–11733.

Niman HL, Houghten RA, Walker LE, Reisfeld RA, Wilson IA, Hogle JM, Lerner RA (1983): Generation of protein-reactive antibodies by short peptides is an event of high frequency: Implications for the structural basis of immune recognition. Proc Natl Acad Sci USA 80:4949–4953.

Prickett KS, Amberg DC, Hopp TP (1989): A calcium-dependent antibody for identification and purification of recombinant proteins. Biotechniques 7:580–589.

Robeva AS, Woodard R, Luthin DR, Taylor HE, Linden J (1996): Double tagging recombinant A1- and A2A-adenosine receptors with hexahistidine and the FLAG epitope. Development of an efficient generic protein purification procedure. Biochem Pharmacol 51:545–555.

Romano C, Yang WL, O'Malley KL (1996): Metabotropic glutamate receptor 5 is a disulfide-linked dimer. J Biol Chem 271:28612–28616.

Sambrook J, Maniatis T, Fritsch EF (1992): Molecular Cloning: A Laboratory Manual, 2nd ed. Cold Spring Harbor, NY: Cold Spring Harbor Laboratory.

Schmidt TG, Skerra A (1993): The random peptide library–assisted engineering of a C-terminal affinity peptide, useful for the detection and purification of a functional Ig Fv fragment. Protein Eng 6:109–122.

Schmidt TG, Skerra A (1994): One-step affinity purification of bacterially produced proteins by means of the "Strep tag" and immobilized recombinant core streptavidin. J Chromatogr A 676:337–345.

von Zastrow M, Kobilka BK (1992): Ligand-regulated internalization and recycling of human β_2-adrenergic receptors between the plasma membrane and endosomes containing transferrin receptors. J Biol Chem 267:3530–3538.

von Zastrow M, Link R, Daunt D, Barsh G, Kobilka B (1993): Subtype-specific differences in the intracellular sorting of G protein–coupled receptors. J Biol Chem 268:763–766.

Wilson IA, Cox NJ (1990): Structural basis of immune recognition of influenza virus hemagglutinin. Annu Rev Immunol 8:737–771.

Wilson IA, Niman HL, Houghten RA, Cherenson AR, Connolly ML, Lerner RA (1984): The structure of an antigenic determinant in a protein. Cell 37:767–778.

CHAPTER 2

IN SITU PHOSPHORYLATION OF G PROTEIN–COUPLED RECEPTORS

MICHEL BOUVIER, STÉPHANE ST-ONGE,
and MONIQUE LAGACÉ

---

1. INTRODUCTION

Constraints on signaling networks provide a high degree of plasticity to signal transduction pathways and ensure proper adaptation to various stimulatory conditions. As the initial point of interaction with transmitters, receptors are uniquely positioned to be the target of multiple regulatory processes that modulate their responsiveness. These regulatory processes play a major role in the

---

Regulation of G Protein–Coupled Receptor Function and Expression,
Edited by Jeffrey L. Benovic.
ISBN 0-471-25277-8   Copyright © 2000 Wiley-Liss, Inc.

sorting and integration of the information detected at the receptor level, allowing the cell to adapt to its environment. Alterations in normal regulation underlie certain pathological conditions related to hyper- or hyposensitivity and the undesirable effects of certain drug therapies. In the last few years, the molecular processes regulating the signaling efficacy of G protein–coupled receptors (GPCR) have begun to be unraveled.

Among various processes, numerous studies have demonstrated the central role played by phosphorylation in the regulation of GPCR (for review, see Benovic et al., 1988). In particular, phosphorylation of serine and threonine residues in cytosolic domains of many GPCR leads to the functional uncoupling that results from sustained stimulation of the receptor itself (agonist-promoted desensitization) or from the activation of a distinct receptor (cross-talk regulation).

For the β_2-adrenergic receptor (β_2AR), which can be considered as prototypical of GPCR, two distinct protein kinases, the cAMP-dependent protein kinase (PKA) and the β-adrenergic receptor kinase (βARK or GRK2), have been implicated in the functional uncoupling leading to rapid agonist-promoted desensitization. Whereas PKA has broad substrate specificity, GRK2 demonstrates a much more restricted specificity and phosphorylates only agonist-occupied receptor (Benovic et al., 1986). GRK2-mediated phosphorylation of serine and threonine residues in the distal portion of the C terminus of the receptor has been shown to promote the association of the protein β-arrestin with the β_2AR, thus inhibiting functional coupling of the receptor to Gs (Lohse et al., 1990; Pitcher et al., 1992b). In contrast, phosphorylation of β_2AR by PKA on sites located in the third cytoplasmic loop and the C tail does not favor the interaction of the receptor with β-arrestin, and the mechanisms by which it promotes uncoupling remain poorly understood. In addition to its role in functional uncoupling, phosphorylation of the receptor by GRK2 has been shown to contribute to its rapid internalization by promoting the formation of a complex between the receptor, β-arrestin, and clathrin (Goodman et al., 1996). Recently, phosphorylation by PKA has been proposed to change the selectivity of coupling of the β_2AR from Gs to Gi and by doing so to favor the activation of the MAP kinase cascade at the expense of the adenylyl cyclase pathway (Daaka et al., 1997).

Unlike PKA, GRK2 is not activated by a second messenger but belongs to the family of G protein–coupled receptor kinases (GRK) (Chen et al., 1993) that are translocated to the plasma membrane and become activated upon receptor stimulation. Interaction between the dissociated active G$\beta\gamma$ complex and GRK2 is responsible for the translocation of the enzyme to the plasma membrane and directly increases the activity of the enzyme (Kameyama et al., 1993; Pitcher et al., 1992a). Prenylation of the γ subunit is believed to play a crucial role in this activation mechanism (Inglese et al., 1992). In the case of other members of the GRK family, palmitoylation of the kinases themselves (GRK4 and GRK6) is responsible for their attachment to the plasma membrane (Stoffel et al., 1994; Premont et al., 1996).

Activation of distinct signaling pathways can also lead to the phosphorylation and desensitization of specific GPCR. Phosphorylation of the β_2AR by the calcium- and phospholipid-dependent protein kinase (PKC) has been shown to promote uncoupling of the receptor upon direct activation of the kinase by phorbol esters or following activation of phospholipase C–coupled receptors such as the α_1-adrenergic receptor (Bouvier et al., 1991). Also, several studies have documented that activation of tyrosine kinase receptors modulates GPCR functions as a result of tyrosine phosphorylation of the receptor (Valiquette et al., 1995; Hadcock et al., 1992). However, in contrast to the abundant literature concerning the phosphorylation of serine and threonine residues in GPCRs, very little is known about phosphorylation of these receptors on tyrosine residues.

As can be concluded from the above introduction, our understanding of the role of phosphorylation in GPCR signaling has evolved considerably in the last decade or so. However, most of what we know was learned from studies carried out on the β_2AR, rhodopsin, and a few other receptors, and the phosphorylation of many GPCRs remains unstudied. Given the fantastic diversity that exists among GPCRs, one can predict that significant knowledge could be gained by extending these studies to additional members of the family. Moreover, even for those receptors for which phosphorylation has been documented, the exact sites that are phosphorylated by the various kinases, their relative contribution to the regulatory actions, and the potential interactions between them remain largely unknown. For example, despite the numerous studies addressing the phosphorylation state of the β_2AR, the precise sites that are phosphorylated by GRK2 *in vivo* have still not been identified. Indeed, although truncation of the distal portion of the β_2AR C tail has been shown to abolish GRK2-mediated phosphorylation and desensitization, this domain of the receptor contains 11 serine or threonine residues that could theoretically all serve as substrates for GRK2. However, the identity of those, which are indeed phosphorylated, remains unknown. This contrasts with the identification of the receptor sites that are phosphorylated *in vitro* by GRK2 and GRK3 (Fredericks et al., 1996). However, caution is warranted when extrapolating these data to the *in vivo* situation. Indeed, for rhodopsin, different sites were found to be phosphorylated *in vivo* and *in vitro* by rhodopsin kinase (GRK1) (Ohguro et al., 1993, 1995).

The relatively slow progress accomplished in the study of GPCR phosphorylation *in situ* is in part due to the technically challenging nature of studying the post-translational modification of proteins that are expressed at very low abundance in most cells. These studies imply the development of purification schemes that allow the isolation of sufficient quantities of receptor to carry out their biochemical characterization. The following sections review the various technical approaches that can be used to efficiently study the phosphorylation of most GPCRs. While some techniques have readily been used for some time and can easily be applied with success to most GPCRs, others are fairly recent and are still in the development phase.

2. DETECTION OF RECEPTOR PHOSPHORYLATION

2.A. Expression systems

As mentioned above, one of the major difficulties in directly studying GPCR phosphorylation is the low amount of receptor that can be isolated from tissues or cell lines naturally expressing these receptors. One approach that we and others have selected to circumvent this problem has been to use heterologous expression, which allows various levels of overexpression. In addition to facilitating purification of sufficient quantities of receptor, these systems permit the study of selected receptor subtypes in a chosen genetic background.

2.A.a. Transient Versus Stable Expression Systems.

Heterologous expression systems can be divided into two classes: transient and stable. The idea behind the two systems is to express sufficient quantity of receptor in a surrogate cell line that normally expresses very little or, when possible, no receptor. For this purpose, the gene or cDNA of the receptor under study is cloned in an expression vector so that its expression is under the control of a strong constitutive or inducible promoter. The expression vector encoding the receptor is then transferred into the selected cell line. This transfer of genetic information, also referred to as *transfection,* can be accomplished with a variety of simple techniques that include pinocytotic internalization of DNA precipitate (calcium-phosphate precipitation) or colloidal suspension (DEAE-dextran), electroporation, or liposome fusion. The choice of the most appropriate technique depends largely on the cell type chosen and on the objectives of the experiment.

Indeed, the proportion of cells effectively harboring the expression vector can vary from 5% to greater than 80% depending on the cell type, the technique, and the quality of the DNA preparations used. Even within a single experiment, the level of expression varies significantly between individual cells because the uptake of the plasmids follows a stochastic distribution. The cells studied therefore represent a heterogeneous population. This does not present a problem if the level of analysis one seeks is qualitative. However, it is a serious handicap if detailed mechanistic and quantitative studies that require a good control of the stoichiometry of the protein studied are essential. In most cases, cells can be harvested 48 to 72 hours after the transfection and the receptor studied. Under this set of circumstances, the expression vector is not inserted in the cell's genome and functions as an independent transcription unit that will not be replicated during cell division. It follows that after many division cycles the expression vector will become diluted and that the level of expression of the receptor will rapidly decline. Also, the episomal plasmid will often be excluded from the cells with time. The expression is therefore transient, and new transfections must be performed for each experiment.

Stable expression takes advantage of a relatively rare recombination of events that leads to random insertion of the expression vector in the genome of the cells. Such recombination events occur in a very small proportion of the cells that have internalized the expression vector. The idea is then to select those cells whose genome have stably incorporated one or more copies of this vector and thus constitutively express the receptor. This is achieved by virtue of selectable markers, such as an antibiotic resistance gene, that are either part of the expression vectors or are co-transfected as independent vectors. Selection of the expressing cells is then realized by treatments with high concentrations of the antibiotic. Cells that did not integrate the resistance gene will die as a result of the toxic effect of the antibiotic. Over subsequent passages, only those cells that have stably integrated the plasmids will survive. At this point, the resistant cells can be treated as a pool and used as a heterologous population, or, alternatively, individual cellular clones originating from a single cell can be isolated. This last option presents the advantage of generating cell lines that are homogeneous and maintain a stable number of receptors over time. It is recommended, when possible, to select several clones that express different numbers of receptors (100 fmol/mg to 10 pmol/mg of protein would represent a good distribution) so that the effects of the level of expression can be controlled. Having to assess more than one clone also allows verification that the observations made are not unique to the clone selected.

The major advantage of transient over stable expression systems is that one does not need to go through the tedious process of selection and cellular cloning (4–6 weeks) and can readily use the cells 48 hours after the transfection. Moreover, higher levels of expression are generally obtained with transient systems. However, stable expression systems provide a homogeneous population of cells that all express the receptor at the same level and allow for repeated experiments under identical conditions. These characteristics should be kept in mind when designing and interpreting experiments with these systems. In general, we believe that stable cell lines offer a better model system to study phosphorylation of GPCRs. Moreover, once obtained, stable cell lines heterologously expressing a given receptor represent valuable biological reagents with which to study various aspects of GPCR function.

In the last few years, the availability of virus-based expression systems has offered an alternative to the classic transfection systems. Vaccinia (Bose et al., 1997) and baculovirus (Luckow and Summers, 1988) have been used to transfer genes of interest to mammalian and insect cells, respectively. Because gene expression occurs through the infectious cycle of the virus, it is transient, the cells dying from the infection several days later. Defining a time window that allows for high levels of expression while maintaining close to 100% cell viability is therefore of primary importance for these systems. The major advantage compared with traditional transient expression systems lies in the fact that the entire cell population can be infected if proper virus titers are used. Also, if the virus stock is properly maintained, the level of expression reached is very reproducible from one experiment to the next.

The baculovirus/Sf9 cell system has been used to study several aspects of GPCR function (Bouvier et al., 1998). Although the precise nature of the insect kinases involved has not been determined, agonist-dependent phosphorylation of several mammalian GPCRs expressed in Sf9 cells have been reported (Ng et al., 1993, 1994; Richardson and Hosey, 1992; Loisel et al., 1997). Figure 2.1 shows isoproterenol-stimulated phosphorylation of the human $\beta_2 AR$ expressed in Sf9 cells with the baculovirus system. This phosphorylation can be mimicked only in part by stimulating the cells with a membrane-permeable cAMP analogue, indicating that the insect PKA is only one of the kinases involved and that an additional kinase (most likely an insect GRK) also contributes to the phosphorylation. Although the baculovirus/Sf9 system cannot be seen as the best system to study phosphorylation as it occurs normally in mammalian cells, it may offer an acceptable alternative to study those receptors for which sufficiently high expression levels cannot be obtained in mammalian cells. Indeed, receptor numbers reaching 50–100 pmol/mg of protein have been reported for many GPCRs with the insect system. Co-infections with several recombinant

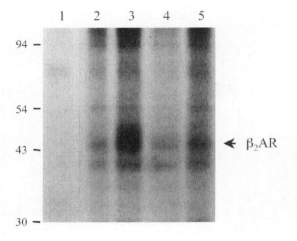

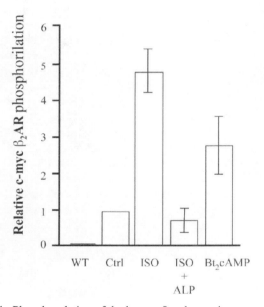

Figure 2.1. Phosphorylation of the human β_2-adrenergic receptor (β_2AR) in Sf9 cells. Sf9 cells were infected with a recombinant baculovirus encoding the human β_2AR and metabolically labeled with ^{32}Pi. The cells were then treated with the β-adrenergic agonist isoproterenol (ISO), with ISO in the presence of an excess of the β-adrenergic antagonist alprenolol (ISO + ALP), with the PKA activator dibutyryl-cAMP (Bt$_2$cAMP), or the vehicle alone (Ctrl). WT indicates cells that were infected with a baculovirus that did not encode the β_2AR. The receptor was isolated by immunoprecipitation with an anti-c-myc antibody directed against an N-terminal epitope-tagged β_2AR. The autoradiogram of an SDS-PAGE of the samples indicates that agonist stimulation promotes phosphorylation of the receptor that is blocked by pretreatment with an antagonist. The agonist-promoted phosphorylation is partially mimicked by the PKA activator.

viruses encoding different proteins of interest (receptors, kinases, G protein subunits, and so forth) also offer interesting alternatives (Bouvier et al., 1998).

2.A.b. *Choice of Cell Lines.* African green monkey kidney fibroblasts (COS-7 cells, ATCC CRL 1651; and COS-1 cells, ATCC CRL 1650) are certainly the most widely used mammalian cell lines for transient expression. These cells are transformed by an origin-defective mutant of SV40 that codes for the wild-type T antigen. The line thus contains the T antigens and can support the replication of SV40 viruses. Expression vectors encoding an SV40 origin of replication will therefore be replicated to high copy number in these cells. This accounts for the very high level of expression obtained in these cells. For example, with the expression vector pBC12BI encoding the human β_2AR, levels of expression reaching 15 pmol/mg membrane protein can easily be attained in COS cells. However, although COS cells remain a very popular system, the development of various vectors that place the expression of the protein of interest under the control of very strong promoters allows the use of any cell that is amenable to transfection. Examples include CHW-1102, NIGMS GM0459; CHO, ATCC CRL 9096; Ltk⁻, ATCC CCL 1.3; and HEK 293, ATCC CRL 1573. For the β_2AR, expression levels reaching 10 pmol/mg of protein can be obtained in HEK 293 cells with the pCDNA3 plasmid (Invitrogen). It should be remembered, however, that in most transient expression systems only a small portion of the total cell population will indeed internalize the vector and will be responsible for the massive expression. It follows that, although transient expression systems may be used to determine if phosphorylation does occur, they might not represent the best systems with which to estimate the normal stoichiometry of phosphorylation or to study the normal regulation of this phosphorylation.

As is the case for transient expression, a large number of cell lines can be used to generate stable expression systems. Considerations in selecting a cell line should include how easy it is to cultivate large quantities of these cells, the absence of endogenous expression for the receptor to be transfected, the presence of the signaling pathway(s) to which the receptor should be functionally coupled, the presence of other signaling molecules that could be considered in the course of the study, and so forth. Chinese hamster fibroblasts (CHW or CHO), mouse fibroblasts (Ltk⁻), and human kidney fibroblasts (HEK 293) have been used with success to stably express various GPCRs. Finding a cell that does not endogenously express the receptor to be studied is sometimes difficult, and the presence of such a receptor can complicate the interpretation, in particular in studies involving site-directed mutagenesis of the transfected receptor. To circumvent this problem, epitope-tagged receptor constructs have been used to allow the purification (see below) of the transfected receptor only.

Although heterologous expression is a tool of choice with which to study posttranslational modifications of receptors, it should be emphasized that using such systems is always a compromise between having access to cells that will allow purification of a large quantity of receptor for biochemical studies and studying a system that is not entirely physiologically relevant. In particular, problems of relative stoichiometries between signaling molecules are always a concern when using overexpression systems. When available, natural cell lines expressing sufficient quantities of a well-characterized receptor subtype is always preferable.

2.B. Cell Culture and Metabolic Labeling

The approach most widely used to study receptor phosphorylation consists of directly monitoring the covalent attachment of radiolabeled phosphate molecules to the receptor after isotopic labeling of the cellular ATP pool. The following discussion describes the methods used to adequately perform such metabolic labeling.

Once the cell line of choice has been generated or selected, the culture conditions should be maintained as usual until 1 day before the experiments. In our case, CHW, Ltk⁻, or HEK cells are grown as monolayers in 600 cm^2 dishes in Dulbeco's minimum eagle medium (MEM) supplemented with 10% fetal bovine serum, penicillin (100 units/ml), streptomycin (100 µg/ml), fungizone (0.25 µg/ml), and glutamine (1 mM) in an atmosphere of 95% air/5% CO_2 at 37°C. In the case of stably transfected cell lines, we found that maintaining the selective pressure with the selectable antibiotic (in most cases neomycin) helps to maintain the level of receptor expression at stable levels for longer periods of time (between 30 and 60 passages, depending on the clones considered). The use of 600 cm^2 instead of 75 cm^2 flasks allows the preparation of sufficient material without the increased radioactive hazards caused by manipulating a large number of flasks.

When the baculovirus/insect cell system is used, Sf9 cells are grown in Grace's insect medium (Gibco) supplemented with fetal bovine serum (10% v/v) at 27°C. Although Sf9 cells can also be grown as monolayers in flasks, metabolic labeling is more easily carried out when they are grown in suspension using Erlenmeyer flasks placed in an orbital shaker in the presence of pluronic acid to prevent cell tearing due to agitation. To permit expression of the receptor at sufficient levels in the entire population, cells should be infected at a multiplicity of infection (MOI) of approximately five recombinant baculovirus molecules per Sf9 cell when they are in a logarithmic phase of growth at a density of 1×10^6 to 2×10^6 cells per milliliter. The optimum time of infection should be determined for each recombinant baculovirus, but, in most cases, maximal levels of expression are found at 48 hours postinfection, a time when the proportion of viable cells, as assessed by their ability to exclude trypan blue, is usually greater than 90%. Additional details about baculovirus and Sf9 cells have been published by O'Reilly et al. (1992).

At 30 hours postinfection for Sf9 cells or when mammalian cells reach 90% confluency, cells are transferred to serum- and phosphate-free media for 18 hours to synchronize them in G_0 and to reduce the endogenous levels of ATP to facilitate equilibrium-isotopic labeling. Following this period, cells are incubated with 5–15 mCi of ^{32}P-orthophosphoric acid in culture media containing 50 µM NaH_2PO_4, 100 µM Na_3VO_4, and 1% serum for 2 hours at 37°C. Inorganic phosphate and sodium vanadate act as nonspecific phosphatase inhibitors, whereas the serum allows re-entry of the cells in G_1 growth phase. These labeling conditions allow isotopic equilibrium of the ATP pools for most cell types as directly assessed by measuring the specific activity of total cellular ATP (see below). When using a new cell type, the time of labeling required to achieve isotopic equilibrium should be determined. Once equilibrium labeling has been reached, the effects of various drugs (receptor agonist, selective kinase or phosphatase activators or inhibitors, and so forth) on the phosphorylation state of the receptor can be assessed by adding them or the vehicle to the cells for various periods of time. When studying the effects of selective phosphatase inhibitors or activators, sodium vanadate and high concentrations of inorganic phosphate should be omitted during the metabolic labeling. Following the labeling and treatment periods, cells are harvested and processed and the receptor purified as described below.

2.B.a. Membrane Preparation. Following metabolic labeling, cells are rinsed twice with ice-cold PBS and disrupted by sonication in ice-cold buffer containing 5 mM TrisHCl, pH 7.4, 2 mM EDTA, 5 µg/ml leupeptin, 10 µg/ml benzamidine, 5 µg/ml soybean trypsin inhibitor, 100 µM Na_3VO_4, and 50 µM NaH_2PO_4. Thereafter, all buffers should contain the protease and phosphatase inhibitors to prevent degradation of the receptor or its dephosphorylation during the solubilization and purification steps. Lysates are then centrifuged at 500g for 5 minutes at 4°C, the pellets are sonicated once more and spun again, and the supernatants are pooled. The resulting supernatant is then centrifuged at 45,000g, and the pelleted membranes are washed twice in the same buffer.

2.B.b. Solubilization. Solubilization represents an important and delicate step in the isolation of the receptors. Indeed, for most GPCRs, only a few detergents (e.g., digitonin and *n*-dodecyl maltoside) preserve their ligand-binding activity. This is an important issue because this biological activity may be used in an affinity purification scheme. Moreover, maintaining the ligand-binding activity allows measuring the number of functional receptors throughout the purification steps. Such quantification is a prerequisite for comparing the phosphorylation states observed in different conditions and determining the stoichiometry of phosphorylation. The best detergent (or detergent mixture) and optimum solubilization conditions should be determined for each receptor in a given cell type. For example, whereas 2% (w/v) digitonin is required to efficiently solubilize the human β_2AR expressed in CHW cells, only 0.3% of the same detergent is required to solubilize the receptor from Sf9 cell membranes using in both cases membrane preparations at 2 mg protein/ml.

To solubilize the human β_2AR from CHW cell membranes, pelleted membrane preparations are resuspended to a final concentration of 2 mg protein/ml in a buffer containing 100 mM NaCl, 10 mM Tris-HCl, 5 mM EDTA, pH 7.4, 2.0% digitonin (or 0.3% *n*-dodecyl maltoside), and protease and phosphatase inhibitors (as described above). The suspension is mildly stirred with a magnetic stirring bead at 4°C for 90 minutes. Nonsolubilized material is then removed by centrifugation at 100,000g for 20 minutes at 4°C. The solubilized material can then be use immediately for the purification procedures or quickly frozen and kept at −80°C.

2.C. Receptor Purification

The various approaches that are generally used to purify GPCRs can be classified into three categories: (1) methods that take advantage of the ligand-binding properties of the receptor and use an affinity resin, (2) immunoprecipitation techniques with polyclonal or monoclonal antibodies raised against the receptor sequence, and (3) methods based on the construction of fusion recombinant proteins between the receptor and epitopes specifically designed to allow purification. In the following section, methods based on ligand binding and on the addition of epitopes that were successfully used to purify the β_2AR are described.

2.C.a. Affinity Purification. Solubilized β_2AR can be purified by affinity chromatography with a sepharose matrix coupled to the β-adrenergic antagonist alprenolol (Benovic et al., 1984; Caron et al., 1979). The alprenolol-sepharose resin (6 ml of gel) is equilibrated with 50 ml of buffer A (100 mM NaCl, 10 mM Tris-HCl, 2 mM EDTA, pH 7.4, 0.05% digitonin [or 0.02% *n*-dodecyl maltoside], and protease and phosphatase inhibitors). Solubilized receptor (2–4 ml) is then added to the affinity resin and shaken gently for 2 hours at room temperature to allow binding of the receptor to the matrix. After this incubation, the resin is transferred to a 15-ml column and the supernatant allowed to flow through.

Following this batchwise loading procedure, the columns are placed at 4°C and washed with 15 ml of an ice-cold buffer containing 500 mM NaCl, 50 mM Tris-HCl, 2 mM EDTA, pH 7.4, and 0.05% digitonin (or 0.02% *n*-dodecyl maltoside) that is allowed to flow through the column by gravity. The original ionic strength is restored by washing with 30 ml of buffer A. The columns are then returned to room temperature and the receptor biospecifically eluted with buffer A containing 60 µM alprenolol at a rate of 5ml/hr.

Although the elution is carried out at room temperature to facilitate the dissociation of the receptor from the alprenolol resin (increasing its off-rate), the eluate is collected on ice to prevent proteolytic degradation. It is also of primary importance to maintain protease inhibitors in the elution buffer. As indicated above, the cocktail that we usually use includes soybean trypsin inhibitor, leupeptin, benzamidine and EDTA. However, adding PMSF, bestatin, and aprotinin can also be necessary in some cases. The eluted receptor is then concentrated to a final volume of 50 µl by ultrafiltration with Centriprep and Centricon cartridges (Amicon).

After desalting the purified sample on a G-50 gel filtration column to remove alprenolol, β_2AR recovery can be measured by soluble binding with ^{125}I-cyanopindolol (CYP) as the radioligand (Bouvier et al., 1988). Nonspecific binding is assessed in the presence of (–)-alprenolol, 10 µM. Receptor-bound ^{125}I-CYP is routinely separated from free ligands through gel filtration over sephadex G-50 columns. Alternatively, free ligand can be separated from the receptor–ligand complex by filtration over GF/B fiberglass filters (Whatman) pre-soaked in a solution of Tris, 25 mM (pH. 7.4), including polyethylenimine (0.3%) and BSA (0.1%). The purified samples can then be prepared immediately for electrophoretic analysis or kept at –80°C for a few days following quick-freeze in liquid nitrogen. If longer storage is required, it is preferable to freeze the samples before removing the alprenolol by desalting because the presence of the antagonist stabilizes the receptor and protects it against proteolytic degradation.

Affinity chromatography purification schemes similar to the one described above has been extremely useful to study the phosphorylation state of receptors such as the β_2AR and the dopamine receptors for which good affinity resins have been developed. However, such affinity matrixes are relatively tedious to prepare and are not available for all GPCRs. This leads to the use of alternative approaches, which are described below.

2.C.b. *Immunopurification of the c-myc/β₂AR.* Although immunoprecipitation with antibodies raised against receptors would be a technique of choice to purify GPCRs, attempts to raise antibodies against these receptors have generally failed to produce high-affinity antibodies that can be used satisfactorily in immunoaffinity purification procedures. To date only a few such antibodies have been generated. To circumvent this problem, many investigators have constructed recombinant receptors fused in frame with sequences encoding specific epitopes that can be recognized by commercially available antibodies.

Epitopes derived from parts of the c-myc and the hemaglutinin coding sequence as well as the Flag epitope are among the most popular. Antibodies that have been used successfully to immunoprecipitate epitope-tagged receptors include the anti-c-myc 9E10 antibody (Santa Cruz Biotechnology), the anti-hemaglutinin 12CA5 antibody (Boehringher), and the anti-Flag M2 antibody (Eastman Kodak). In addition to immunoreactive epitopes, stretches of six or more histidines have also been used. These poly-histidine epitopes interact via a chelation reaction with high affinity to nickel and thus allow purification of proteins bearing this epitope using an agarose resin coupled to nickel (Ni-NTA-agarose; Qiagen). The poly-His–tagged receptors can be eluted from the resin with neutral imidazole, which preserves the binding activity of the receptor.

Theoretically, the epitopes can be introduced anywhere within the coding sequence of the receptor. However, positioning them either at the N or C terminus has generally been favored over an insertion within the core of the receptors with the idea that (1) it would result in a greater accessibility of the epitopes to the antibodies or nickel and (2) it should have less detrimental effects on the three-dimensional structure of the receptors. Successful purification schemes were developed using both N- and C-terminally positioned epitopes. Because phosphorylation sites are found in the C tail of many GPCRs, we and others have chosen to use N-terminal epitopes to study this post-translational modification to avoid any potential interference between the epitope and the phosphorylation sites. In addition, to permit immunopurification of the receptors, the presence of the epitope allows visualization of the receptor by immunoblotting throughout the purification steps.

Characterization of the human β₂AR phosphorylation in Sf9 cells was easily carried out with recombinant receptor tagged at its NH2 termini with the c-myc epitope with the following procedure. After metabolic labeling of Sf9 cells expressing the c-myc/β₂AR, receptors are solubilized as described above and the tagged receptor immunoprecipitated using the mouse anti-c-myc monoclonal antibody (9E10). A crucial factor for optimal immunoprecipitation is to reduce the concentration of the detergent to the minimum required to maintain the receptor in solution (e.g., ~0.03% digitonin). Removal of detergent and concentration of the solubilized receptor can be done by dialysis with centriprep cartridges (Amicon). This is accomplished by successive concentration and dilution of the samples in a detergent-free buffer containing 100 mM NaCl, 10 mM Tris-HCl, pH 7.4, 2 mM EDTA, and protease inhibitors (buffer B).

The immunoprecipitation is initiated by adding purified anti-c-myc 9E10 antibody (at a 7:1 antibody to receptor molar ratio) to the concentrated solubilized receptor (classically, a final volume of less than 1 ml is used for the immunoprecipitation) and agitated for 2 hours at 4°C. The secondary antimouse IgG antibody coupled to agarose (Sigma; at an 11:1 secondary to primary antibody molar ratio) is then added and the mixture incubated overnight at 4°C under agitation. The immunoprecipitate is centrifuged at 12,000 rpm in a microfuge for 10 minutes; the pellet is rinsed extensively with buffer B and resuspended in SDS-PAGE loading buffer. The denaturation of the proteins under reducing conditions is allowed to proceed at room temperature for 30 minutes. It is not recommended to boil the sample as it may cause aggregation of many GPCR. The relative yields of the immunoprecipitation carried out for the different conditions studied cannot be directly assessed by radioligand-binding assay because the immunoprecipitated receptor loses its ability to bind but can be semiquantitatively estimated by Western blot analysis with the anti-epitope antibody.

To purify poly-His–tagged β_2AR, solubilized membrane preparations that do not contain EDTA are loaded onto a 2.5-cm inner diameter × 10 cm column already packed with 10 ml of Ni-NTA-agarose (Qiagen) at a rate of 0.5 ml/min in the presence of 10 mM imidazole to reduce the nonspecific binding of proteins to the resin. After the loading procedure, the column is washed with 20 volumes of a buffer containing 25 mM Hepes (pH 7.4), 500 mM NaCl, 10 mM imidazole, 0.02% n-dodecyl-maltoside, and the protease inhibitor cocktail at a rate of 1 ml/min. The column is then re-equilibrated with 5 volumes of a lower ionic strength buffer in which the NaCl concentration has been reduced to 100 mM. Finally, the poly-His–tagged receptor is eluted at 0.5 ml/min with 3 volumes of a buffer containing 25 mM Hepes (pH 7.4), 100 mM NaCl, 100 mM imidazole, 0.02% n-dodecyl-maltoside, 20% glycerol (v/v), and the protease/phosphatase inhibitor cocktails. The eluted receptor is then concentrated by ultrafiltration with centriprep and centricon cartridges and the amount of receptor purified measured by radioligand binding as described above. All the chromatographic steps are carried out at 4°C to prevent proteolytic degradation.

Immunopurification and nickel-based chromatography methods usually lead to overall purification yields that are higher than those attained by affinity chromatography (can easily reach 85% of the solubilized receptor compared with ~50% for the alprenolol-sepharose affinity method described above). Generally, however, more contaminating bands accompany the receptor following these epitope-tag–based procedures (Fig. 2.2). Identification of the phosphorylated band(s) that correspond to the receptor therefore becomes particularly important. Western blot analysis of the same samples with either an anti-tag antibody or an antibody directed against the receptor itself can be used to directly identify the receptor band(s). Such analysis can be carried out on the same nitrocellulose membrane that is used for the autoradiographic detection of the phosphorylated proteins (see below). Alternatively, a duplicate polyacrylamide gel can be prepared specifically for the Western blot analysis.

Another approach that permits the identification of the receptor band is the photoaffinity labeling of the receptor protein with radiolabeled photoreactive ligands. Indeed, for several receptors, selective radioligands that can be covalently linked to the receptor after photoactivation have been developed. For the

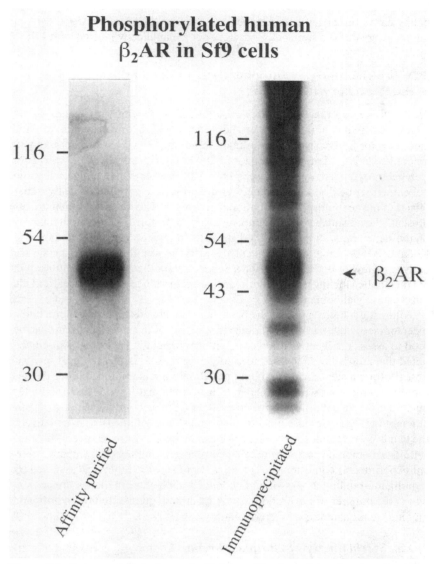

Figure 2.2. Immunoprecipitation and affinity purification of the phosphorylated β_2-adrenergic receptor (β_2AR) in Sf9 cells. Metabolically labeled c-myc–tagged β_2AR were purified by alprenolol-sepharose affinity chromatography or immunoprecipitation with the 9E10 antibody. The identity of the receptor band was confirmed by Western blot analysis with the anti-c-myc antibody and by photoaffinity radiolabeling (data not shown).

β_2AR, photolabeling with ^{125}I-cyanopindolol-diazerine (Dupont-NEN radiochemicals) has been used successfully to identify the protein bands corresponding to the receptor (Bouvier et al., 1988; Mouillac et al., 1992; Valiquette et al., 1995). Because autoradiography does not allow one to easily distinguish between proteins that are ^{32}P-phosphorylated or ^{125}I-iodinated, the photoaffinity labeling has to be carried out on solubilized samples that were treated in parallel but that were derived from cells that were not metabolically labeled with ^{32}Pi.

2.D. Stoichiometry of Phosphorylation and Identification of the Phosphorylation Sites

2.D.a. Assessing the Phosphorylation Level. Following purification of the receptors with one of the techniques described above, a known amount of purified receptor as assessed by radioligand binding (when possible) or a known amount of total protein is prepared for SDS-PAGE (Laemmli, 1970) with 10% slab gels. After electrophoretic separation of the proteins, the polyacrylamide gel can either be dried immediately or the proteins electrophoretically transferred to nitrocellulose or immobilon membranes. The transfer to membranes presents the advantage of allowing one to easily perform Western blot analysis on the same preparations. The dried gel or the nitrocellulose is then exposed to Kodak XAR-5 film or the equivalent at −70°C for periods of time that can vary between a few hours and a few days depending on the amount of ^{32}Pi used in the metabolic labeling, the level of receptor expression in the cells used, and the purification yield obtained.

Following identification of the phosphorylated bands that correspond to the receptor (see above), laser densitometric scanning of the autoradiograms can be used to assess the incorporation of ^{32}P into the receptor. However, it should be noted that autoradiography is linear only over a small range of signal intensities. Blotting a standardized amount of ^{32}P onto the membrane before the autoradiography allows for the generation of a standard curve that facilitates quantitation. Phosphorimager, when available, can also provide a good quantitation of the signal because it is linear over many logs of intensity. Alternatively, the bands corresponding to the receptor can be cut out from the gel or the nitocellulose membrane and prepared for scintillation counting. To compare different experimental conditions, an identical number of receptors as assessed by radioligand-binding assays should be loaded in neighboring lanes. In cases in which the number of purified receptors cannot be quantitated by radioligand binding, semiquantitative Western blotting can be used.

2.D.b. Stochiometry of Phosphorylation. An important aspect that concerns the potential functional importance of GPCR phosphorylation is the stoichiometry of such phosphorylation. Conventional wisdom predicts that a very low sub-stoichiometric phosphorylation would have little effect on receptor function. The phosphorylation stoichiometry can be rather easily assessed by determining the specific activity of the ATP pools in equilibrium metabolic labeling experiments. The amount of phosphate incorporated is then determined by direct scintillation counting of the purified receptor following SDS-PAGE electrophoresis and the stoichiometry of phosphorylation calculated based on the amount of receptor (as assessed by radioligand binding) loaded on the gel.

To determine the specific activity of the ATP pools, cells are metabolically labeled and lysed and the membrane fraction processed for receptor purification as described above. In parallel, the proteins of the cytosolic fraction are extracted using 2 volumes of acetonitrile and centrifuged at 45,000g for 20 minutes at 4°C. The supernatant is then concentrated by lyophilization (speed-vac) and the ATP isolated by HPLC with an absosphere nucleotide–nucleoside column (Allteck). The peak corresponding to ATP is then collected and counted by liquid scintillation. The ATP concentration is determined by ultraviolet absorbance with a cold ATP standard curve. Typically, with the metabolic labeling conditions described above, the specific activity of the radiolabeled ATP reaches 0.05–0.1 Ci/ mmol, depending on the cell line used.

This type of analysis allows one to determine the overall phosphorylation stoichiometry of a given receptor under different sets of conditions but does not provide information on the identity of the phosphorylated residues or on the relative stoichiometries of individual phosphorylation sites. It follows that the number of moles of phosphate per mole of receptor obtained cannot be equated to the number of phosphorylation sites present in a given receptor. It represents the minimum number of sites assuming a stoichiometry of one for each of the sites.

2.D.c. *Phosphoamino Acid Analysis.* A thorough characterization of the phosphorylation state of a protein includes the identification of the phosphorylation sites. A first estimate of such sites can be obtained by phosphoamino acid analysis. Such a procedure is based on the ability to separate labeled phosphoamino acids following complete hydrolysis of the protein of interest. Various high-voltage electrophoresis and chromatographic procedures have been described. Here we describe a two-dimensional separation procedure based on sequential high-voltage electrophoresis and ascending chromatography. This technique allows an excellent resolution of the phosphotyrosines from the other phosphorylated species and does not require any sophisticated electrophoresis apparatus.

Acid hydrolysis of the purified and electrophoretically separated β_2AR with 5.7 M HCl for 1 hour at 110°C can be performed directly on excised pieces of immobilon membranes. Nonradioactive pSer, pThr, and pTyr standards (1 µg) are then added to the hydrolysate. The mixture is subjected to separation on cellulose-coated thin layer sheets (Sigma) by electrophoresis at 500 V for 1 hour at 4°C in a solution consisting of glacial acetic acid:pyridine:H_2O (50:156:1,794), pH 1.9, followed by ascending chromatography in isobutyric acid:0.5 M ammonium hydroxide (5:3) (Duclos et al., 1991; Cooper et al., 1983). The plates are then dried, and phosphoamino acid standards are visualized by colorimetric reaction with sprayed ninhydrin upon drying with a hot air blower. The radiolabeled phosphoamino acids are then revealed by autoradiography of the chromatographic plates. An example of phosphoamino acid determination for the agonist-promoted phosphorylation of human β_2AR in Sf9 cells is presented in Figure 2.3.

Semiquantitative assessment of the phosphoamino acid species can be obtained by laser scanning densitometry of the autoradiograms. However, such assessment is not absolute because none of the steps involved are entirely quantitative and because of differential stability and recoveries of the phosphoamino acids. Indeed, the three phosphoamino acid species show distinct sensitivities to various conditions. For example, phosphotyrosines are relatively stable to alkali, whereas phosphoserine and phosphothreonine are degraded by such treatment due to base-catalyzed β-elimination.

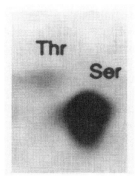

Figure 2.3. Phosphoamino acid analysis of the phosphorylated β_2AR. Phosphorylated β_2AR isolated by affinity chromatography from Sf9 cells was hydrolyzed in the presence of HCl and the amino acid mixture separated by sequential high-voltage thin-layer electrophoresis and ascending chromatography. The positions of the phosphoserine, phosphothreonine, and phosphotyrosine were determined by the addition of nonradioactive standards and ninhydrin coloration. The autoradiogram shown indicates that serines and threonines but not tyrosines are phosphorylated. Interestingly, phosphotyrosines are found when the phosphorylation experiments are carried out in mammalian cells instead of the insect Sf9 cells, indicating the existence of cell-specific differences.

An alternative approach to the phosphoamino acid determination requiring chemical hydrolysis of the receptor is the use of anti-phosphoamino acid antibodies. Theoretically, the use of these antibodies in Western blot analysis of the purified receptors would allow one to determine the effect of specific treatments on the extent of phosphorylation on phosphoserines, phosphothreonines, and phosphotyrosines. Unfortunately, the poor selectivity and affinity of the phosphoserine and phosphothreonine antibodies developed thus far have limited their usefulness. In contrast, the availability of high-affinity antibodies specifically directed against phosphotyrosine residues is an important asset with which to study protein tyrosine phosphorylation.

In many instances, immunoprecipitation of a protein with an antiphosphotyrosine antibody has been interpreted as evidence that this protein is indeed phosphorylated on tyrosine residues. Similarly, specific immunoblotting of an immunoaffinity-purified protein with an antiphosphotyrosine has been routinely used to demonstrate and quantitate tyrosine phosphorylation. For the β_2AR, such an approach was successfully used to demonstrate that this receptor was phosphorylated on at least one tyrosine residue in mammalian cells and that this phosphorylation could be promoted by stimulation of the cells with insulin (Valiquette et al., 1995; Hadcock et al., 1992).

Following electrophoresis of the purified β_2AR, proteins are electrophoretically transferred to a nitrocellulose membrane (Schleicher and Schuell). Membranes are incubated for 3 hours at 25°C in a blocking buffer consisting of 4% BSA, 10 mM Tris, pH 7.4, 150 mM NaCl, 0.05% Nonidet P-40, and 0.05% Tween 20. The polyclonal phosphotyrosine antibody (Upstate Biotechnology Inc., No. 06-123) is then added at a final concentration of 2 μg/ml and incubated for 3 hours. Selectivity of the immunoreactivity is assessed by preincubating the antibody with 80 mM pTyr or pSer/pThr at room temperature for 1 hour. After five washes of 5 minutes each in rinse buffer (blocking buffer without BSA), the membranes are incubated for 1 hour with 1 µCi/ml ^{125}I-protein A, washed, dried, and exposed for autoradiography. A peroxidase-coupled secondary (antirabbit) antibody can also be used to reveal the immunoreactivity by chemiluminescence.

2.D.d. Sites of Phosphorylation. Once phosphorylation of a GPCR has been established and shown to be modulated by various treatments, identifying the sites that are phosphorylated becomes an important step in determining the effects that such phosphorylation may have on receptor function. A direct way to determine the phosphorylation sites of the receptors would be to directly sequence the phosphopeptides generated following chemical or enzymatic digestion of the purified phosphorylated receptor. Such an approach has been successfully used to determine the β_2AR sites that are phosphorylated *in vitro* by GRK-2 and GRK-5 with purified receptor and kinase preparations (Fredericks et al., 1996).

[32]P-phosphorylated receptor was digested with the protease Glu-C, and the phosphopeptides generated were separated by reverse-phase HPLC with a C-18 column and detected by scintillation counting. The isolated peptides were then subjected to amino acid sequence analysis. The sequences obtained allowed the identification of the phosphopeptides as specific receptor sequences, and the determination of the amount of radioactivity released at each Edman degradation permitted the identification of the exact phosphorylation sites on the receptor. Although this is the most direct approach, it is not readily feasible for *in situ* phosphorylation conditions. Indeed, 400–1,000 pmol of purified β_2AR was used in the *in vitro* experiments, whereas only 2–10 pmol of purified receptor can be obtained with the metabolic labeling systems described above. Attempts at using phosphopeptide mapping in the absence of subsequent amino acid sequence analysis for determining the sites of *in situ* phosphorylation have been reported (Hadcock et al., 1992). However, this approach has generally not been used successfully for GPCRs.

Due to the difficulty of directly determining the identity of the sites that are phosphorylated in a whole-cell setting, site-directed mutagenesis is still the most popular approach used today. Indeed, the use of heterologous expression systems allows one to express, in the same genetic background, mutant forms of the receptor that lack potential phosphorylation sites (Hausdorff et al., 1989; Bouvier et al., 1989; Valiquette et al., 1995). Systematic scanning of mutations of all the potential phosphorylation sites, mutations of clusters of potential sites, as well as mutations of specific consensus sequences that represent putative sites for a given kinase is widely used. Eukaryotic expression plasmids encoding the receptors of choice are mutated by site-directed mutagenesis to substitute selected serine or threonine residues (with alanines or glycines) or tyrosine residues (with phenylalanine). After confirming that the appropriate mutations were introduced into an otherwise normal receptor by direct DNA sequencing, plasmids expressing the mutant and wild-type receptors are transfected in the chosen cell type. Stable lines expressing equivalent numbers of receptors are then selected to carry out the phosphorylation experiments (transient expression systems can also be used, but the variability in the transfection efficiency from one experiment to the next may complicate data interpretation). The effects of the mutations on the extent (stoichiometry) of phosphorylation observed are assessed as described above.

A decrease in the phosphorylation level following mutation of specific sites is interpreted as meaning that these sites do indeed represent *in situ* phosphorylation sites, whereas no effect suggests that they are not phosphorylated. This site-directed mutagenesis approach has the advantage of allowing the simultaneous assessment of the effects of the mutations on the functions of the receptor and thus permits one to propose a link between the lack of phosphorylation at a given site and a specific receptor function or attribute. On the negative side, one cannot exclude the possibility that the effect on the phosphorylation and/or functional state could be an indirect consequence of the mutation resulting from a conformational change that does not involve phosphorylation of this site. It follows that, although site-directed mutagenesis is a very powerful tool, results obtained with this technique alone do not represent definitive proof of the phosphorylation of a specific site. Additional direct assessment of this phosphorylation should be attempted.

As mentioned previously, direct determination of the phosphorylation sites within a GPCR *in situ* remains a challenging problem. The recent development of more sensitive mass spectrometry analysis techniques has opened the way to the use of this approach to study GPCR phosphorylation. Indeed, identification of the sites that are phosphorylated *in situ* have been reported for rhodopsin (Papac et al., 1993) and the endothelin B receptor (Roos et al., 1998). The approach used in these two studies is based on the fact that the addition of a phosphate molecule to a polypeptide leads to an 80-Da increase in its molecular mass that can be easily detected by mass spectrometry.

With purified receptor, an initial fingerprinting of the proteolytic fragments (carried out either on the complex mixture of peptides generated or following separation of the peptides by HPLC) is done by liquid secondary ion mass spectrometry (LSIMS) or matrix-assisted laser desorption ionization time of flight mass spectrometry (MALDITOF). By comparing the experimentally determined peptide mass with those predicted by the primary sequence, this first step allows the identification of the peptides that are phosphorylated (the experimentally determined mass being 80 Da greater than the predicted one). In a second step, the presence of a phosphate residue and the identity of the specific amino acid that is bearing this phosphate is confirmed by electrospray ion trap mass spectrometry coupled with fragment analysis by tandem (MS/MS) mass spectrometry. However, in some cases, the complexity of the spectra obtained is such that the precise phosphorylation site(s) in peptides that contain multiple serines and/or threonines and/or tyrosines cannot be established unambiguously. In such cases, comparison of the spectra obtained from the fragmented peptides with those generated by synthetic phosphopeptides of identical compositions can be used to assign the phosphorylation site(s) (Papac et al., 1993).

Although the use of mass spectrometry requires very specialized and expensive equipment and should still be considered in a developmental phase for studying GPCR phosphorylation, it is a very promising approach that presents several important advantages. First, experiments can be carried out with relatively small quantities of receptor (less than 4 pmol of endothelin B receptor was required for the study described above). This is particularly important as it would allow one to study phosphorylation directly in tissues that endogenously express GPCRs and would not require the use of overexpression systems. Second, these experiments can be performed without manipulating large quantities of radioactivity that present non-negligible safety hazards in laboratories. One potential caveat of the mass spectrometric approach is that it is not clear at this point in time if this technique will be appropriate to precisely determine the overall and site-specific stoichiometry of phosphorylation.

3. CONCLUSION

Over the last few years, the study of GPCR phosphorylation *in situ* has been greatly facilitated by the advance in heterologous overexpression systems and by the development of relatively easy and efficient receptor purification methods that are based on the use of epitope tagging. These techniques make it relatively straightforward to determine, by metabolic labeling with ^{32}Pi, if a given receptor can become phosphorylated in a specific cell and whether various treatments (e.g., agonist stimulation, direct activation, or inhibition of specific kinases or phosphatases) can affect this level (stoichiometry) of phosphorylation. Combination of these approaches with site-directed mutagenesis has led to the identification of specific residues or receptor subdomains that most likely represent functionally important phosphorylation sites. There is no doubt that these techniques will continue to be widely used to study GPCR phosphorylation and that they will bring new information on the specific phosphorylation patterns of individual receptors as well as on the general functional relevance of this post-translational modification for this class of receptors.

Nevertheless, one should keep in mind that results obtained in a specific heterologous overexpression system cannot be automatically extrapolated to other cell types or tissues that normally express these receptors. Also, one should not underestimate the possible consequences of overexpression when carrying out quantitative assessment of the phosphorylation state of a receptor. Indeed, the relative stoichiometry of the receptor versus the kinases and other regulatory proteins that are expressed in these cells may have a significant impact on the results obtained. These systems should therefore be considered for what they are: models that make possible the study of a post-translational modification that still cannot be investigated readily in native tissues. However, the generalization of the use of mass spectrometry for the study of GPCR phosphorylation may offer some solutions to the problems described above. Indeed, it may allow the characterization of the phosphorylation state of receptors in their natural site of expression and permit the direct identification of the phosphorylation sites within the receptor structure.

REFERENCES

Benovic JL, Bouvier M, Caron MG, Lefkowitz RJ (1988): Regulation of adenylyl cyclase–coupled β-adrenergic receptors. Annu Rev Cell Biol 4:405–427.

Benovic JL, Shorr RGL, Caron MG, Lefkowitz RJ (1984): The mammalian β_2-adrenergic receptor: Purification and characterization. Biochemistry 23:4510–4518.

Benovic JL, Strasser RH, Caron MG, Lefkowitz RJ (1986): β-Adrenergic receptor kinase: Identification of a novel protein kinase that phosphorylates the agonist-occupied form of the receptor. Proc Natl Acad Sci USA 83:2797–2801.

Bose A, Saha D, Gupta NK (1997): Viral infection. I. Regulation of protein synthesis during vaccinia viral infection of animal cells. Arch Biochem Biophys 342:362–372.

Bouvier M, Collins S, O'Dowd BF, Campbell PT, De Blasi A, Kobilka BK, MacGregor C, Irons GP, Caron MG, Lefkowitz RJ (1989): Two distinct pathways for cAMP-mediated down-regulation of the β_2-adrenergic receptor: Phosphorylation of the receptor and regulation of its mRNA level. J Biol Chem 264:16786–16792.

Bouvier M, Guilbault N, Bonin H (1991): Phorbol-ester induced phosphorylation of the β_2-adrenergic receptor decreases its coupling to G_s. FEBS Lett 279:243–248.

Bouvier M, Hnatowich M, Collins S, Kobilka BK, De Blasi A, Lefkowitz RJ, Caron MG (1988): Expression of a human cDNA encoding the β_2-adrenergic receptor in Chinese hamster fibroblast (CHW): Functionality and regulation of the expressed receptors. Mol Pharmacol 33:133–139.

Bouvier M, Ménard L, Dennis M, Marullo S (1998): Expression and recovery of functional G-protein–coupled receptors using baculovirus expression systems. Curr Opin Biotechnol 9:522–527.

Caron MG, Srinivasan Y, Pitha J, Kociolek K, Lefkowitz RJ (1979): Affinity chromatography of the β-adrenergic receptor. J Biol Chem 254:2923–2927.

Chen C-Y, Dion SB, Kim CM, Benovic JL (1993): β-Adrenergic receptor kinase. Agonist-dependent receptor binding promotes kinase activation. J Biol Chem 268:7825–7831.

Cooper JA, Sefton BM, Hunter T (1983): Detection and quantification of phosphotyrosine in proteins. Methods Enzymol 99:387–405.

Cullen BR (1987): Use of eukaryotic expression technology in the functional analysis of cloned genes. Methods Enzymol 152:684–704.

Daaka Y, Luttrell DK, Lefkowitz RJ (1997): Switching of the coupling of the beta$_2$-adrenergic receptor to different G proteins by protein kinase A. Nature 390:88–91.

Duclos B, Marcandier S, Cozzone AJ (1991): Chemical properties and separation of phosphoamino acids by thin-layer chromatography and/or electrophoresis. Methods Enzymol 201:10–27.

Fredericks ZL, Pitcher JA, Lefkowitz RJ (1996): Identification of the G protein–coupled receptor kinase phosphorylation sites in the human beta$_2$-adrenergic receptor. J Biol Chem 271:13796–13803.

Goodman OB, Krupnick JG, Santini F, Gurevich VV, Penn RB, Gagnon AB, Keen JH, Benovic JL (1996): Beta-arrestin acts as a clathrin adaptor in endocytosis of the beta$_2$-adrenergic receptor. Nature 383:447–450.

Hadcock JR, Port JD, Gelman MS, Malbon CC (1992): Cross-talk between tyrosine kinase and G-protein–linked receptors. Phosphorylation of β_2-adrenergic receptors in response to insulin. J Biol Chem 267:26017–26022.

Hausdorff WP, Bouvier M, O'Dowd BF, Irons GP, Caron MG, Lefkowitz RJ (1989): Phosphorylation sites on two domains of the β_2-adrenergic receptor are involved in distinct pathways of receptor desensitization. J Biol Chem 264:12657–12665.

Inglese J, Koch WJ, Caron MG, Lefkowitz RJ (1992): Isoprenylation in regulation of signal transduction by G protein–coupled receptor kinases. Nature 359:147–150.

Kameyama K, Haga K, Haga T, Kontani K, Katada T, Fukada Y (1993): Activation by G protein βgamma subunits of β-adrenergic and muscarinic receptor kinase. J Biol Chem 268:7753–7758.

Laemmli UK (1970): Cleavage of structural proteins during the assembly of the head of the bacteriophage T4. Nature 227:680–686.

Lohse MJ, Benovic JL, Codina J, Caron MG, Lefkowitz RJ (1990): β-Arrestin: A protein that regulates β-adrenergic receptor function. Science 248:1547–1550.

Loisel TP, Ansanay H, St-Onge S, Gay B, Boulanger P, Strosberg AD, Marullo S, Bouvier M (1997): Recovery of homogeneous and functional beta-2 adrenergic receptors from extracellular baculovirus particles Nat Biotechnol 15:1300–1304.

Luckow VA, Summers MD (1988): Trends in the development of baculovirus expression vectors. Biotechnology 6:47–55.

Mouillac B, Caron M, Bonin H, Dennis M, Bouvier M (1992): Agonist-modulated palmitoylation of β₂-adrenergic receptor in Sf9 cells. J Biol Chem 267:21733–21737.

Ng GY, George SR, Zastawny RL, Caron M, Bouvier M, Dennis M, O'Dowd BF (1993): Human serotonin$_{1B}$ receptor expression in Sf9 cells: Phosphorylation, palmitoylation and adenylyl cyclase inhibition. Biochemistry 32:11727–11733.

Ng GY, Mouillac B, George S, Caron M, Dennis M, Bouvier M, O'Dowd B (1994): Desensitization, phosphorylation and palmitoylation of the human D1 dopamine receptor. Eur J Pharmacol Mol Pharmacol Sec 267:7–19.

O'Reilly DR, Miller LK, Luckow VA (1992): Baculovirus Expression Vectors: A Laboratory Manual. New York: WH Freeman and Co.

Ohguro H, Palczewski K, Ericsson L, Walsh K, Johnson R (1993): Sequential phosphorylation of rhodopsin at multiple sites. Biochemistry 32:5718–5724.

Ohguro H, Van Hooser JP, Milam AH, Palczewski K (1995): Rhodopsin phosphorylation and dephosphorylation *in vivo*. J Biol Chem 270:14259–14262.

Papac DI, Oatis JE, Crouch RK, Knapp DR (1993): Mass spectrometric identification of phosphorylation sites in bleached bovine rhodopsin. Biochemistry 32:5930–5934.

Pitcher JA, Inglese J, Higgins JB, Arriza JL, Casey PJ, Kim C, Benovic JL, Kwatra MM, Caron MG, Lefkowitz RJ (1992a): Role of βgamma subunits of G proteins in targeting the β-adrenergic receptor kinase to membrane-bound receptors. Science 257:1264–1267.

Pitcher JA, Lohse MJ, Codina J, Caron MG, Lefkowitz RJ (1992b): Desensitization of the isolated β₂-adrenergic receptor by β-adrenergic receptor kinase, cAMP-dependent protein kinase, and protein kinase C occurs via distinct molecular mechanisms. Biochemistry 31:3193–3197.

Premont RT, Macrae AD, Stoffel RH, Chung N, Pitcher J, Ambrose C, Inglese J, MacDonald ME, Lefkowitz RJ (1996): Characterization of the G protein–coupled receptor kinase GRK4. J Biol Chem 271:6403–6410.

Richardson RM, Hosey MM (1992): Agonist-induced phosphorylation and desensitization of human m2 muscarinic cholinergic receptors in Sf9 insect cells. J Biol Chem 267:22249–22255.

Roos M, Soskic V, Poznanovic S, Godovac-Zimmerman J (1998): Post-translational modifications of endothelin receptor B from bovine lungs analyzed by mass spectrometry. J Biol Chem 273:924–931.

Stoffel RH, Randall RR, Premont RT, Lefkowitz RJ, Inglese J (1994): Palmitoylation of G protein–coupled receptorkinase, GRK6. J Biol Chem 269:27791–27794.

Valiquette M, Parent S, Loisel TP, Bouvier M (1995): Mutation of tyrosine-141 inhibits insulin-promoted tyrosine phosphorylation and increased responsiveness of the human β₂-adrenergic receptor. EMBO J 14:5542–5549.

CHAPTER 3

RECONSTITUTION AND *IN VITRO* PHOSPHORYLATION OF G PROTEIN–COUPLED RECEPTORS

JUDITH A. PTASIENSKI and M. MARLENE HOSEY

Regulation of G Protein–Coupled Receptor Function and Expression,
Edited by Jeffrey L. Benovic.
ISBN 0-471-25277-8   Copyright © 2000 Wiley-Liss, Inc.

I. INTRODUCTION

One of the most fascinating aspects of G protein–coupled receptors (GPCRs) is the agonist-dependent regulation of the receptors by protein phosphorylation (Freedman and Lefkowitz, 1996). Many GPCRs are believed to be regulated via this process; however, very few GPCRs actually have been analyzed directly as substrates for protein kinases that are believed to regulate their activity. The biggest obstacle to studying the regulation of GPCRs has been that their low density in native cells precludes detailed biochemical studies of the phosphorylation of the receptors. However, the advent of heterologous expression systems has allowed for generation of higher levels of protein and the direct characterization of many more receptors as substrates for regulation by phosphorylation (Freedman and Lefkowitz, 1996).

It is informative to perform two types of analyses in parallel in characterizing the phosphorylation of a protein. *In vivo* studies with ^{32}P-labeled cells (as discussed in Chapter 2) allow for characterization of cellular events associated with changes in receptor phosphorylation. In parallel, *in vitro* studies with purified protein kinases are very useful to dissect the molecular events that occur *in vivo*. This chapter details methods that we have used extensively to characterize GPCRs as substrates *in vitro* for various types of protein kinases.

Historically, such phosphorylation studies have required that the receptors first be purified and reconstituted into detergent-free lipid vesicles so that effective phosphorylation could be achieved. Why is this so? Why not use membrane fractions as the source of the receptors in protein phosphorylation assays? As introduced above, the most important limiting factor in studies of phosphorylation of membrane-bound GPCRs is the availability of sufficient quantities of protein. Only the visual GPCR rhodopsin is present in sufficient quantity to allow studies of phosphorylation in native membranes. For all other GPCRs, the extensive phosphorylation of other membrane proteins by protein kinases endogenous to the membranes creates a "high background" and consequently a low "signal-to-noise" ratio, making it extremely difficult, if not impossible, to visualize the phosphorylation of the GPCRs.

Thus, the option that has been successfully employed for multiple types of GPCRs has been to isolate the proteins by purification strategies and then to insert the purified proteins back into a phospholipid vesicle for subsequent analysis in phosphorylation studies. This approach has been extremely useful in that it has provided much insight into the molecular events associated with the regulation of various GPCRs, and, importantly, was critical for the discovery and characterization of the G protein–coupled receptor kinases (GRKs; see Chapter 7). The methods that we have found to most consistently yield reproducible results in the studies of the regulation of purified and reconstituted GPCRs are detailed below.

With another approach and with the advent of heterologous expression systems, several successful attempts have been made to study phosphorylation of the GPCRs directly in membrane fractions (DebBurman et al., 1995a; Pei et al., 1994). However, in both cases, it was still necessary to enrich the membranes in the GPCRs, by either stripping peripheral proteins (DebBurman et al.,

1995a) or preparing purified membrane fractions with sucrose density gradients (Pei et al., 1994). The success of both methods requires rather high expression of the GPCRs. As the strategies may be of use for other GPCRs, a brief description of the former method is provided in this chapter. For all the work described in this chapter, we provide information that we have gained working with subtypes of muscarinic cholinergic receptors (mAChR), particularly the M_2 subtype. Although every GPCR does not behave exactly the same, we anticipate that the methods, in general, will be applicable for other GPCRs. We detail the exact procedures used for the mAChR and indicate where deviations may or should be made for other GPCRs.

2. REAGENTS AND EQUIPMENT

Studies of reconstitution of receptors are simple enough provided that the proper reagents are used, and that "fatal" impurities do not result in receptor inactivation. We will list the sources of reagents we use when they are first mentioned in this chapter.

Of particular importance is to utilize detergents of excellent quality. Many GPCRs have been solubilized with CHAPs, octylglucoside, or digitonin under conditions that retain ligand-binding activity of the GPCRs. We have had extensive experience with digitonin in studies of GPCRs. For our work on muscarinic cholinergic receptors, "good" digitonin is a requirement. Digitonin from many sources is a mixture of soluble and insoluble glycosides (see Cremo et al., 1981). Digitonin acquired from most sources is rather insoluble but can be brought into solution by boiling or other approaches (Cremo et al., 1981). However, digitonin solutions that were initially insoluble also have a high tendency to come out of solution over time. In contrast, highly purified digitonin, such as is available from Gallard-Schlessinger, is readily soluble and will stay in solution for prolonged periods of time. Nevertheless, it is advisable not to use digitonin solutions that are over 2 weeks old, as we have observed loss of ligand-binding activity with "aged" solutions of digitonin. In our experiences, it is useful to test one or more available lots from Gallard-Schlessinger and then purchase a useful lot in large quantities in order to have a stable source of detergent.

Cholic acid is also a product that contains impurities that can be harmful. In our studies, we use cholic acid that has been boiled in ethanol, charcoal treated, and recrystallized. We make a working stock solution of 20% cholate, pH 8, from the purified cholic acid. This cholate is used throughout the entire purification and reconstitution protocol. Other approaches to purify cholic acid have been described (see Brandt et al., 1983).

The reconstitution studies detailed here do not require any special equipment except a high-power sonicator, although others have used bath-type sonicators with success (Haga et al., 1986). We use a Heat Systems Ultrasonics W-385 sonicator that has a maximum power output of 475 W. The system we use is fitted with a cup horn sonicator so that the probe does not directly access the solution being sonicated.

3. GETTING READY: PURIFICATION OF THE RECEPTORS

3.A. Membrane Isolation and Receptor Solubilization

To obtain reconstituted receptors suitable for studies of protein phosphorylation, it is usually necessary to purify the GPCRs so that the ^{32}P incorporated
into the receptor protein can be readily detected with autoradiography or phosphorimaging. For these studies, it is not always essential to purify the GPCRs
to homogeneity. However, we have found that the purer the preparation, the
more stable the receptor will be over time. For each GPCR, a unique strategy
can be worked out if affinity chromatography with a specific receptor ligand is
a viable approach (e.g., Parker et al., 1991). Otherwise, tagging receptors with
a His^6 or an epitope tag might provide an alternative approach.

One paper has outlined an approach to purify mAChR via a His^6 tag; however, the results required the design of a mutant receptor lacking glycosylation
sites, as well as most of the third intracellular loop, and other fairly radical
changes (Hayashi and Haga, 1996). What we outline below is a very brief summary of the steps involved in the purification of M_2 mAChR according to a procedure that was first described by Haga and colleagues (see Haga and Haga,
1985) and subsequently used in a number of other laboratories (Parker et al.,
1991; Richardson et al., 1993). Although some of the discussion below is specific for muscarinic receptors, the general approach, and the tips provided with
regard to receptor preparation *following* elution from the affinity column, can
be broadly applied.

Briefly, Sf9 insect cells (Invitrogen) are grown in spinner flasks in serum-
free Sf-900II media (Gibco/BRL). A 1-liter flask containing 500 ml of media
and 1×10^7 cells is infected with recombinant virus (Parker et al., 1991;
Richardson et al., 1993) directing expression of the receptors, and the cells are
harvested 3 days later. All subsequent steps are performed at 4°C unless otherwise noted.

Cells are centrifuged for 20 minutes at 5,000 rpm in a Beckman JA-10 rotor.
The medium is discarded, and the cells are resuspended in 70 ml of homogenization buffer plus protease inhibitors (buffer A, formulas below) and homogenized with a Tri-R Homogenizer at setting 7.5 for six to seven strokes on ice
in a glass dounce homogenizer fitted with a Teflon pestle. The homogenate is
centrifuged at 40,000 rpm (125,000g) for 30 minutes in a Beckman Ti45 rotor.
The supernatant is discarded, and the pellet is resuspended in 35 ml of buffer
A with one to two strokes of the Tri-R homogenizer. The material is centrifuged
a second time at 40,000 rpm for 30 minutes. The supernatant is discarded and
the pellet is finally resuspended with one to two strokes with the Tri-R homogenizer in 15 ml of ice-cold buffer A. The protein concentration of the resulting
mixture is usually 6–10 mg protein/ml.

To achieve solubilization of the receptors, we use a 4:1 mixture of digi-
tonin/cholate as described by Schimerlik and colleagues (see Cremo et al.,
1981). Typically, an equal volume of 2× solubilization buffer (buffer B) is
added to the resuspended membrane fraction. The mixture is placed on a rotator (Glas-Col) at 4°C for 30 minutes and then centrifuged at 40,000 rpm for 1
hour in the Ti45 rotor. The supernatant contains the solubilized receptors.

3.B. Purification of the Receptors by Affinity and Hydroxylapatite Chromatography

The approach used is to couple an affinity chromatography step directly to a hydroxylapatite (HTP, Bio-Rad HTP Bio-Gel) column, which significantly improves the purification (Haga and Haga, 1985). The solubilized receptors described in Section 3.A are added batchwise to 7 ml of affinity resin; for the M_2 mAChR, this is 3-(2'-amino-benz-hydryloxy)-tropane (ABT) coupled to epoxy-activated Sepharose-6B (Pharmacia) (Haga and Haga, 1983) that had been pre-equilibrated with 10–20 ml of a 1:1 solution of buffers A and B. The mixture of receptors and affinity resin is placed on a rotator overnight at 4°C. The next day the mixture is allowed to settle, and the supernatant ("breakthrough") is removed.

Wash buffer (buffer C) is used to transfer the gel to a disposable chromatography column (Bio-Rad, Econopac columns). The column is then washed (using gravity flow) with 35 ml of wash buffer containing salt (buffer D). Meanwhile, a slurry of HTP is rehydrated according to the manufacturer's specifications except that the decantation is repeated until there are virtually no "fines" and the slurry settles very rapidly (in 1 minute or less); this step can be performed several days ahead, but it is critical for the success of this column to make sure that all "fines" are removed; otherwise the columns will clog during use.

The HTP (1 ml) is then transferred to a column and pre-equilibrated with at least 5 ml of buffer D (allow the washing to occur by gravity; the column should flow at a rate of ~0.25–0.5 ml/min). Once the HTP column is prepared and the affinity column washed, the outlet of the affinity column is then connected to the HTP column. The receptor is eluted directly from the ABT-Sepharose column onto the HTP column with 40 ml of buffer D containing 2 mM atropine. High concentrations of antagonist are routinely necessary to elute GPCRs from the affinity matrices.

When the elution is complete, the columns are disconnected and the HTP column is washed with 5 ml of ice-cold buffer D and brought to room temperature. (The purpose of performing the HTP chromatography at room temperature is to prevent the high concentrations of phosphate buffer used in the elution from crystallizing). The HTP column is eluted sequentially with a step gradient of $NaKPO_4$, pH 7.4, with 3 ml of 0.1 M, 4 ml of 0.25 M, and finally 2 ml of 0.4 M of the phosphate buffer at room temperature. The 0.4-M fraction is eluted directly into a tube containing 2 ml of buffer C to prevent crystallization of the $NaKPO_4$. The eluates are placed on ice and assayed in ligand-binding studies; for the M_2 mAChR, this is achieved with the antagonist ^3H-quinuclidinyl benzilate (QNB, New England Nuclear). Normally all of the receptor elutes in the 0.25 M $NaKPO_4$ fraction.

3.C. Concentrating the Purified Receptors Before Reconstitution

It is critical to concentrate the receptors before the reconstitution steps described below because the reconstitution step works poorly with dilute samples. Additionally, it should be noted that not only must the volume be small, but the specific activity (concentration) of receptor must be high, as dilute receptor solutions are unstable during the reconstitution and the yield will be poor. We usually find that the amount of receptor reconstituted is usually one half of the amount assayed in the concentrated receptor.

In the concentration steps, it is *critical* not to concentrate detergent (in our case, digitonin) to levels that are toxic to the receptors. Typically, we use devices such as the Centricon centrifugation concentrators (Amicon) to concentrate the receptors. Low concentrations of detergents such as digitonin are not retained by the Centricon concentrators; however, above a critical concentration, the detergent is retained by the concentrators, and a gel-like solution forms that has unwanted consequences on receptor activity. Thus, several additional steps are used to dilute the detergent and reconcentrate the receptor until the digitonin concentration is reduced to acceptable levels.

First, the fraction(s) from the HTP column that contain the receptor are divided into two Amicon concentrators (30,000 molecular weight cutoff) and concentrated by centrifugation until the volume is approximately 0.5 ml in each. After the first concentration step, 1 ml of 0.05% digitonin in 10 mM Tris-HCl, pH 7.4, is added to each concentrator, and the receptor fractions are concentrated to ~0.1 ml. Finally, the contents of both concentrators are combined into one concentrator, and an additional 1 ml of the 0.05% digitonin/Tris buffer is added and the receptor solution is finally concentrated down to a volume of 0.05–0.1 ml. This approach provides concentrated receptors in a low concentration of detergent.

3.D. Buffers for Receptor Purification

1. Homogenization buffer (buffer A): 20 mM Tris-HCl, pH 7.4; 1 mM EDTA, pH 7.4; 2 mM $MgCl_2$; protease inhibitors (PIs; see below)
2. 2× Solubilization buffer (buffer B): 20 mM KPO_4, pH 7; 0.1 mM EDTA, pH 7.4; 0.8% digitonin/0.16% cholate; PIs
3. Wash buffer (buffer C): 20 mM KPO_4, pH 7; 0.1 mM EDTA, pH 7.4; 0.1% digitonin/0.02% cholate; PIs
4. Wash/salt buffer (buffer D): 20 mM KPO_4, pH 7; 0.1 mM EDTA, pH 7.4; 0.2 M NaCl; 0.1% digitonin/0.02% cholate; PIs
5. HTP elution buffers: 0.1, 0.25, or 0.45 M $NaKPO_4$, pH 7; 0.1 mM EDTA, pH 7.4; 0.1% digitonin/0.02% cholate; PIs
6. Protease inhibitors (PIs): phenylmethylsulfonyl fluoride, 17.4 μg/ml (Sigma); aprotinin, 1 μg/ml (ICN Biomedicals); leupeptin, 10 μg/ml (BACHEM); pepstatin A, 1.4 μg/ml (BACHEM); soybean trypsin inhibitor, 10 μg/ml (type 1-S, Sigma); iodoacetamide, 184 μg/ml (Sigma).

3.E. Detergent Assay for Solubilized Receptors

When isolating receptors by affinity chromatography, it is likely that ligand associated with the receptors that are eluted from the column will carry over into the ligand-binding assay. Thus, it is easy to underestimate the amount of receptor present in fractions containing the purified receptors. To ensure against this, it is prudent to use several different amounts of purified receptors in the binding assays and/or to vary the volume of the assay.

Typically we assay 1–5 μl in a 1-ml assay. The exact conditions for assaying different GPCRs will vary with each receptor. Those used for mAChR include 20 mM KPO$_4$, pH 7, radioligand of appropriate concentration (e.g., 2 nM ^{3}H-QNB) ± excess cold ligand (e.g., 10–100 μM atropine), and 1–5 μl receptor. Depending on the receptor and ligand, the assays are incubated for sufficient time to achieve equilibrium. For the M$_2$ mAChR, assays are incubated in a 37°C shaking water bath for 1 hour.

At the end of the incubation, 300 μl of 30% polyethylene glycol (PEG, 8K, Sigma) and 100 μl of bovine serum albumin (BSA, fraction V, Sigma; 1 mg/ml) are added to each assay tube. The reactions are mixed by vortexing and placed at 4°C for 10 minutes. Afterwards, the reactions are filtered over Whatman GF/F filters that were presoaked in 0.3% polyethylenimine (PEI) (Aldrich) solution for 1 hour. The tubes and filters are quickly rinsed with 4 × 4 ml ice cold 5 mM KPO$_4$ buffer. The filters are dried, placed in scintillation cocktail, and counted after the filters are thoroughly saturated with fluor.

4. RECONSTITUTION

Several different protocols are possible. We only describe one method in detail here and refer the reader to other sources for approaches that have been used with success for a number of different GPCRs and other proteins (Brandt et al., 1983; Cerione et al., 1985, 1986; Florio and Sternweis, 1985; Haga et al., 1986; Maloney and Ambudkar, 1989). The one used successfully in our laboratory has been to mix the purified receptors with lipids and pass the resulting mixture over an Extracti-Gel D (Pierce) column to remove the detergent and allow for vesicle formation. Depending on the intended use of the reconstituted receptors, the choice of lipid is important as the lipid can have important consequences in terms of receptor activity and in protein phosphorylation reactions where the protein kinases are sensitive to lipids.

For example, it is well known that isoforms of protein kinase C are lipid dependent, while it is less well recognized that the GRKs, especially GRK2 and GRK3, are lipid sensitive (DebBurman et al., 1995b). For a neutral background (i.e., one that does not appear to have much effect on PKCs or GRKs), it is desirable to use highly purified phosphatidylcholine (PC) (from Avanti) as the lipid source in reconstitution assays (see DebBurman et al., 1995b). However, if one is not interested in studying the lipid dependence of phosphorylation per se, a convenient approach is to use vesicles formed from phosphatidylserine (PS). Receptors (at least the M$_2$ mAChR) in PS vesicles are stable, and the PS will act as the necessary lipid activator in studies with GRK2 and GRK3 (DebBurman et al., 1995b) or PKC. Vesicles can also be prepared from crude PC (such as Sigma product P-9513), which also will possess kinase-stimulating activity. In addition, phospholipid vesicles formed from lipids extracted from chick heart provide stable receptor preparations that work well in phosphorylation studies (see Richardson et al., 1992, 1993).

4.A. Lipid Preparation for L-α-Phosphatidyl-L-Serine Vesicles

Use purified L-α-phosphatidyl-L-serine (PS) from Avanti or Sigma. Take 400 μl of lipids (10 mg/ml in chloroform/methanol) and dry thoroughly under a stream of nitrogen gas. Add 200 μl of the deoxycholate/cholate in HEN buffer (see below) and sonicate in a covered glass vial placed in an ice bath in a Heat Systems Ultrasonics W-385 sonicator using 20% maximum power output and continuous operation for 20 minutes. Replace ice as needed during the sonication procedure. The lipids will appear milky white, and there will be no visible "chunks" when prepared correctly. Sonicate for a longer time if this is not the case. Watch closely during the entire procedure, and adjust power output as needed so that it never goes over 20% maximum. It is imperative to prepare fresh lipids on the day of use.

4.B. Preparation of Purified Chick Heart Lipids

The method we have used to prepare crude chick heart lipid preparations is adapted from a procedure of Haga and colleagues (1986), who developed this method for purifying lipids from porcine cerebrum based on the method of Folch et al. (1957). Our experience suggests that this procedure works well if fresh tissue is used, while similar approaches using frozen tissue have been notably less successful.

Two grams of fresh (not frozen) chick hearts are washed with ddH$_2$O to get rid of excess blood, minced with scissors, and homogenized with a Polytron homogenizer (Brinkman) fitted with a PT20 generator in 38 ml of chloroform:methanol (2:1). The homogenate is filtered through Whatman No. 1 paper in a glass funnel, and the volume of the filtrate is measured. Add water equal to one fourth the measured volume (e.g., to a measured volume of 32 ml, add 8 ml of water). This mixture is shaken vigorously in a separatory flask. Note that proportions are critical throughout.

The separatory flask is placed at 4°C overnight to allow the phases to separate. The bottom yellow, organic layer is collected, and the top milky layer is discarded. The organic extract is placed in a clean separatory flask, and to this 8 ml of 0.1 M KCl saturated with chloroform:methanol (2:1) is added. The mixture is shaken, and the phases are allowed to separate for approximately 30 minutes at 4°C. The bottom yellow organic phase is collected and filtered through No. 1 Whatman filter paper. Add 0.5 gm of sodium sulfate (anhydrous) to the mixture and filter through glass wool. The lipids are stored layered with nitrogen gas in glass tubes with polypropylene caps at –20°C.

4.C. Reconstitution of Purified Receptors

The method described below has been repeatedly performed with a high degree of success over a period of ~10 years in our laboratory. It was originally developed using a combination of methods first described by others (Cerione et al., 1985; Haga et al., 1986).

Prepare a column of 1 ml of Extracti-Gel D and wash extensively overnight or for several hours with 0.2% bovine serum albumin (BSA) in HEN buffer (see below) and then pre-equilibrate with 5 ml of HEN/40 buffer. Just before use, allow the Extracti-Gel column to come to room temperature, and perform the loading of receptor and elutions at room temperature. After the appropriate lipid is ready (as discussed above), mix 50–100 μl of purified, concentrated receptor (optimally in the range of 100–400 pmol) with 20 μl of 100 mM agonist (e.g., oxotremorine) prepared in HEN buffer. (It is important to use very high concentrations of agonist as this appears to protect the receptor from denaturation during the removal of detergent in the reconstitution procedure.) Let stand at room temperature for 20 minutes. Alternatively, the mixture may be kept at 4°C overnight.

Next, add 80 μl of the sonicated lipids and 50 μl of HEN buffer. Mix by vortexing thoroughly and place on ice for 5–10 minutes. Load onto the Extracti-Gel column. Elute with HEN/40 buffer and collect the following fractions: (1) flow through, (2) 150 μl, (3) 200 μl, (4) 250 μl, and (5) 200 μl. Assay the fractions for receptors (procedure is given below); the receptors are usually in fractions 3 and 4. It is necessary to pre-equilibrate and elute the Extracti-Gel column with a reduced salt buffer (HEN/40) because kinases that may later be used to phosphorylate the receptors, such as GRK2 and GRK3, are extremely sensitive to the ionic strength of the buffering solution (Benovic, 1991).

4.D. Buffers used for Reconstitution

1. HEN buffer: 20 mM Hepes, pH 8.0; 1 mM EDTA, pH 8; 160 mM NaCl
2. HEN/40 buffer: 20 mM Hepes, pH 8.0; 1 mM EDTA, pH 8; 40 mM NaCl
3. Deoxycholate/cholate in HEN buffer: 0.18% deoxycholate (deoxycholic acid–sodium salt monohydrate); 0.04% cholate (cholic acid that has been charcoal purified and recrystallized; see above)

4.E. Assay for Reconstituted Receptors

Because there is generally a carry over of agonist (e.g., oxotremorine) into the reconstituted receptors, it is important to obtain a faithful readout in the ligand-binding assays used to determine the concentration of the reconstituted receptors. It is essential to use a very small amount of receptor in a large assay volume (e.g., 1–2 μl of receptor in a 2-ml assay is usually acceptable). It may be necessary to use even less receptor (i.e., diluting receptors and using 0.5 μl or so). Conditions should be worked out individually for each type of GPCR by trying various concentrations of reconstituted receptors such that ligand-binding activity increases linearly as a function of receptor concentration.

The assay contents will vary for different GPCRs. The conditions used to assay mAChR are 20 mM KPO_4, pH 7; 10 mM $MgCl_2$; 50 mM NaCl; radioligand (e.g., 2 nM ^{3}H-QNB); ± excess unlabeled ligand (e.g., 10–100 μM atropine); and reconstituted receptor. Mix all ingredients, adding receptor last. Incubate in a 37°C shaking water bath for 30–45 minutes (longer assay times sometimes result in loss of binding activity). At the end of this time, add 600 μl of 30% PEG and 200 μl of BSA (1 mg/ml) to each 2-ml assay tube. Vortex well and place at 4°C for 10 minutes. Filter over GF/F filters that have been presoaked for 1 hour in 0.3% PEI solution. Rinse tubes and filters with 4×4 ml of ice-cold 5 mM KPO_4 buffer. Dry filters, add scintillation cocktail, and wait until filters are thoroughly saturated to count. In initial studies one should characterize the affinity of the reconstituted receptors for antagonist ligands to ensure that ligand binding is occurring as expected and that severe denaturation has not occurred.

5. PHOSPHORYLATION OF PURIFIED RECONSTITUTED RECEPTORS

5.A. Phosphorylation of the Reconstituted Receptors with GRKs

5.A.a. General Considerations. Agonist-dependent phosphorylation of GPCRs by GRKs can readily be studied with reconstituted GPCRs provided that the reconstitution process has successfully removed all traces of detergent and that the GPCRs remain in an active state capable of binding agonists. In addition, the ionic strength of the reaction needs to be low to avoid inhibition of GRKs (see Chapter 7). A consideration with *in vitro* phosphorylation studies in general is that one hopes to be able to faithfully reproduce phosphorylation as it occurs in the intact cell (i.e., to observe phosphorylation at the same sites that are phosphorylated *in vivo* and at the same stoichiometries). Thus, it is important that the protein substrate not be denatured so that phosphorylation occurs *in vitro* on sites that are normally exposed in the native protein and not on sites that might have become exposed as a result of partial denaturation.

In studies with the GRKs that appear to have the widest substrate specificities (GRK2 and GRK3), a built-in feature of the reactions requires that the substrates not be denatured. Very little phosphorylation of the GPCRs is observed in the absence of agonist; stoichiometric phosphorylation of the receptors is only observed upon agonist binding to the receptors (e.g., see Figs. 1 and 2 of Richardson et al., 1993). If significant phosphorylation of reconstituted GPCRs by GRKs occurs in the absence of agonist, it is likely due to carry over of agonist from the reconstitution reactions that contain very high levels of agonist. A check for this is to add a reaction containing antagonist; if antagonists reduce phosphorylation of the GPCRs in the absence of added agonist, this is a fairly good indication that there is agonist associated with the reconstituted receptors (again, see Fig. 1 of Richardson et al., 1993). (Alternative explanations might come into play with inverse agonists, but effects of these agents have not been extensively explored.)

Another consideration for *in vitro* phosphorylation reactions is to attempt to use concentrations of reagents that approximate *in vivo* concentrations. This is slightly more difficult to achieve because the GPCRs are present in cells at low concentrations, and it is necessary to use sufficient amounts of GPCRs in the *in vitro* reactions so that the level of phosphorylation can be accurately measured. In addition, the GRKs are unstable when diluted to low concentrations. For optimal results under the conditions tested, it is necessary to maintain concentrations of GRKs above 10 nM in the *in vitro* phosphorylation reactions to prevent loss of enzyme activity. On the other hand, to achieve a high specific radioactivity, the concentration of ATP that is used *in vitro* is considerably lower than the intracellular concentration of ATP (50 μM vs ~3 mM, respectively).

In our experience, the stoichiometries of phosphorylation achieved *in vitro* with the M_2 mAChR and GRK2 or GRK3 when the *in vitro* reactions are carried out in the absence of Gβγ are in the same range as those achieved *in vivo* in response to agonist stimulation (Pals-Rylaarsdam et al., 1995, 1997). When *in vitro* reactions are carried out in the presence of Gβγ, the stoichiometries approach twice that observed in intact cells (Richardson et al., 1993). For the M_2 mAChR, agonist stimulation of cells results in phosphorylation at multiple sites and a stoichiometry of 3–5 mol phosphate incorporated/mol receptor (Kwatra et al., 1987; Richardson and Hosey, 1992; Pals-Rylaarsdam et al., 1995). In *in vitro* studies with GRKs, we observe stoichiometries of 4–5 mol P/mol receptor with wild-type M_2 mAChR, while a mutant M_2 mAChR that does not undergo phosphorylation in intact cells still can be phosphorylated at one site by GRKs *in vitro* (Pals-Rylaarsdam et al., 1997). Overall, the results we have obtained suggest that the *in vitro* phosphorylation conditions result in a phosphorylation profile that is quite similar to that observed *in vivo,* although under certain conditions additional sites may be phosphorylated.

5.A.b. Suggested Protocol for a Typical Assay for Phosphorylation With GRK2 or GRK3.
First, make sure that the concentration of monovalent salts is not greater than 40 mM, and avoid excess detergents. Prepare [γ-^{32}P]ATP by mixing the radiolabeled ATP (Amersham) with nonradioactive ATP such that the final concentration of ATP in the assay will be 50 μM and the specific activity will be ~1,000 cpm/pmol (to determine the specific activity, count an aliquot containing less than 1×10^6 cpm and divide cpm by fmol or pmol added). Prepare 10× phosphorylation buffer: 200 mM Tris, pH 7.4; 50 mM $MgSO_4$; 20 mM EDTA. Prepare a phosphorylation reaction (final volume, 0.1 ml) in a 1.5-ml plastic microfuge tube containing reconstituted receptor, 0.3–1 pmol/100 μl reaction; ± agonist, and/or ± antagonist; GRK (GRK2 or GRK3 is typically used at a final concentration of 30 nM), ± G-protein βγ subunits; 10× phosphorylation buffer (10 μl in a 100-μl reaction); 10 μl of 0.5 mM (10×) [γ-^{32}P]ATP; and ddH$_2$O to adjust the final volume to 100 μl. (Reactions can also be performed in larger or smaller quantities by adjusting the reagents appropriately.)

All the ingredients except the Gβγ subunits, GRK, and ATP are placed into the assay tube. Then the Gβγ subunits are added; the tube is vortexed and placed at room temperature for 5–10 minutes. (This preincubation with Gβγ appears to be necessary to observe consistent effects with this reagent and may reflect a need for the Gβγ to associate with the phospholipid vesicle; the most plausible explanation of the stimulatory effects of Gβγ is that it helps to orient the GRKs to the phospholipids that are required for activity.) Next, the GRK is added, and finally the ATP is added to start the reaction. The assay is incubated at 37°C for the required length of time. The reactions are stopped by the addition of 5× SDS gel electrophoresis sample buffer.

Alternatively, if it is desirable to not have excess ^{32}P in the samples, the following procedure is used: To a 100-μl reaction, add 2.5 μl of 2% digitonin to lyse the vesicles. Place the reaction at 4°C for 20 minutes and then transfer it to a Centricon 30 and centrifuge at 5,000g for 4 minutes. Next, add 200 μl of HEN40/0.05% digitonin, and centrifuge the reaction again down to the desired final volume. Measure the volume and add an appropriate amount of 5× SDS gel electrophoresis sample buffer. The samples are electrophoresed on SDS gels and processed for autoradiography or phosphorimaging, the latter being the more desirable for its greater speed, sensitivity, and quantitative analysis.

Stoichiometries of phosphorylation are calculated after calculating the amount of radioactivity present in the receptor band on the gel and relating it to the amount of receptor that was loaded onto the gel. If reactions do not work according to expectations, it is useful to determine if the reconstituted receptors contain unsuspected inhibitory agents; this can be performed by testing how the reconstituted receptors affect the phosphorylation of a positive control, such as rhodopsin in urea stripped rod outer segments.

5.A.c. Reagents for GRK Reactions. GRK2 and GRK3 enzymes are purified as described in Chapter 7 and diluted in TE buffer (20 mM Tris-HCl, pH 7.4; 2 mM EDTA) or "GRK buffer" (20 mM Hepes, pH 7.4; 5 mM EDTA; 200 mM NaCl; 0.02% Triton X-100, 15% glycerol) if necessary, just before addition to the assay. It is critical that the dilutions not be done until the last minute.

Gβγ subunits can be prepared or obtained from other investigators. For studies with Gβγ from transducin, it is usually necessary to dilute the Gβγ in TE buffer before adding to the assay.

5.B. Phosphorylation With PKC

Generally, the same principles discussed above apply. Reactions are similar to those with GRKs except that $CaCl_2$ (0.5 mM final) and PS (0.1 mg/ml final) are added to the assays (Richardson et al., 1992). In general, the phospholipids present with the reconstituted receptors are sufficient to activate PKC. PKC can be purified from brain or purchased.

6. PHOSPHORYLATION OF MEMBRANE-BOUND GPCRs

A universal method for performing such studies has yet to be developed. For studies of expressed subtypes of mAChRs, we found the following approach to be useful (DebBurman et al., 1995a). The receptors were expressed in Sf9 cells to levels of 15–35 pmol/mg protein using recombinant baculoviruses. Receptors expressed at lower levels were difficult to analyze. Membranes were prepared as described above (see Section 3.A.). Sf9 cell membranes (2–5 mg protein/ml) containing the expressed mAChRs were treated with varying concentrations of urea (1–5 M) according to procedures modified from those used for stripping rod outer segments of various peripheral proteins, including rhodopsin kinase, with urea (Shichi and Somers, 1978). The urea-treated membranes were sonicated on ice for 10–15 minutes, followed by centrifugation at 100,000g for 30 minutes. The resulting pellets were rinsed repeatedly to remove excess urea and finally resuspended in 20 mM Hepes, pH 7.4, 2 mM EDTA, 1 mM $MgCl_2$ plus PIs (see above) to the original volume. Recoveries of receptors and total membrane protein were determined by ligand-binding and protein assays.

Phosphorylation of membrane-bound mAChRs by GRKs was carried out in reaction mixtures (100 µl) very similar to those used for the reconstituted receptors. The reactions contained 0.3–1.0 pmol of membrane-bound GPCRs in 20 mM Tris-HCl, pH 7.4, 2 mM EDTA, 5 mM $MgCl_2$, ± agonist or antagonist, and different GRKs (30–50 nM). The reactions were started by adding 0.05 mM [γ-^{32}P]ATP (500–1,000 cpm/pmol) and were incubated at 37°C for 15 minutes or for the times indicated. The reactions were stopped by adding 25 µl 5× SDS sample buffer and electrophoresed on SDS gels. Phosphoproteins were visualized and quantified by phosphorimaging with a BAS 2000 Fuji Bioimage analyzer. In some studies, G$\beta\gamma$ subunits were used in concentrations from 5 to 200 nM and were added directly to the reactions. The results obtained can be seen in DebBurman et al. (1995a).

REFERENCES

Benovic JL (1991): Purification and characterization of beta-adrenergic receptor kinase. Methods Enzymol 200:351–362.

Brandt DR, Asano T, Pedersen SE, Ross EM (1983): Reconstitution of catecholamine-stimulated guanosinetriphosphatase activity. Biochemistry 22:4357–4362.

Cerione RA, Staniszewski C, Benovic JL, Lefkowitz RJ, Caron MG, Gierschik P, Somers R, Spiegel AM, Codina J, Birnbaumer L (1985): Specificity of the functional interactions of the beta-adrenergic receptor and rhodopsin with guanine nucleotide regulatory proteins reconstituted in phospholipid vesicles. J Biol Chem 260:1493–1500.

Cerione RA, Staniszewski C, Gierschik P, Codina J, Somers RL, Birnbaumer L, Spiegel AM, Caron MG, Lefkowitz RJ (1986): Mechanism of guanine nucleotide regulatory protein–mediated inhibition of adenylate cyclase. Studies with isolated subunits of transducin in a reconstituted system. J Biol Chem 261:9514–9520.

Cremo CR, Herron GS, Schimerlik MI (1981): Solubilization of the atrial muscarinic acetylcholine receptor: A new detergent system and rapid assays. Anal Biochem 115:331–338.

DebBurman SK, Kunapuli P, Benovic JL, Hosey MM (1995a): Agonist-dependent phosphorylation of human muscarinic receptors in *Spodoptera frugiperda* insect cell membranes by G protein–coupled receptor kinases. Mol Pharmacol 47:224–233.

DebBurman SK, Ptasienski J, Boetticher E, Lomasney JW, Benovic JL, Hosey MM (1995b): Lipid-mediated regulation of G protein–coupled receptor kinases 2 and 3. J Biol Chem 270:5742–5747.

Florio VA, Sternweis PC (1985): Reconstitution of resolved muscarinic cholinergic receptors with purified GTP-binding proteins. J Biol Chem 260:3477–3483.

Folch J, Lees M, Stanley HS (1957): A simple method for the isolation and purification of total lipids from animal tissues. J Biol Chem 226:497–509.

Freedman NJ, Lefkowitz RJ (1996): Desensitization of G protein–coupled receptors. Recent Prog Horm Res 51:319–351.

Haga K, Haga T (1983): Affinity chromatography of the muscarinic acetylcholine receptor. J Biol Chem 258:13575–13579.

Haga K, Haga T (1985): Purification of the muscarinic acetylcholine receptor from porcine brain. J Biol Chem 260:7927–7935.

Haga K, Haga T, Ichiyama A (1986): Reconstitution of the muscarinic acetylcholine receptor. Guanine nucleotide–sensitive high affinity binding of agonists to purified muscarinic receptors reconstituted with GTP-binding proteins (Gi and Go). J Biol Chem 261:10133–10140.

Hayashi MK, Haga T (1996): Purification and functional reconstitution with GTP-binding regulatory proteins of hexahistidine-tagged muscarinic acetylcholine receptors (m2 subtype). J Biochem (Tokyo) 120:1232–1238.

Kwatra MM, Leung E, Maan AC, McMahon KK, Ptasienski J, Green RD, Hosey MM (1987): Correlation of agonist-induced phosphorylation of chick heart muscarinic receptors with receptor desensitization. J Biol Chem 262:16314–16321.

Maloney PC, Ambudkar SV (1989): Functional reconstitution of prokaryote and eukaryote membrane proteins. Arch Biochem Biophys 269:1–10.

Pals-Rylaarsdam R, Gurevich VV, Lee KB, Ptasienski JA, Benovic JL, Hosey MM (1997): Internalization of the m2 muscarinic acetylcholine receptor. Arrestin-independent and -dependent pathways. J Biol Chem 272:23682–23689.

Pals-Rylaarsdam R, Xu Y, Witt-Enderby P, Benovic JL, Hosey MM (1995): Desensitization and internalization of the m2 muscarinic acetylcholine receptor are directed by independent mechanisms. J Biol Chem 270:29004–29011.

Parker EM, Kameyama K, Higashijima T, Ross EM (1991): Reconstitutively active G protein–coupled receptors purified from baculovirus-infected insect cells. J Biol Chem 266:519–527.

Pei G, Tiberi M, Caron MG, Lefkowitz RJ (1994): An approach to the study of G-protein–coupled receptor kinases: An *in vitro*–purified membrane assay reveals differential receptor specificity and regulation by G beta gamma subunits. Proc Natl Acad Sci USA 91:3633–3636.

Richardson RM, Hosey MM (1992): Agonist-induced phosphorylation and desensitization of human m2 muscarinic cholinergic receptors in Sf9 insect cells. J Biol Chem 267:22249–22255.

Richardson RM, Kim C, Benovic JL, Hosey MM (1993): Phosphorylation and desensitization of human m2 muscarinic cholinergic receptors by two isoforms of the beta-adrenergic receptor kinase. J Biol Chem 268:13650–13656.

Richardson RM, Ptasienski J, Hosey MM (1992): Functional effects of protein kinase C–mediated phosphorylation of chick heart muscarinic cholinergic receptors. J Biol Chem 267:10127–10132.

Shichi H, Somers RL (1978): Light-dependent phosphorylation of rhodopsin: Purification and properties of rhodopsin kinase. J Biol Chem 253:7040–7046.

CHAPTER 4

IDENTIFICATION OF RESIDUES THAT ARE PHOSPHORYLATED WITHIN A RECEPTOR

KRZYSZTOF PALCZEWSKI, J. PRESTON VAN HOOSER, and HIROSHI OHGURO

Regulation of G Protein–Coupled Receptor Function and Expression,
Edited by Jeffrey L. Benovic.
ISBN 0-471-25277-8   Copyright © 2000 Wiley-Liss, Inc.

I. INTRODUCTION

Protein phosphorylation is one of the most versatile mechanisms involved in the regulation of enzymes, channels, and receptors. Understanding the mechanism by which phosphorylation regulates the activity of G protein–coupled receptors (GPCRs) requires the identification of specific residues within a given receptor that are modified by protein kinases. Although several classic chemical methods are available for the analysis of phosphorylation sites, recent advances in mass spectrometric (MS) techniques have the potential to revolutionize the field of protein chemistry. Although newer MS applications will emerge as a consequence of technical innovation, the progress will still depend on successes in practical aspects of phosphopeptide preparations. This chapter explores the critical aspects of phosphopeptide preparation in the studies of GPCR phosphorylation, with a major emphasis on the identification of phosphorylation sites within rhodopsin.

2. GENERAL CONCEPTS IN THE IDENTIFICATION OF PHOSPHORYLATION SITES

2.A. Limited Proteolysis

The most straightforward method for identifying phosphorylation sites is limited proteolysis. This technique relies on proteolysis of a ^{32}P-labeled protein in native conditions, occasionally in the presence of a ligand that protects the core of the protein against complete digestion. Only exposed regions of the protein accessible to the protease will be cleaved. In addition, the cleavage will be restricted to only a subset of these specific sites in which unfavorable amino acids (e.g., Pro or Gly) are not present. Each protease also has a kinetic preference for a particular peptide sequence. Frequently, the cleavage of the polypeptide chain will not disrupt the folding of the remaining parts of the protein, thus limiting the number of peptides generated. In the case of GPCRs, digestion may be carried out with a membrane-embedded receptor. The presence of membranes prevents the release of highly hydrophobic transmembrane fragments and simplifies the analysis of peptides released from the cytoplasmic surface of the receptor. For example, digestion of membrane-bound rhodopsin by Asp-N endoproteinase results in a single cleavage at the C terminus, releasing a soluble peptide that contains all of the phosphorylation sites catalyzed by rhodopsin kinase (Palczewski et al., 1991) (Fig. 4.1). Initial phosphoamino acid analysis (Boyle et al., 1991)

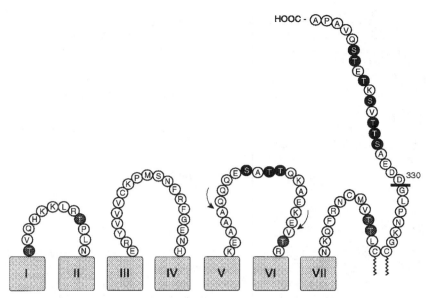

Figure 4.1. The cytoplasmic surface of bovine rhodopsin. Potential phosphorylation sites are shown in black. Threonine residues shown in gray are in close proximity to the membrane surface and are therefore not likely to be phosphorylated. D^{330} and a bar denote an endoproteinase Asp-N cleavage site. Arrows denote *Staphylococcus aureus* V8 endoproteinase and thermolysin cleavage sites in the V–VI loop.

may be useful in selecting proteolytic enzymes. Use of proteolytic enzymes with restricted specificity, such as Asp-N, Arg-C, and Asn-C endoproteinases,[1] is often the most informative approach. There are also many examples of the use of trypsin, which cleaves specifically at the C side of Arg and Lys residues.

SDS-PAGE and autoradiography are employed to prove that small radiolabeled fragments are generated in conditions of limited proteolysis while the core of the protein is no longer radiolabeled. If all of the phosphorylated peptides are soluble, radioactive peptides can be isolated from a mixture of relatively few peptides by one of the standard methods, such as reversed-phase high performance liquid chromatography (HPLC), thin-layer chromatographies, or two-dimensional electrophoresis/electroblotting onto PVDF (poly[vinylidene fluoride]) membranes. If the peptides are insoluble in aqueous solutions, initial gel filtration in concentrated formic acid may be necessary (e.g., Hargrave et al., 1980). Once peptides are isolated, they are identified by amino acid sequencing. If the analysis relies solely on sequencing, there is a possibility that the radioactivity could come from a peptide that is not accessible for Edman degradation, while a contaminating peptide is sequenced. Quantitative analysis of phosphate and peptides may be necessary to assess the purity of the peptide. This approach is further strengthened if it is used in connection with MS. Peptide fragments

[1]The nomenclature stands for the enzyme that cleaves the polypeptide chain at the N or C terminus of the indicated amino acid, although, occasionally, cleavage is observed at unpredictable peptide bonds.

with unusual properties (small, highly hydrophobic, or highly charged) pose additional challenges, and steps must be taken (such as the selection of another protease that will generate more manageable peptides) to ensure that these are not lost.

To prevent dephosphorylation, which can be a significant problem in some reconstitution systems, use of γ-^{35}S-ATP instead of γ-^{32}P-ATP may be advantageous. Most, if not all, kinases utilize this ATP analogue, yielding thiophosphorylated proteins that are resistant to dephosphorylation (Ohguro et al., 1994). Disadvantages of thiophosphorylation include reactivity with potent reducing and oxidizing agents and hydrolysis. Furthermore, thiophosphorylated protein should not be carboxymethylated due to high reactivity of thiophosphate with iodoacetic acid.

2.B. Peptide Mapping

Peptide mapping is a widely used method for the identification of phosphorylated peptides. The protein is typically radiolabeled and digested completely in the presence of detergents or chaotropic salts. Substitution of γ-^{32}P-ATP in the phosphorylation reaction by γ-^{33}P-ATP is advantageous because of the slower decay of ^{33}P. This substitution allows the radiolabeled sample to be used over a prolonged period. An additional advantage of this radionuclide is its lower radiation energy, allowing the diluted sample to be handled without protective shielding.

The generation of phosphorylated peptides is highly reproducible if certain preliminary steps are followed. The Cys residues of radiolabeled protein should be blocked to prevent S–S bond formation (e.g., by carboxymethylation or pyridylation). Recently, *in gel* digestion has been frequently used in cases where purification of a protein of interest is incomplete. In this technique, the mixture of proteins is separated by SDS-PAGE, the gel is washed to remove SDS, the proteins are silver stained, and the band of interest is cut out and mixed with the proteolytic enzyme. Generated peptides are typically extracted with aqueous solutions of CH_3CN and purified by reversed-phase HPLC. The detailed procedures are described by Matsudira (1993) and in a series entitled *Techniques in Protein Chemistry* published under the auspices of the Protein Society (Academic Press).

Critical steps in phosphopeptide analysis are the isolation of individual phosphopeptides and their sequencing. Two-dimensional paper electrophoresis/chromatography or cellulose thin-layer separation of peptides is used to rapidly generate the entire map for the investigated protein (van der Geer and Hunter, 1994). More modern methods of separation are based on reversed-phase HPLC (Hunter and Games, 1994) or capillary electrophoresis (Watts et al., 1994) or a combination of both (Hynek et al., 1997). The detection of peptides and their sequencing should be done with one of the mass spectrometric methods. Peptide mapping may be extremely difficult for certain mixtures of peptides, such as those generated from GPCRs, due to the extremely hydrophobic nature of many peptides.

Fe^{3+}-metal ion affinity chromatography is frequently used to selectively enrich phosphopeptides from a complex mixture of peptides or even to purify them. This type of chromatography is an example of immobilized metal affinity chromatography (IMAC) of which the best known is the purification of His tag proteins on an Ni^{2+}-immobilized column. Ferric ions are strongly chelated to iminodiacetic acid–substituted agarose or Sepharose (Chelex-Sepharose from Pharmacia). The immobilized Fe^{3+} complex acts as a selective immobilized metal affinity adsorbent for phosphopeptides, as well as phosphoproteins (Muszynska et al., 1986). Although the Fe^{3+}-Chelex chromatography is selective, some acidic peptides and those that are capable of chelating Fe^{3+} will also bind to the resin. Overall purification is highly efficient (~90% yield), making IMAC a dependable choice for phosphopeptide purification. Michel et al. (1988) carried out pioneering studies using this approach. Tryptic digests of spinach photosystem II cores, containing four phosphoproteins (8.3, 32, 34, and 44 kD), were loaded onto Fe^{3+}-chelating Sepharose in acidic conditions, and phosphopeptides were eluted at high pH. Peptides were further cleaned up (to remove salt and separate individual components) by reversed-phase HPLC. MS/MS was used to identify the phosphorylation sites.

As useful as IMAC is in the isolation of phosphopeptides, it also illustrates an important problem that arises when working with these peptides, namely, the adhesion of phosphopeptides to the steel components of the HPLC (columns, tubings, pumps, and so forth). Multiple phosphorylated peptides may be completely lost or their yield greatly reduced upon contact with metallic surfaces of the equipment. Appropriate selection of Teflon parts (or other organic solvent-resistant plastics) may alleviate this problem.

Recently, a novel procedure for micropurification of phosphopeptides was presented by Posewitz and Tempst, 1999. Employing IMAC-Ga^{3+}, phosphopeptides are retained in a near-quantitative and highly selective manner, to yield a concentrated sample for mass spectrometry. Ga (III) ions offer distinct advantages over the most often used Fe^{3+}.

2.C. Protein Sequencing

Proteins of interest are phosphorylated by a protein kinase in the presence of γ-^{32}P-ATP. Next, the radiolabeled protein is fragmented by enzymatic or chemical methods. ^{32}P-phosphopeptides are purified to homogeneity and sequenced by Edman degradation performed on solid phase or gas/liquid phase instruments. PTH (phenylthiohydantoin)-phospho-Ser and PTH-phospho-Thr cannot be measured directly because they are unstable during Edman degradation and poorly extracted into an organic phase (e.g., butyl chloride). Because phosphate (^{32}P) is also poorly extracted into an organic phase, a phosphorylation site may not be identified. This is particularly a problem with multiple phosphorylations or phosphorylation of a region enriched in Ser and Thr residues. After each cycle, PTH derivatives of amino acids are removed, and total amounts of ^{32}P-labeled phosphopeptides or hydrolyzed free phosphate associated with a sequencing membrane are measured by scintillation counting (Roach and Wang, 1991). Shannon and Fox (1995) describe solid-phase Edman degradation methods for phosphopeptide analysis that offer several advantages over gas-phase instruments.

To avoid the problem of ^{32}P hydrolysis from phosphopeptides, several protocols were developed to convert phospho-Ser to dehydroalanine by β-elimination of the phosphate group, without breaking peptide bonds. Next, a nucleophile, such as an amine or sulfite, is added to form an adduct with the double bond of dehydroalanine, and conventional sequencing is conducted. The identification of the site of phosphorylation is accomplished by specific elution of derivatized de-hydroalanine. β-Elimination and nucleophylic addition are poorly controlled reactions; thus, quantitative analysis of these results is not feasible. Under the conditions of β-elimination, phospho-Tyr is unaffected, while phospho-Thr undergoes β-elimination to α-aminodehydrobutyric acid that is resistant to nucleophilic addition (Meyer et al., 1991). Recently, Fischer et al. (1997) developed a new procedure that takes advantage of a manual Edman degradation protocol and allows the identification of ^{32}P-labeled peptide phosphorylation sites at the sub-picomole level by using both a volatile reagent, trifluoroethyl isothiocyanate, and volatile buffers for the extraction steps. This new protocol identifies the sites of phosphorylation in phospho-Ser– and phospho-Tyr–containing peptides. Further use of this method will permit more rigorous evaluation, and its applicability for various sequences containing phosphoamino acids must be tested.

2.D. Mutagenesis and Consensus Sequences

Protein kinases catalyze transfer of the terminal phosphate of ATP to a protein or peptide substrate. Three-dimensional structures of several protein kinases have been solved, giving precise insight into the mechanisms of the catalysis at atomic levels of resolution. Due to the recognition of specific peptide sequences, many protein kinases are characterized by "consensus sequences": the protein or peptide sequence that is phosphorylated by the kinase (Pearson and Kemp, 1991). For example, cAMP-dependent protein kinase phosphorylates sequences containing RRXS motifs, where X is any amino acid. Within the protein, the consensus sequence can only be phosphorylated if it is exposed. All of the most abundant kinases are well characterized in terms of the sequence motifs that are phosphorylated. Occasionally, some of the substrates fall outside well-defined consensus sequences; there are also kinases, including G protein–coupled receptor kinases (GRKs), that do not have clearly defined phosphorylation sequences (Palczewski, 1997). The analysis of phosphorylation sites within GPCRs will help identify kinases involved in receptor regulation and expand our understanding of the features that influence kinase receptor interactions.

Site-directed mutagenesis is a powerful method of probing protein structure and has been useful in identifying sites of phosphorylation. The analysis of potential sites of phosphorylation is limited to those residues that fall into a consensus sequence of a particular kinase, or all putative sites of phosphorylation are systematically eliminated. Typically, the residues that are putative sites of phosphorylation are changed to Ala, abolishing phosphorylation. For Ser/Thr phosphorylation, another approach would be to replace Ser with Thr or vice versa. Because the chemistry of phosphotransfer to Ser and Thr residues by protein kinases is almost identical, kinases use both residues for phosphorylation. Simple phosphoamino acid analysis would confirm the change from phosphorylation of Ser (Thr) to Thr (Ser), suggesting a phosphorylation site (Kennelly, 1994). Site transformation gives more evidence for the accuracy of identified sites of phosphorylation by mutagenesis. A pitfall of this method is the formal possibility that the introduced mutation will change the secondary structure of the protein substrate, thereby hindering phosphorylation and leading to misinterpretation of putative phosphorylation sites. Another cautionary note is that some kinases, including GRKs, have their primary binding site separated from the phosphorylation region (Palczewski, 1997) and have poorly defined consensus sequences. Under these circumstances, elimination of one site could lead to phosphorylation of an alternative position (Zhang et al., 1997; reviewed by Palczewski, 1997). Thus, for a given kinase sequence, analysis and mutagenesis may not always yield useful information.

Alternative approaches include the creation of fusion proteins containing a consensus sequence for phosphorylation for typical protein kinases coupled with a carrier protein that assists with high expression levels and efficient purification. After the initial studies have revealed the optimal conditions for phosphorylation and separation of relevant peptides, it is imperative that parallel studies are carried out on the full-length protein and, ideally, *in vivo,* as has been done for the *N*-methyl-D-aspartate receptor (Omkumar et al., 1996).

2.E. Mass Spectrometry

With a combination of electric and magnetic fields, the mass spectrometer separates charged compounds in their gaseous forms according to their mass-to-charge ratios. Recent progress in ionization and vaporization techniques (such as electrospray [ES] and matrix-assisted laser desorption/ionization [MALDI]) makes possible the analysis of proteins and peptides without their decomposition. The accuracy of mass determination is on the order of 0.01%, allowing precise determination of the mass for proteins and peptides with an accuracy not achievable by other methods. Even unusual modifications can be elucidated because in principle, these techniques do not require standards.

For an example, the time course of bovine rhodopsin phosphorylation *in vitro* is shown in Figure 4.2. With time, the C-terminal fragment of rhodopsin is successively phosphorylated, and formation of ions with m/z that correspond to mono-, di-, and triphosphorylated (each phosphate group increases the mass by 80 Da) species is observed. If detergent and salt were used during the phosphorylation procedures, the peptides have to be desalted before MS measurements. Ionization of the peptide by ES is particularly sensitive to low amounts of salts compared with MALDI/MS.

The incredible power of this technique is illustrated by application of tandem mass spectrometers, with the first quadrupole of the spectrometer serving to isolate the ion of interest (a mass separator), so a mixture of peptides can be applied (5–10 peptides, even if they are N-terminally blocked or post-translationally modified by unknown groups). Once a peptide is selected, typically no larger than 25 amino acids long, it is collided with a neutral gas in the second quadrupole, causing collision-induced dissociation, breaking it down to ions missing one or more amino acids containing the N-terminal or C-terminal ends (the "b" and "y" ions, respectively). These ions are separated in the third quadrupole of the mass spectrometer, producing a distinctive, unique fingerprint-like pattern. Deconvolution of the fragmentation patterns, with information on the precise mass of the fragments, leads to the identification of the amino acid sequence.

For an example, Figure 4.3 shows the MS/MS spectrum of the [330]DDDAS(P)ATASKTE ion obtained from the C-terminal region of mouse rhodopsin, generated first by Asp-N protease cleavage of rhodopsin followed

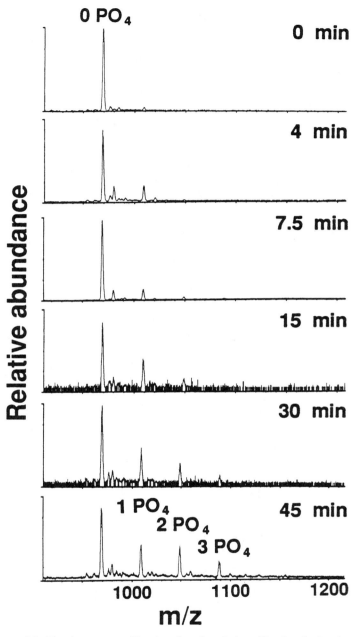

Figure 4.2. The time course of *in vitro* phosphorylation of bovine rhodopsin as observed by the analysis of the C-terminal peptide by ES/MS. Rhodopsin C-terminal peptide obtained by Asp-N digestion of samples of rod outer segment phosphorylated at different time points (0, 4, 7.5, 15, 30 or 45 minutes) was further purified by a reversed-phase C18 HPLC column as described by Ohguro et al. (1993, 1994) and directly subjected to analysis by ES/MS. The ions marked with 0 PO_4, 1 PO_4, 2 PO_4, and 3 PO_4 represent the corresponding peptides with zero, one, two, and three phosphates, respectively. (From Ohguro et al. [1993], with permission from the American Chemical Society.)

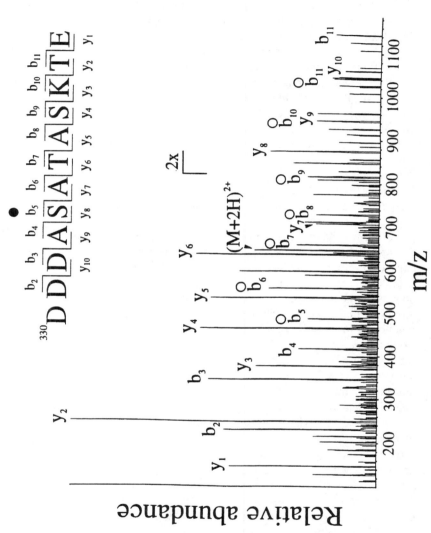

Figure 4.3. MS/MS spectrum of monophosphorylated [330]DDDASPATASKTE from mouse rhodopsin. The MS/MS spectrum of the monophosphorylated $(M + 2H)^{2+}$ ion of m/z is 645.8. The ion nomenclature is that proposed by Biemann (1990), where the γ and β ions are numbered from the C and N terminus, respectively. Those ion designations with open circles indicate additional β-elimination of a phosphate. The site of phosphorylation is indicated by a closed circle above the sequence. From Ohguro et al. [1995], with permission from the American Society for Biochemistry and Molecular Biology.)

by *Staphylococcus aureus* V8 protease proteolysis of the peptide. The mass of the $(M + 2H)^{2+}$ ion suggests that the peptide is phosphorylated on a single residue, and from the fragmentation pattern the position of phosphate could only be assigned to [334]Ser. No other assignment of the phosphate group would give the observed MS/MS pattern. Note that Ser (as well as Thr, Glu, and Asp) may undergo dehydration, creating peptide ions reduced by 18 Da from corresponding peptide ions. Phospho-Ser or Phospho-Thr may undergo β-elimination, losing a phosphate group. A detailed description of these techniques is outside the scope of this chapter and can be found in recent reviews (Mann and Wilm, 1995; Siuzdak, 1994; Resing and Ahn, 1997).

MS techniques also have inherent problems that should not be overlooked: (1) In general, MS should not be considered as a quantitative method, because the efficiency of ionization depends on the nature of the peptides. Phosphate groups suppress ionization of the peptide by ~35%; thus alternative analytic techniques (such as amino acid analysis or ultraviolet light [UV]) are helpful in more precise quantitative analysis. (2) Interpretation of the MS/MS spectra for long peptides (~20 amino acids long) is often complex, and further subcleavage may be necessary. Synthetic phosphopeptides, customized and available commercially, may help in interpreting the MS/MS spectrum. Alternatively, dephosphorylation of phosphopeptides by alkaline phosphatases or protein phosphatases and comparison with the phosphorylated peptide may help with the interpretation of MS/MS spectra. Of the two, protein phosphatases are perhaps less useful because they do not efficiently dephosphorylate short phosphopeptides. A recent approach allows one to simultaneously monitor the peptide map using the mass spectrometer (LC-ESMS) in two modes: one for full scan monitoring of all peptides in positive ion mode and the other for selective ion monitoring for negative phosphate ion (J. Crabb, personal communication). (3) Complete desalting of peptides may be a problem if the HPLC is routinely used for other purposes. The HPLC may require extensive washing to remove residual amounts of KCl before an efficient ionization employing ES/MS can be obtained. (4) Ambiguities can be created if a phosphopeptide contains more than one phosphate or if one phosphate is distributed among several residues. Thus, it is imperative that MS/MS techniques are employed in conjunction with complementary methods, such as separation of different forms of phosphopeptides, further subcleavage, or phosphoamino acid analysis. Proximity of phosphoamino acids will typically inhibit proteases from cleavage at their predicted site, thus allowing selective cleavages of unphosphorylated peptides and separation from phosphorylated peptides. (5) Unsuccessful isolation of a phosphopeptide is the most common cause of the failure to identify phosphorylation sites. The most difficult peptides are those that are either too small or hydrophilic and those that do not bind to the column or peptides that are too large or hydrophobic and therefore do not elute with high CH_3CN concentrations. Peptides that are phosphorylated to low stoichiometry may also be overlooked. In addition, incomplete cleavage of a phosphoprotein will introduce unnecessary heterogeneity. Other methods, in conjunction with mass spectrometry, should assist in quantitative analysis of phosphorylation, including phosphoamino acid analysis, UV detection of separated peptides, ^{32}P-labeling determination of the peptide concentration from amino acid analysis, peptide mapping, and mutagenesis.

2.F. Immunological Techniques

The idea is to develop a specific antibody that will recognize the phosphorylated forms of the GPCR. For example, bovine rhodopsin has seven Ser and Thr residues at the C terminus, and, ideally, the antibody should recognize one of several combinations among the multiple phosphorylation sites. Potentially, such methods would detect *in vivo* phosphorylation of a protein without use of radioactive tracers. However, in contrast to direct chemical methods, immunological approaches have limited use in study of GPCRs.

In general, this approach was successful in the analysis of receptor phosphorylation on Tyr residues (White and Backer, 1991), although, typically, the antibody was unsuccessful in discriminating different positions of phosphorylation within the primary sequence of the protein. Identification of sites of phosphorylation on Ser and Thr residues has been much more difficult (Heffetz et al., 1991). Several problems were encountered during the analysis by immunological methods: (1) time-consuming procedures for the preparation of the sample for detection by immunocytochemistry; (2) cross-reactivity with unphosphorylated protein that depends on the concentrations of unphosphorylated protein and antibodies and time of their incubation, thus not allowing the quantitative analysis of phosphorylation sites; and (3) cross-reactivity with several phosphorylated proteins.

An antiphosphorylated neurofilament antibody has been developed that recognizes phosphorylated rhodopsin in photoreceptors (Balkema and Drager, 1985). A second antibody was developed by Hargrave and colleagues (see Adamus et al., 1988, 1991) that reacted preferentially with phosphorylated rhodopsin. These antibodies were useful in the preliminary analysis of light-dependent phosphorylation in the retina (Li et al., 1995; Ohguro et al., 1995).

3. METHODS FOR DETERMINING THE SITES OF PHOSPHORYLATION WITHIN RHODOPSIN

3.A. General Strategy

Photoreceptor membranes can be readily purified from frog and bovine retinas. We have adopted these techniques for the purification of mouse disk membranes for determining the sites of rhodopsin after *in vivo* phosphorylation (see Fig. 4.5A, below). In this technique, mouse retinas are dissected and transferred to water, causing the cells to burst due to osmotic pressure. Phosphorylation and dephosphorylation are stopped by a combination of low temperature, immense dilution, and a mixture of inhibitors. In a reconstituted system composed of bovine components, phosphorylation and dephosphorylation can be stopped in an analogous manner.

The first step in determining the sites of phosphorylation within phosphorylated rhodopsin by Ser/Thr-specific rhodopsin kinase was selection of a method for specific cleavage of this transmembrane molecule. Only those Ser and Thr residues within photolyzed rhodopsin that are well exposed were considered as potential phosphorylation sites. Such residues are located within the V–VI loop and at the C terminus (Fig. 4.1). Residues in the I–II loop (^{251}Thr) and an extra loop formed between helix VII and palmitoylated Cys residues are too close to the membranes, and consequently phosphorylation of these residues is unlikely. This constraint arises due to the structure of the active site of protein kinase, which must accommodate bound ATP. These predictions are further supported by experimental observations (described below) and are consistent with results obtained with other enzymes, such as soluble proteases that have no access to the immediate vicinity of the membranes. In the case of rhodopsin, Asp-N endoproteinase removes a highly soluble C-terminal fragment (bovine, [330]DDEASTTVSKTETSQVAPA, Fig. 4.1; mouse, [330]DDDASA-TASKTETSQVAPA) that can be separated from the insoluble remainder of the rhodopsin molecule by centrifugation. The C terminus has almost all, if not all, sites that are phosphorylated by rhodopsin kinase as deduced by SDS-PAGE and autoradiography. A peptide from loop V–VI can be removed by thermolysin and *S. aureus* V8 protease, and it does not contain any phosphorylated residues, confirming that all of the sites of phosphorylation are located at the C terminus.

The C-terminal peptide can be readily purified with a C18 column (Palczewski et al., 1991). With trifluoroacetic acid (TFA) and a steep CH_3CN gradient, bovine unphosphorylated and phosphorylated [330]DDEASTTVSK-TETSQVAPA co-eluted. The progress of phosphorylation can be monitored by specific radioactivity measurements (analysis of ^{32}P-phosphate incorporation vs peptide concentration as measured by amino acid analysis). To separate phosphorylated peptides from an excess of unphosphorylated peptides, a separate run with a C18 column with a perfluorinated carboxylic acid as a counterion was employed (the concept is shown in Figure 4.4 with a hypothetical mixture of peptide d that is phosphorylated in different positions shown as peptides a, b, and c). We, and others, obtained the best separation of phosphorylated (eluted earlier) from nonphosphorylated peptides (eluted later) in the presence of heptabutyric acid, with differences in retention times as large as 20 minutes. The chromatographic method is applicable for all phosphopeptides tested to date. For example, Daas et al. (1994) discovered that the incorporation of the phosphate group reduces retention time in proportion to the resulting change in hydrophobicity of the peptide and that all peptides exhibited an increase in retention time with an increase in the counterion hydrophobicity. Taking advantage of these properties, we were able to separate the same peptides phosphorylated at different positions, acidic or basic phospho-Ser/Thr– or phospho-Tyr–containing peptides (Dass et al., 1994; Ohguro and Palczewski, 1995; Pearson and McCroskey, 1996).

The isolated 19 amino acid long peptide contains seven putative phosphorylation sites, and precise analysis of the phosphorylation sites involves additional subdigestion with *S. aureus* V8 protease, thermolysin, or trypsin, as shown in Figure 4.4. If a phosphorylated residue is in the vicinity of the site that normally is cleaved in the unphosphorylated peptide, some proteases will not cleave at this site. This phenomenon can be used to generate diagnostic fragments specific for a particular site of phosphorylation. After digestion, peptides are analyzed by LC-MS/MS for precise identification of the sites of phosphorylation. For bovine rhodopsin, phosphorylated residues were found on [334]Ser, [338]Ser, [343]Ser and [336]Thr.

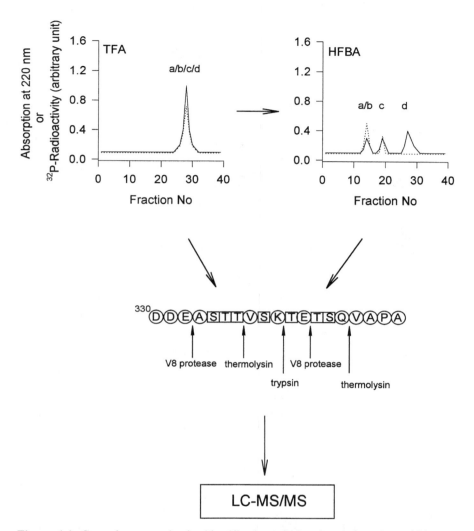

Figure 4.4. General strategy in the identification of phosphorylation sites within rhodopsin. The C terminus of rhodopsin is cleaved with endoproteinase Asp-N and the rest of the receptor removed by centrifugation. The C-terminal peptide is purified with a C18 column and a steep CH_3CN gradient. A mixture of 19-amino acid long peptides, unphosphorylated and phosphorylated at different positions (indicated as a, b, c, and d), was subjected to either further subdigestion or separation with heptafluorobutyric acid (HFBA) as a counterion and a C18 column. An unphosphorylated peptide was readily separated from phosphorylated forms of the same peptide. In some cases, peptides phosphorylated at different positions also exhibited the same elution profile. This additional step of separation may be necessary to simplify a mixture of peptides that differ in the position of phosphorylation. A mixture of long peptides phosphorylated at different positions but with the same stoichiometry may be difficult (or impossible) to analyze by MS/MS, particularly if one site is phosphorylated to a lesser degree. Subdigestion of the peptide by *S. aureus* V8 protease, thermolysin, and/or trypsin followed by liquid chromatography is necessary for precise MS/MS mapping in these cases. TFA, trifluoroacetic acid.

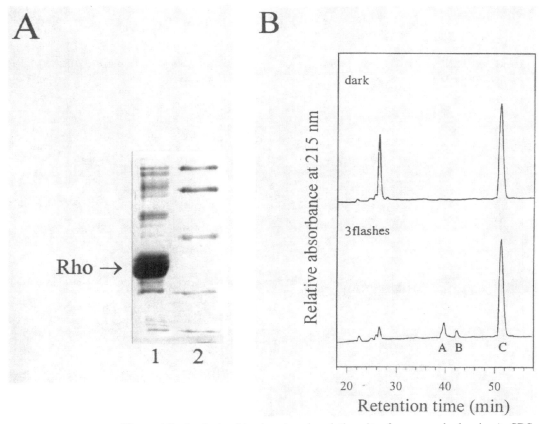

Figure 4.5. Analysis of *in vivo* phosphorylation sites for mouse rhodopsin. **A:** SDS-PAGE analysis of ROS preparation from mouse retina. Lane 1 contains 10 μg of protein, with the prominent band corresponding to rhodopsin. Lane 2 contains molecular weight standards (from the top, 92, 67, 43, 30, 21 kD; from Pharmacia Biotech). **B:** One hundred mice kept in the dark for 2 hours were sacrificed by cervical dislocation. Each mouse was sacrificed either in the dark or immediately after three consecutive flashes at 20-second intervals. Exposure to the flashes led to bleaching of ~43% rhodopsin. Retinas were promptly dissected and put into 0.8 ml of hypotonic solution containing rhodopsin kinase and phosphatase inhibitors (20 mM potassium fluoride, 2 mM EDTA, and 5 mM adenosine). The disk membranes were isolated by sucrose density gradient centrifugation and treated overnight with 40 ng of endoproteinase Asp-N at room temperature. The centrifugation supernatant was injected into a reversed-phase narrow-bore C18 HPLC column (2.1 × 250 mm, Vydac 218TP52), and the C-terminal rhodopsin peptide (^{330}DDDASATASKTETSQVAPA) was obtained by a CH$_3$CN linear gradient from 0% to 52% in 0.05% TFA for 30 minutes. The flow rate was 0.3 ml/min. The peptide was rechromatographed by a reversed-phase microbore C18 HPLC column (1.0 × 250 mm, Vydac 218TP51) employing a CH$_3$CN linear gradient from 0% to 20% for 80 minutes in the presence of 0.04% HFBA. The flow rate was 0.15 ml/min. Peptides were monitored by absorption at 215 nm. Upper and lower traces show microbore HPLC elution profiles of the sample obtained from dark-adapted mice and those mice exposed to three flashes, respectively. Peptides designated by A, B and C were identified by mass spectrometry and compared with synthetic phosphopeptides as follows. A, monophosphorylated peptide at ^{338}Ser (a major component) or ^{343}Ser (a minor component, less than 15%); B, monophosphorylated peptide at ^{334}Ser; C, unphosphorylated peptide. The peptide eluted at 27 minutes was identified as a peptide derived from the α subunit of G$_t$, transducin. (From Ohguro et al. [1995], with permission from the American Society for Biochemistry and Molecular Biology.)

There are newer modifications of the MS/MS techniques that require smaller amounts of sample and allow analysis of a complex mixtures (e.g., the precursor ion scan technique as described by Annan and Carr, 1997; Carr et al., 1996) or negative-ion orifice-potential stepping and capillary liquid chromatography/ES ionization MS as described by Ding et al. (1994).

Rapid isolation and even partial purification of investigated protein, proteolysis, HPLC or capillary electrophoresis, and a combination of MS techniques, enable the identification of *in vivo* phosphorylation sites without radioactive labeling. Only a few picomoles (tens of femtomoles in favorable cases) of phosphopeptide are needed for more modern MS analyses (Annan and Carr, 1997) to identify the sequence. Ambiguity arising from multiple phosphorylation can be minimized by initial separation of different forms of the same peptides as depicted in Figure 4.4 using HFBA as a counter ion (Fig. 4.5). MS techniques are not quantitative methods, and rigorous separation of peptides by amino acid analysis or use of UV or fluorescence for detection is necessary for precise quantification of phosphorylation. It is useful before analysis of *in vivo* phosphorylation to perform experiments employing an *in vitro* purified system. Synthetic standard phosphorylated and unphosphorylated peptides are valuable in setting up the analysis.

It is important to note that, in addition to our work on the sites of phosphorylation within rhodopsin, complementary and independent studies of Papac et al. (1993) and McDowell et al. (1993) were carried out. In general, these studies are in reasonable agreement with one another for the *in vitro* conditions. Other studies on the sites of phosphorylation that involve GPCR include, for example, the most recent work of Diviani et al. (1997) for α_{1B}-adrenergic receptor, Fredericks et al. (1996) for β_2-adrenergic receptor, Pak et al. (1997) for μ opioid receptor, and Roth et al. (1997) for somatostatin receptor subtype 3.

3.B. Purification of Rod Outer Segment Membranes and Rhodopsin Kinase From Bovine Retina

Bovine rod outer segments (ROS) were prepared from fresh retinas according to the procedure of Papermaster (1982). Retinas were dissected, and ROS were isolated under dim red illumination at 0°–5°C. Rhodopsin kinase was purified using heparin-Sepharose in the presence of Tween 80 as described by Palczewski (1993) or, alternatively, using a detergent-free method on immobilized recoverin as described by Chen et al. (1995).

3.C. Purification of ROS Membranes from Mouse Retina

3.C.a. *Material Required*

Buffer A: 67 mM phosphate buffer, pH 7.0, containing 5% sucrose

Curved dissecting scissors, 5 1/8

Deionized H_2O

Dry ice

Eppendorf test tube

Ethanol

Freshly neutralized 10 mM hydroxylamine with 100 mM NaOH

Kontes dual tissue grinder, 3 ml

Solution containing 20 mM potassium fluoride, 2 mM EDTA, and 5 mM adenosine

Sucrose solution in buffer A (1.11 gm/ml density)

Sucrose solution in buffer A (1.15 gm/ml density)

Two 20-W fluorescent bulbs

3.C.b. Equipment

SDS-PAG electrophoretic system

Centrifuge J2-HS, Beckman, or equivalent

Ultracentrifuge TLX, Beckman, or equivalent

3.C.c. Methods. All procedures are performed at 4°C unless otherwise stated. This procedure may be applied to dark adapted animals or animals exposed to different lighting conditions.

Mice are dark adapted or exposed to light (e.g., 54 fc produced by exposing animals to two 20-W fluorescent bulbs from a distance of 60 cm). Animals are sacrificed via cervical dislocation. Immediately following death, the eyes are removed and rinsed with deionized H_2O. The muscles, optic nerve, and lens are carefully removed. The retinas are isolated and collected into an Eppendorf test tube containing 80 μl/retina of a solution consisting of 20 mM potassium fluoride, 2 mM EDTA, and 5 mM adenosine to inhibit protein kinases and phosphatases. The procedure at this point takes less than 25 seconds. The samples are frozen on dry ice/ethanol immediately after removal. Next, 1.5 ml of 67 mM phosphate buffer, pH 7.0, containing 5% sucrose and 10 mM hydroxylamine is added to the frozen sample (1 ml per 10 retinas). The suspension is transferred to a 3-ml glass/glass homogenizer (Kontes dual tissue grinder, 3 ml) and homogenized 10 times. The suspension is transferred to a 2-ml Eppendorf tube and shaken on an Eppendorf shaker at room temperature for 15 minutes. The suspension (total volume = 2.0 ml) is layered on top of a continuous sucrose gradient (1.11–1.15 gm/ml density) and centrifuged for 150 minutes at 13,000 rpm (J2-HS, Beckman). The ROS disc membranes are collected and transferred to a 25-ml cylinder and 1:1 vol of the above phosphate buffer is added and mixed gently. The ROS suspension is transferred to four 1.5-ml Beckman microcentrifuge tubes and centrifuged for 15 minutes at 40,000 rpm (TLX, Beckman). Pelleted ROS are analyzed with SDS-PAGE (Fig. 4.5) and stored at −80°C until use.

3.D. Asp-N Endoproteinase Digestion

3.D.a. Material Required

Acetonitrile (CH_3CN)

Buffer A: 10 mM Hepes, pH 7.5, containing 200 mM KCl

Buffer B: 10 mM Hepes, pH 7.5

Endoproteinase Asp-N from Boehringer Mannheim (Cat. No. 1054 589)

Eppendorf tubes

Heptafluorobutyric acid (HFBA)

Narrow-bore C18 reversed-phase HPLC column (Vydac 218TP52, 2.1 × 250 mm)

Microbore C18 reversed-phase HPLC column (Vydac 218TP51, 1.0 × 250 mm)

Trifluoroacetic acid (TFA)

3.D.b. Equipment

SDS-PAG electrophoretic system

Centrifuge J2-HS, Beckman, or equivalent

HPLC suitable for narrow-bore and micro bore columns, HP1050, or equivalent

3.D.c. Methods.

Endoproteinase Asp-N (EC 3.4.24.2), from a mutant strain of *Pseudomonas fragi,* is a 27-kD protease of high purity and specificity (Noreau and Drapeau, 1979). It cleaves many, but not all, peptide bonds that involve N-terminal Asp or cysteic acid residues. For example, with myoglobin as the substrate, only four of the six aspartyl bonds present were cleaved (Drapeau, 1980). The enzyme can be purchased from Boehringer Mannheim (Cat. No. 1054 589) as a lyophilized powder. A solution of the protease is stable for 1 week at 4°C. The enzyme is effective in various buffers (pH 6.0–8.5); however, chelating agents should be avoided as endoproteinase Asp-N is a metalloprotease that utilizes Zn^{2+} or Co^{2+} as a cofactor. This property can be used to effectively inhibit protease activity with 2 mM EDTA. To avoid autolysis, the temperature of the digestion should not exceed 37°C.

Endoproteinase Asp-N is active (>50%) in 0.01% SDS, 2 M urea, 1 M guanidine hydrochloride, or 15% CH_3CN (Hagmann et al., 1995). In general, proteases exhibit strong preference for specific peptide bonds; however, at high enzyme concentration over extended periods of time, other less kinetically favorable bonds will be cleaved as well. For example, although X-Asp is cleaved 2,000-fold faster than most X-Glu bonds, under certain conditions both bonds may be cleaved (Tetaz et al., 1990; Hagmann et al., 1995).

Endoproteinase Asp-N cleaves rhodopsin specifically at the C terminus, removing a highly soluble 19 amino acid long peptide that contains all of the phosphorylation sites. The original description of this cleavage has been published (Palczewski et al., 1991). Bovine or mouse rhodopsin in ROS or disc membranes (0.5–1 mg rhodopsin) was extensively washed with 10 mM Hepes, pH 7.5, containing 200 mM KCl (two to three times) followed by 10 mM Hepes, pH 7.5 (two to three times), using 1.5-ml Eppendorf tubes, each shaken at maximal speed for 5–10 minutes and then centrifuged at 14,000 rpm for 5 minutes at room temperature. This procedure was omitted in the case of purified rhodopsin. The membranes were washed as above and mixed with 0.5–1 ml of 10 mM Hepes, pH 7.5, containing 50–100 ng of endoproteinase Asp-N at the enzyme substrate ratio of 1:10,000, and incubated at room temperature overnight. The cleavage rate was faster in hypotonic buffer than in buffer containing salt. After the incubation, the extent of the cleavage was assessed with SDS-PAGE, and the rhodopsin C-terminal peptide was separated from membranes by centrifugation at 16,000 rpm for 10–20 minutes (in this step, the peptide recovery was higher with added salt, which hardened the membrane pellet).

The peptides were further purified with a narrow-bore C18 reversed-phase HPLC column (Vydac 218TP52, 2.1 × 250 mm) employing a linear gradient of 0%–52% CH_3CN and from 0.08% to 0.1% TFA for 30 minutes. The flow rate was 0.3 ml/min, and the peptides were detected at 215 nm. A mixture of unphosphorylated and phosphorylated rhodopsin C-terminal 19 amino acid peptide (bovine, [330]DDEASTTVSKTETSQVAPA; mouse, [330]DDDASATASKTETSQ-VAPA) was eluted at ~37% CH_3CN. The unphosphorylated and phosphorylated peptides can be separated by re-chromatography with a microbore C18 reversed-phase HPLC column (Vydac 218TP51, 1.0 × 250 mm) employing a CH_3CN gradient from 0% to 5% for 5 minutes and from 5% to 24% for 95 minutes in 0.04% HFBA. The flow rate was 0.12 ml/min, and the peptides were detected at 215 nm. The unphosphorylated, monophosphorylated, and multiple phosphorylated bovine peptides were eluted at 60 minutes (16% CH_3CN), 50 minutes (14% CH_3CN), and 40 minutes (12% CH_3CN), respectively. Similarly, unphosphorylated and two monophosphorylated mouse peptides at [334]Ser or [338]Ser were separated from each other (eluted at 52 minutes [11.4% CH_3CN], 43 minutes [9.6% CH_3CN], and 38 minutes [8.8% CH_3CN], respectively). This HPLC chromatographic method with HFBA as a counterion was also effective in the separation of basic or acidic phospho-Ser/Thr peptides or phospho-Tyr peptides and their parent unphosphorylated peptides.

3.E. Subdigestion of Unphosphorylated and Phosphorylated Asp-N–Treated Rhodopsin C-Terminal Peptides

3.E.a. Material Required

Acetonitrile (CH_3CN)

Buffer A: 100 mM Tris-HCl, pH 8.0

Eppendorf tubes

Narrow-bore C18 reversed-phase HPLC column (Vydac 218TP52, 2.1 × 250 mm)

S. aureus V8 protease (Boehringer Mannheim)

Thermolysin (Boehringer Mannheim)

Trypsin-TPCK (Worthington),

Trifluoroacetic acid (TFA)

3.E.b. Equipment

HPLC suitable for narrow-bore columns, HP1050, or equivalent

3.E.c. Methods.
The Asp-N–treated C-terminal bovine rhodopsin peptide ([330]DDEASTTVSKTETSQVAPA, ~5 nmol) was subdigested with trypsin-TPCK (0.2 nmol, Worthington), thermolysin (15 pmol, Boehringer Mannheim), or *S. aureus* V8 protease (0.2 nmol, Boehringer Mannheim) in 0.1 ml of 100 mM Tris-HCl, pH 8.0, at 30°C for 1 hour, 2 hours, or 16 hours, respectively (Fig. 4.4). Trypsin and thermolysin cleaved the peptide into two fragments

(^{330}DDEASTTVSK, ^{340}TETSQVAPA) and three fragments (^{330}DDEASTT, ^{337}VSKTETSQ, ^{345}VAPA), respectively. These digestions were completely inhibited by the phosphorylation near the cleavage sites (trypsin; phosphorylation at ^{338}Ser, perhaps at ^{340}Thr, thermolysin; phosphorylation at ^{338}Ser, ^{343}Ser). However, *S. aureus* V8 protease always cleaved the peptides into three fragments (^{330}DDE, ^{333}ASTTVSKTE, ^{342}TSQVAPA) regardless of phosphorylation at any sites. Subdigestions of mouse rhodopsin peptide (^{330}DDDASATASKTETSQVAPA) by these proteolytic enzymes were similar to those of the bovine peptide.

Using a combination of these subdigestions and analysis by MS, we have found *in vitro* that photolyzed rhodopsin is initially phosphorylated at Ser residues (^{334}Ser, ^{338}Ser, and ^{343}Ser). For simple determination of the initial sites of rhodopsin phosphorylation, we performed tryptic digestion of a radioactive sample followed by HPLC separation. Briefly, Asp-N–treated rhodopsin C-terminal peptides were subdigested by trypsin-TPCK as described above, and the mixture was directly subjected to a narrow-bore C18 reversed-phase HPLC column (Vydac 218TP52). The peptides were eluted by a linear gradient of 0% to 24% CH_3CN and 0.08% to 0.1% TFA for 60 minutes. Three singly phosphorylated (^{330}DDEAS[P]TTVSK, ^{340}TETS[P]QVAPA, and ^{330}DDEASTTVS[P]KTETSQVAPA) and multiphosphorylated rhodopsin peptides were eluted at ~30 minutes (12% CH_3CN), 35 minutes (14% CH_3CN), 44 minutes (17.6% CH_3CN), and 41 minutes (16.4% CH_3CN), respectively. Stoichiometries of phosphorylation at specific sites were estimated by radioactivity counting.

3.F. Carboxypeptidase and Aminopeptidase Digestions

Carboxypeptidase and aminopeptidase digestions are also useful for confirming the sites of phosphorylation determined by MS. Their application is particularly beneficial in cases of multiple phosphorylation sites adjacent to the N or C terminus. Two examples are described below.

3.F.a. Material Required

Acetonitrile (CH_3CN)

Buffer A: 50 mM sodium acetate buffer, pH 4.6

Buffer B: 80 mM Tris-HCl, pH 8.0, containing 2 mM $MgCl_2$

Carboxypeptidase P (Boehringer Mannheim)

Eppendorf tubes

Leucine aminopeptidase (Sigma)

Narrow-bore C18 reversed-phase HPLC column (Vydac 218TP52, 2.1 × 250 mm)

Trifluoroacetic acid (TFA)

3.F.b. Equipment

HPLC suitable for narrow bore columns, HP1050, or equivalent

3.F.c. Carboxypeptidase P Digestion.
MS/MS analysis of monophosphorylated ^{333}ASTTV(P)SKTE suggested ^{338}Ser phosphorylation but did not eliminate

the possibility of [340]Thr phosphorylation. The peptide (~2 nmol) was treated with 0.2 nmol of carboxypeptidase P (Boehringer Mannheim) in 50 mM sodium acetate buffer, pH 4.6, at 30°C for 0.5, 2, or 16 hours. At each time point, the sample was injected into a C18 reversed-phase HPLC column (Vydac, 218TP52), and peptides were eluted by a linear gradient of CH_3CN from 0% to 24% in 0.05% TFA at a flow rate of 0.3 ml/min. After 0.5 and 2 hours of digestion, two peptides eluted at earlier times (11 and 16 minutes) than the original peptide (19 minutes), and a peptide eluting at 11 minutes was obtained after 16 hours of digestion. As identified by ES/MS, peptides that eluted at 11 and 16 minutes were monophosphorylated [333]ASTTVS and monophosphorylated [333]ASTTVSK, respectively. Further digestion beyond the phospho-Ser was not observed.

3.F.d. *Leucine Aminopeptidase Digestion.* MS/MS analysis of monophosphorylated [342]T(P)SQVAPA did not assign the site of phosphorylation because no collision-induced dissociation between [342]Thr and [343]Ser was evident in the spectrum. The peptide (~2 nmol) was treated with 120 pmol of leucine aminopeptidase (Sigma) in 80 mM Tris-HCl, pH 8.0, containing 2 mM $MgCl_2$ at 30°C for 2 hours, and the mixture was directly applied to a reversed-phase HPLC. Two peptides obtained were monophosphorylated [343]SQVAPA and monophosphorylated [343]SQVAP, respectively, by ES/MS and MS/MS. Evidently, the aminopeptidase contained carboxypeptidase activity and the aminopeptidase activity was greatly reduced at phospho-Ser similarly to carboxypeptidase.

3.G. Electrospray (ES/MS) and Tandem Mass Spectrometry (MS/MS)

ES/MS and MS/MS were acquired on a triple-quadrupole mass spectrometer fitted with a nebulization-assisted ES ionization source (Sciex API III, PE/Sciex, Thronehill, Ontario). Fractions containing peptides were lyophilized by a Speed vac concentrator and solubilized in 1:1 methanol/H_2O (v/v) and 0.1% formic acid. Samples were introduced into the ES source via a silica transfer line (50 mm inner diameter) using a Harvard Apparatus pump. The mass spectrometer was scanned with a mass step of 0.1 Da at the rate of 1 ms/step. Resolution was adjusted to a 20% valley between adjacent isotope peaks in a singly charged cluster.

Precursor ions were selected with the first of three quadrupoles (Q1) for collision-induced dissociation (at ~30 eV) with argon in the second quadrupole (Q2) for MS/MS. The third quadrupole (Q3) was scanned with a mass step of 0.25 Da and 1 ms/step. Parent-ion transmission was maximized by reducing the resolution of Q1 to transmit a 2–3 m/z window around of the selected ion, and Q3 resolution was adjusted to ~50% valley between peaks 3 Da apart.

ACKNOWLEDGMENTS

This research was supported by grants from NIH EY08061; Core Facilities grant EY01730; an award from Research to Prevent Blindness, Inc. (RPB) to the Department of Ophthalmology at the University of Washington; Japanese

Ministry of Health (Surveys and Research on Specific Diseases); Naito Memorial Foundation; Ciba-Geigy Foundation for the Promotion of Science; The Machida Memorial Foundation for Medical and Pharmaceutical Research; and Uehara Memorial Foundation. K.P. is a recipient of a Jules and Doris Stein Professorship from RPB. The authors thank Dr. K. Walsh (University of Washington School of Medicine) for help with mass spectrometric methods, members of their laboratory, and Drs. J. Saari, P. Hargrave, and J. Crabb for helpful comments.

REFERENCES

Adamus G, Zam ZS, Arendt A, Palczewski K, McDowell JH, Hargrave PA (1991): Anti-rhodopsin monoclonal antibodies of defined specificity: Characterization and application. Vis Res 31:17–31.

Adamus G, Zam ZS, McDowell JH, Shaw GP, Hargrave PA (1988): A monoclonal antibody specific for the phosphorylated epitope of rhodopsin: Comparison with other anti-phosphoprotein antibodies. Hybridoma 7:237–247.

Annan RS, Carr SA (1997): The essential role of mass spectrometry in characterizing protein structure: Mapping posttranslational modifications. J Protein Chem 16:391–402.

Balkema GW, Drager UC (1985): Light-dependent antibody labelling of photoreceptors. Nature 316:630–633.

Biemann K (1990): Mass values for amino acid residues in peptides. Methods Enzymol 193:886–887.

Boyle WJ, van der Geer P, Hunter T (1991): Phosphopeptide mapping and phospho-amino acid analysis by two-dimensional separation on thin-layer cellulose plates. Methods Enzymol 201:110–149.

Carr SA, Huddleston MJ, Annan RS (1996): Selective detection and sequencing of phosphopeptides at the femtomole level by mass spectrometry. Anal Biochem 239:180–192.

Chen CK, Inglese J, Lefkowitz RJ, Hurley JB (1995): Ca^{2+}-dependent interaction of recoverin with rhodopsin kinase. J Biol Chem 270:18060–18066.

Dass C, Mahalakshmi P, Grandberry D (1994): Manipulation of ion-pairing reagents for reversed-phase high-performance liquid chromatographic separation of phosphorylated opioid peptides from their non-phosphorylated analogues. J Chromatogr 678:249–257.

Ding J, Burkhart W, Kassel DB (1994): Identification of phosphorylated peptides from complex mixtures using negative-ion orifice-potential stepping and capillary liquid chromatography/electrospray ionization mass spectrometry. Rapid Commun Mass Spectrom 8:94–98.

Diviani D, Lattion AL, Cotecchia S (1997): Characterization of the phosphorylation sites involved in G protein–coupled receptor kinase– and protein kinase C–mediated desensitization of the α1B-adrenergic receptor. J Biol Chem 272:28712–28719.

Drapeau GR (1980): Substrate specificity of a proteolytic enzyme isolated from a mutant of *Pseudomonas fragi*. J Biol Chem 255:839–840.

Fischer WH, Hoeger CA, Meisenhelder J, Hunter J, Craig AG (1997): Determination of phosphorylation sites in peptides and proteins employing a volatile Edman reagent. J Protein Chem 16:329–334.

Fredericks ZL, Pitcher JA, Lefkowitz RJ (1996): Identification of the G protein–coupled receptor kinase phosphorylation sites in the human beta2-adrenergic receptor. J Biol Chem 271:13796–13803.

Hagmann ML, Geuss U, Fischer S, Kresse GB (1995): Peptidyl-Asp metalloendopeptidase. Methods Enzymol 248:782–787.

Hargrave PA, Fong S-L,McDowell JH, Mas MT, Curtis DR, Wang JK, Juszczak E, Smith DP (1980): The partial primary structure of bovine rhodopsin and its topography in the retina rod cell disc membrane. Neurochem Int 1:231–244.

Heffetz D, Fridkin M, Zick Y (1991): Generation and use of antibodies to phosphothreonine. Methods Enzymol 201:44–53.

Hunter AP, Games DE (1994): Chromatographic and mass spectrometric methods for the identification of phosphorylation sites in phosphoproteins. Rapid Commun Mass Spectrom 8:559–570.

Hynek R, Kasicka V, Kucerova Z, Kas J (1997): Fast detection of phosphorylation of human pepsinogen A, human pepsinogen C and swine pepsinogen using a combination of reversed-phase high-performance liquid chromatography and capillary zone electrophoresis for peptide mapping. J Chromatogr 688:213–220.

Kennelly PJ (1994): Identification of sites of serine and threonine phosphorylation via site-directed mutagenesis—Site transformation versus site elimination. Anal Biochem 219:384–386.

Li T, Franson WK, Gordon JW, Berson EL, Dryja TP (1995): Constitutive activation of phototransduction by K296E opsin is not a cause of photoreceptor degeneration. Proc Natl Acad Sci USA 92:3551–3555.

Mann M, Wilm M (1995): Electrospray mass spectrometry for protein characterization. Trends Biochem Sci 20:219–224.

Matsudira P (1993): A Practical Guide to Protein and Peptide Purification for Microsequencing, ed 2. San Diego: Academic Press.

McDowell JH, Nawrocki JP, Hargrave PA (1993): Phosphorylation sites in bovine rhodopsin. Biochemistry 32:4968–4974.

Meyer HE, Hoffmann-Posorske E, Heilmeyer LM Jr (1991): Determination and location of phosphoserine in proteins and peptides by conversion to S-ethylcysteine. Methods Enzymol 201:169–185.

Michel H, Hunt DF, Shabanowitz J, Bennett J (1988): Tandem mass spectrometry reveals that three photosystem II proteins of spinach chloroplasts contain N-acetyl-O-phosphothreonine at their NH_2 termini. J Biol Chem 263:1123–1130.

Muszynska T, Andersson L, Porath J (1986): Selective adsorption of phosphoproteins on gel-immobilized ferric chelate. Biochemistry 25:6850–6853.

Noreau J, Drapeau GR (1979): Isolation and properties of the protease from the wild-type and mutant strains of *Pseudomonas fragi*. J Bacteriol 140:911–916.

Ohguro H, Johnson RS, Ericsson LH, Walsh KA, Palczewski K (1994): Control of rhodopsin multiple phosphorylation. Biochemistry 33:1023–1028.

Ohguro H, Palczewski K (1995): Separation of phospho- and non-phosphopeptides using reversed phase column chromatography. FEBS Lett 368:452–454.

Ohguro H, Palczewski K, Ericsson LH, Walsh KA, Johnson RS (1993): Sequential phosphorylation of rhodopsin at multiple sites. Biochemistry 32:5718–5724.

Ohguro H, Van Hooser JP, Milam AH, Palczewski K (1995): Rhodopsin phosphorylation and dephosphorylation *in vivo.* J Biol Chem 270:14259–14262.

Omkumar RV, Kiely MJ, Rosenstein AJ, Min KT, Kennedy MB (1996): Identification of a phosphorylation site for calcium/calmodulin dependent protein kinase II in the NR2B subunit of the N-methyl-D-aspartate receptor. J Biol Chem 271:31670–31678.

Pak Y, O'Dowd BF, George SR (1997): Agonist-induced desensitization of the mu opioid receptor is determined by threonine 394 preceded by acidic amino acids in the COOH-terminal tail. J Biol Chem 272:24961–24965.

Palczewski K (1993): Purification of rhodopsin kinase from bovine rod outer segments. Methods Neurosci 15:217–225.

Palczewski K (1997): GTP-binding-protein–coupled receptor kinases—Two mechanistic models. Eur J Biochem 248:261–269.

Palczewski K, Buczylko J, Kaplan MW, Polans AS, Crabb JW (1991): Mechanism of rhodopsin kinase activation. J Biol Chem 266:12949–12955.

Papac DI, Oatis JE Jr, Crouch RK, Knapp DR (1993): Mass spectrometric identification of phosphorylation sites in bleached bovine rhodopsin. Biochemistry 32:5930–5934.

Papermaster DS (1982): Preparation of retinal rod outer segments. Methods Enzymol 81:48–52.

Pearson JD, McCroskey MC (1996): Perfluorinated acid alternatives to trifluoroacetic acid for reversed-phase high-performance liquid chromatography. J Chromatogr A 746:277–281.

Pearson RB, Kemp BE (1991): Protein kinase phosphorylation site sequences and consensus specificity motifs: Tabulations. Methods Enzymol 200:62–81.

Posewitz, MC, Tempst, P (1999): Immobilized gallium (III) affinity chromatography of phosphopeptides. Anal Chem 71:2883–2892.

Resing KA, Ahn NG (1997): Protein phosphorylation analysis by electrospray ionization-mass spectrometry. Methods Enzymol 283:29–44.

Roach PJ, Wang YH (1991): Identification of phosphorylation sites in peptides using a microsequencer. Methods Enzymol 201:200–206.

Roth A, Kreienkamp HJ, Meyerhof W, Richter D (1997): Phosphorylation of four amino acid residues in the carboxyl terminus of the rat somatostatin receptor subtype 3 is crucial for its desensitization and internalization. J Biol Chem 272:23769–23774.

Shannon JD, Fox JW (1995): Identification of phosphorylation sites by Edman degradation. Techniques Protein Chem VI:117–123.

Siuzdak G (1994): The emergence of mass spectrometry in biochemical research. Proc Natl Acad Sci USA 91:11290–11297.

Tetaz T, Morrison JR, Andreou J, Fidge NH (1990): Relaxed specificity of endoproteinase Asp-N: This enzyme cleaves at peptide bonds N-terminal to glutamate as well as aspartate and cysteic acid residues. Biochem Int 22:561–566.

van der Geer P, Hunter T (1994): Phosphopeptide mapping and phosphoamino acid analysis by electrophoresis and chromatography on thin-layer cellulose plates. Electrophoresis 15:544–554.

Watts JD, Affolter M, Krebs DL, Wange RL, Samelson LE, Aebersold R (1994): Identification by electrospray ionization mass spectrometry of the sites of tyrosine phosphorylation induced in activated Jurkat T cells on the protein tyrosine kinase ZAP-70. J Biol Chem 269:29520–29529.

White MF, Backer JM (1991): Preparation and use of anti-phosphotyrosine antibodies to study structure and function of insulin receptor. Methods Enzymol 201:65–79.

Zhang L, Sports CD, Osawa S, Weiss ER (1997): Rhodopsin phosphorylation sites and their role in arrestin binding. J Biol Chem 272:14762–14768.

CHAPTER 5

PALMITOYLATION OF G PROTEIN–COUPLED RECEPTORS

HUI JIN, SUSAN R. GEORGE, MICHEL BOUVIER, and BRIAN F. O'DOWD

Regulation of G Protein–Coupled Receptor Function and Expression,
Edited by Jeffrey L. Benovic.
ISBN 0-471-25277-8   Copyright © 2000 Wiley-Liss, Inc.

I. INTRODUCTION

Cellular proteins may be modified by a variety of biochemical processes, including phosphorylation, glycosylation, acetylation, arginylation, ADP-ribosylation, methylation, and lipidation. Protein lipidations (Towler et al., 1988; Casey, 1995) are classified on the basis of the nature of the attached lipid moieties into palmitoylation (acylation by palmitic acid), myristoylation (acylation by myristic acid), prenylation (acylation by isoprenoids), arachidonoylation (acylation by arachidonic acid), and modification by glycosylphosphatidylinositol (GPI). Myristoylation (Resh, 1996; Johnson et al., 1994; Boutin, 1997), prenylation (Zhang and Casey, 1996; Resh, 1996; Higgins and Casey, 1996), and GPI modification (Udenfriend and Kodukula, 1995; Stevens, 1995; England, 1993) have been extensively studied and reviewed, whereas arachidonoylation is a relatively recent finding (Hallak et al., 1994). These diverse lipid modifications have been shown to enhance the association of the modified proteins with the membrane and may also modulate protein–protein interactions. Accumulating evidence demonstrates that palmitoylation of proteins may also serve such a purpose and influence function in a similar manner (Mumby, 1997; Bouvier et al., 1995; Bizzozero et al., 1994; Bizzozero, 1997).

Palmitoylation is a reversible post-translational modification (Bonatti et al., 1989) whereby a 16-carbon saturated fatty acyl chain is covalently attached to a protein through a thioester bond (Bizzozero et al., 1994). As shown in Table 5.1, an extensive array of proteins have been proven to be palmitoylated, yet in only a few of them have studies been carried out to determine the functional role. In those investigations on cytosolic proteins, palmitoylation seems to increase the degree of membrane association.

Palmitoylated proteins can roughly be classified into three categories (Bizzozero et al., 1994):

1. Single transmembrane glycoproteins (e.g., transferrin receptor).
2. Soluble proteins (e.g., the Ras family, some members of the Src family, and G protein α subunits). As summarized elsewhere (Resh, 1994; Casey, 1995), mounting evidence suggests that palmitoylation is a major determinant of their ability to be associated with the plasma membrane.
3. Multiple transmembrane proteins (e.g., G protein–coupled receptors [GPCRs]). The majority of GPCRs are believed to be palmitoylated at cysteine(s) within the C terminus, although only a few have been investigated experimentally (Table 5.2).

After the original report (Ovchinnikov et al., 1988) describing the palmitoylation of bovine rhodopsin on the two cysteines in its cytoplasmic tail, many GPCRs have been shown to be palmitoylated (Table 5.2). For these receptors, the palmitoylated cysteines were found in their C tail, in a location very similar to that of the palmitoylated cysteines of rhodopsin. Although no strict consensus sequence can be established for receptor palmitoylation, most of the confirmed palmitoylated cysteines are located within the first 20 amino acids of the receptor's C tail where they are flanked by a large proportion of hydrophobic and positively charged residues. Although much effort has been made to clarify the role of palmitoylation in receptor coupling to its cognate

TABLE 5.1. Palmitoylated Proteins and Functions of Palmitoylation

Protein	Function of Palmitoylation	References
A 55-kD membrane protein of human erythrocytes, different from band 3		Das et al. (1992)
An 85-kD protein in rat adipocytes similar to CD36 (platelet membrane glycoprotein IV)		Jochen and Hays (1993)
Acetylcholinesterase	Anchoring to plasma membrane	Randall (1994)
Adenylyl cyclase	Chemical depalmitoylation decreased its activity	Mollner et al. (1995)
Asialoglycoprotein receptor (ASGP-R)		Zeng et al. (1995)
Band 3 protein in the human erythrocyte membrane	Not required for anion exchange	Okubo et al. (1991), Kang et al. (1994)
Cation-dependent mannose 6-phosphate receptor	Essential for the normal trafficking and lysosomal enzyme sorting	Schweizer et al. (1996)
Caveolin	Not necessary for localization of caveolin to caveolae; may stabilize the oligomer	Dietzen et al. (1995), Monier et al. (1996)
CD4, the HIV receptor	Not required for expression of CD4 on the cell surface or for binding of p56lck to its cytoplasmic domain	Crise and Rose (1992)
CD44	May play an active role in receptor interactions and signal transduction in normal human T lymphocytes	Guo et al. (1994)
Chloroplast 32-kD herbicide-binding protein		Mattoo and Edelman (1987)
Cyclic nucleotide phosphodiesterase		Agrawal et al. (1990)
Endothelial nitric oxide synthase (eNOS)	Required for targeting of NOS to caveolae	Robinson et al. (1995), Garcia-Cardena et al. (1996)
Erythrocyte ankyrin		Staufenbiel (1987)
G protein α subunits	Lack of palmitoylation reduces membrane association	Wedegaertner et al. (1993), Veit et al. (1994), Parenti et al. (1993), Linder et al. (1993), Grassie et al. (1993)
GAD65 in pancreatic β cells	Membrane anchorage	Solimena et al. (1994), Christgau et al. (1992)
GAIP (G α-interacting protein)	Membrane anchorage	De Vries et al. (1996)
Glucose transporter GLUT1		Pouliot and Beliveau (1995)
Glutamate kainate receptor GluR6	Kainate-gated currents produced by the unpalmitoylated mutant receptor were indistinguishable from those of the wild type	Pickering et al. (1995)
Glycoprotein G of rabies virus		Gaudin et al. (1991)

(*continued*)

TABLE 5.1. Palmitoylated Proteins and Functions of Palmitoylation (*Continued*)

Protein	Function of Palmitoylation	References
Glycoprotein Ib in megakaryocytes		Schick and Walker (1996)
gp41 of HIV and SIV		Yang et al. (1995)
Gramicidin transmembrane channel		Koeppe et al. (1995)
GRK-6 and GRK-4	Essential for membrane association	Stoffel et al. (1994), Premont et al. (1996)
Hemagglutinin of influenza virus	No effect on fusion activity	Steinhauer et al. (1991), Philipp et al. (1995)
	May regulate the maturation and budding of influenza virus	Veit and Schmidt (1993), Portincasa et al. (1992)
Human CD36		Tao et al. (1996)
Influenza A virus M2 protein		Veit et al. (1991), Sugrue et al. (1990)
L-type calcium channel β_{2a} subunit	Palmitoylation-deficient β_{2a} mutant still localized to membrane particulate fractions and able to target functional channel complexes to the plasma membrane similar to wild type; however, channels formed with a palmitoylation-deficient β_{2a} subunit exhibited a dramatic decrease in ionic current per channel	Chien et al. (1996)
Low-affinity neurotrophin receptor (p75LNTR)		Barker et al. (1994)
Membrane glycoproteins of Semliki Forest virus		Schmidt and Burns (1989), Scharer et al. (1993)
Myelin proteolipid protein		Bizzozero et al. (1990)
Myelin-associated glycoproteins (MAG)		Pedraza et al. (1990)
Na channel α subunit		Schmidt and Catterall (1987)
Neural cell adhesion molecules (NCAM)		Murray et al. (1987)
Neuromodulin (GAP-43)	Palmitoylation is important for membrane targeting, Golgi localization, and neurite transport of neuromodulin	Sudo et al. (1992), Liu et al. (1993), (1994)
Nicotinic ACh receptor		Olson et al. (1984)
P-selectin		Fujimoto et al. (1993)
p21H-ras, p21N-ras, and p21K-ras(A)	Combine with the CAAX motif to target specific plasma membrane localization; the posttranslational processing of ras p21 is critical for stimulation of yeast adenylate cyclase	Jackson et al. (1990), Hancock et al. (1990), Horiuchi et al. (1992)

(*continued*)

TABLE 5.1. Palmitoylated Proteins and Functions of Palmitoylation (*Continued*)

Protein	Function of Palmitoylation	References
p21rhoB		Adamson et al. (1992)
p62	May be important in vesicular transport during mitosis	Mundy and Warren (1992)
p63 of endoplasmic reticulum		Schweizer et al. (1995)
Pertussis toxin	Required for cellular toxicity	Hackett et al. (1995)
Ras-related Rap2 protein		Beranger et al. (1991)
Ras-related YPT/rab proteins in *S. pombe*		Newman et al. (1992)
Rh polypeptides of RBC		Hartel-Schenk and Agre (1992)
SCG10 in the growth cones of developing neurons		Di Paolo et al. (1997)
Sindbis virus E2 glycoprotein	Important for budding	Ivanova and Schlesinger (1993)
SNAP-25		Hess et al. (1992)
Src-family tyrosine kinases (p56lck, p59fyn, p55fgr, and p59hck)	Confer localization to caveolae	Wolven et al. (1997), Shenoy-Scaria et al. (1994), Rodgers et al. (1994), Robbins et al. (1995), Resh (1994), Paige et al. (1993), Koegl et al. (1994)
Synaptotagmin		Veit et al. (1996), Chapman et al. (1996)
Torpedo cysteine string protein (T-Csp)		Gundersen et al. (1994)
Transferrin receptors	Associated with inhibition of the rate of transferrin receptor endocytosis	Alvarez et al. (1990)
Transforming growth factor (TGF)-α	Critical in interaction with associated proteins	Shum et al. (1996)
Transmembrane proteins in murine leukemia virus	Does not affect its transport, processing, surface expression, or cell fusion activity	Yang and Compans (1996), Hensel et al. (1995)
Tubulin		Ozols and Caron (1997), Caron (1997)
Vesicular stomatitis virus (VSV) G protein		Rose et al. (1984)
xlcaax-1 of *Xenopus*	Mutation of the palmitoylation sites inhibited the association of xlcaax-1 with the membrane	Reddy et al. (1991), Kloc et al. (1991, 1993)
Yeast Ras2 proteins	Necessary for membrane association	Kuroda et al. (1993)
Yeast synaptobrevin homologue (Snc1)	Affects protein stability	Couve et al. (1995)

TABLE 5.2. Palmitoylation of G Protein–Coupled Receptors

Receptor	Site	Mutant	G Protein Coupling	Other Functions	References
5-HT1A	C417, C420				Butkerait et al. (1995)
5-HT1B	C388				Ng et al. (1993)
α_2AR	C442	C442A/S	No change	Mut failed to down-regulate	Kennedy and Limbird (1993), Eason et al. (1994)
β_2AR	C341	C341G	Decreased		O'Dowd et al. (1989)
D1DR	C347, C351	C347/351G	Decreased		Ng et al. (1994a) Jensen et al. (1995)
		C347/351A	No change		Jin et al. (1997),
D2DR	C442				Ng et al. (1994b)
ETA	C1215	C1215S, C1215A	No change for Gs; abolished Gq coupling		Horstmeyer et al. (1996)
ETB	C402, C403, C405	C402S; C403/405S; C402/403/405S	No change for C402S; abolished For rest		Okamoto et al. (1997)
LH/CG	C621, C622	C621/622A	No change	Mut increased internalization and down-regulation	Kawate et al. (1997), Kawate and Menon (1994)
M$_2$AChR	C457	C457G	No change decreased		van Koppen and Nathanson (1991), Hayashi et al. (1997)
mGluR4	C890				Alaluf et al. (1995)
Rhodopsin	C322, C323	C322S/C323S	No change		Karnik et al. (1993)
TRH-R	C335, C337	C335/337S/G	No change	Mut decreased internalization	Nussenzveig et al. (1993)
V2	C341, C342	C341/342S	No change		Sadeghi et al. (1997)

5-HT1A, 5-hydroxytryptamine1A receptor; 5-HT1B, 5-hydroxytryptamine1B receptor; α_2AR, α_2-adrenergic receptor; β_2AR, β_2-adrenergic receptor; D1DR, D1 dopamine receptor; D2DR, D2 dopamine receptor; ETA, endothelin A receptor; ETB, endothelin B receptor; LH/CG, luteinizing hormone/chorionic gonadotropin receptor; M$_2$ AChR, M$_2$ acetylcholine receptor; mGluR4, metabotropic glutamate receptor subtype 4; TRH-R, thyrotropin releasing–hormone receptor; V2, V2 vasopressin receptor.

G protein, its phosphorylation state, or its rate of internalization or down-regulation, no consensus on the functional importance of this post-translational modification can be clearly established for these receptors.

2. DETECTION OF RECEPTOR PALMITOYLATION

Systematic studies aimed at investigating the palmitoylation state of GPCRs should try to answer the following questions: (1) Is the receptor acylated? (2) Is the fatty acid attached to the receptor via a palmitate residue? (3) What is the chemical nature of the bond? (4) Where is the palmitoylation site? Once a receptor has been determined to be palmitoylated, the next logical step would be to investigate the functional role played by this modification.

The following sections review the various approaches and methods that are currently used to study palmitoylation of GPCRs.

2.A. Receptor Expression and Cell Culture

Although palmitoylation of GPCRs can theoretically be studied in cells endogenously expressing a given receptor, the typically low level of expression of these proteins together with the low specific activity of the available tracers renders such experiments very difficult. As a consequence, most studies have taken advantage of the relatively high level of expression that can be obtained with recombinant receptors expressed in heterologous surrogate cell lines. DNAs encoding the desired GPCRs subcloned into an expression vector (such as pRC/CMV or pCDNA3 from Invitrogen) are often used to transfect a mammalian cell line, usually a fibroblast such as COS cells for transient expression (e.g., rhodopsin [Karnik et al., 1993], thyrotropin-releasing hormone receptor [TRH-R] [Nussenzveig et al., 1993], and V2 vasopressin receptor [Sadeghi et al., 1997]) and Chinese hamster ovarian (CHO), baby hamster kidney (BHK) or Madin Darby canine kidney (MDCKII) cells for permanent expression (e.g., β_2-adrenergic receptor [β_2AR] [O'Dowd et al., 1989; Moffett et al., 1993] and M_2 muscarinic receptor [van Koppen and Nathanson, 1991; Habecker et al., 1993]).

Increasingly, researchers are using the baculovirus/Sf9 cell system (Summers and Smith, 1987) to achieve high levels of protein expression, which facilitates the analysis of palmitoylation of GPCRs. The receptors examined in this system include the β_2AR (Loisel et al., 1996), 5-HT1B receptor (Ng et al., 1993), D1 (Ng et al., 1994a) and D2L (Ng et al., 1994b) dopamine receptors, and the endothelin ET_A receptor (Horstmeyer et al., 1996).

2.A.a. Transient Expression. COS cells are grown and kept under 75% confluence in appropriate medium (Dulbecco's modified Eagle's medium [DMEM] [Sadeghi et al., 1997] or RPMI medium [Horstmeyer et al., 1996] for COS) supplemented with 10% fetal bovine serum (FBS), penicillin (50 units/ml), and streptomycin (50 μg/ml) at 37°C in a humidified atmosphere containing 5% CO_2. One day before transfection, cells are plated out at a density of 0.5×10^6 cells/100-mm dish. The following day, DNAs are transfected into cells using one of the following common transfection methods: calcium-phosphate co-precipitation (Chen and Okayama, 1987), DEAE-dextran (Luthman and Magnusson, 1983), or lipofectin or lipofectamine (Life Technologies, Inc., Gaithersburg, MD). After 48 hours, the transfected cells are ready for metabolic labeling with radioactive palmitic acid.

2.A.b. Stable Cell Lines. Cells suitable for permanent transfection (e.g., CHO, BHK, and MDCKII) are transfected as above. However, a selection culture medium containing 500 µg/ml geneticin (G418) is used to select the resistant colonies, which are allowed to expand in cell culture flasks, propagated in the selection medium, and screened for positive clones by means of radioligand binding. Ideally, cell lines expressing similar receptor levels are selected for further study. The whole process takes approximately 1–2 months.

2.A.c. Culture of Sf9 Cells. Cloning and purification of recombinant baculovirus are essentially carried out as described elsewhere (Vialard et al., 1990) and has been facilitated by the availability of the Bac-to-Bac Kit from Life Technologies. Recombinant plasmids encoding the GPCR of interest are used with the recombinant baculovirus to create the desired working virus. The viruses are generated by transfection of Sf9 cells grown in 60-mm dishes as monolayers in TC100 medium (Horstmeyer et al., 1996) or Grace's supplemented medium (Ng et al., 1994a), with the appropriate DNA construct by the calcium phosphate precipitation or lipofection method. The resultant baculovirus culture is harvested as the stock and diluted to infect fresh Sf9 cells. Alternatively, the crude stock can be used to purify the recombinant baculoviruses by successive plaque assays using β-galactosidase detection.

For receptor expression, Sf9 cells are grown in suspension culture. Cells at a density of 2×10^6/ml are infected with the appropriate virus at a multiplicity of infection of 2–5 (Ng et al., 1994a; Horstmeyer et al., 1996). Cells are harvested within 48–72 hours (Ng et al., 1994b; Horstmeyer et al., 1996) after infection.

2.B. Metabolic Labeling with Radioactive Palmitic Acid

The transfected cells are washed and cultured in serum-free medium before metabolic labeling. This period of incubation with serum-free medium varies with different reports, ranging from 0 to 20 hours. Serum deprivation is carried out to deplete the stock of cellular palmitoyl-CoA and to reduce metabolic activity (by promoting cell growth arrest) to favor palmitate uptake and to increase the specific activity of the tracer once the radioactive palmitate is added. However, it has been shown that without preincubation with serum-free medium, the COS cells can still take up 60%–70% of label added after 30 minutes incubation with tritiated palmitic acid (Kennedy and Limbird, 1993). Therefore, starving the cells may not be absolutely necessary.

Following the preincubation, cells are incubated with $[9,10-^3H]$palmitic acid in fresh medium containing 1% DMSO in DMEM either free of serum (Okamoto et al., 1997; Moffett et al., 1993; Kennedy and Limbird, 1993) or with low serum levels, such as 1% FBS (O'Dowd et al., 1989; Ng et al., 1994b; Karnik et al., 1993). However, 10% FBS has also been used (Sadeghi et al.,

1997; Kennedy and Limbird, 1994) successfully. The addition of serum allows re-initiation of cell growth in a synchronized manner in those experiments in which the cells were previously starved of serum. The amount of tritiated palmitate added ranged from 0.1 mCi/ml (Karnik et al., 1993), 0.2 mCi/ml (Ng et al., 1994a; Moffett et al., 1993; Horstmeyer et al., 1996), 0.285 mCi/ml (Okamoto et al., 1997; Kawate and Menon, 1994), and 0.4 mCi/ml (O'Dowd et al., 1989) to 1.0 mCi/ml (Sadeghi et al., 1997; Kennedy and Limbird, 1993) and 1.2 mCi/ml (Kennedy and Limbird, 1994). The specific activity of the tritiated palmitate that we purchased from NEN/Dupont was 35.9 Ci/mmol. The incubation time for labeling also may vary from 2 hours (O'Dowd et al., 1989; Moffett et al., 1993; Karnik et al., 1993), 4 hours (Ng et al., 1994a; Horstmeyer et al., 1996), 6 hours (Okamoto et al., 1997; Kawate and Menon, 1994), and 8 hours (Kennedy and Limbird, 1994) to 18 hours (Zhu et al., 1995). The optimum time of labeling varies with the cell type used. For instance, maximum tritiated-palmitate incorporation into the β_2AR was observed following a 1-hour labeling in Sf9 cells, whereas a 3-hour labeling was required to attain maximal incorporation in HEK 293 cells (M. Bouvier, unpublished observation). When possible, labeling longer than 4 hours should be avoided to reduce the incorporation of the label into amino acids following metabolic degradation of the fatty acid.

2.C. Membrane Preparation, Receptor Solubilization, and Purification

Because receptors are not the only proteins expressed on the cell membrane and because the palmitic acids are labeled with a weak radioisotope, it is critical in most cases to purify and concentrate receptors to reduce the background and obtain a strong signal.

2.C.a. Membrane Preparation.
To remove free radioactive palmitate after labeling, adherent cells are rinsed with ice-cold PBS and harvested, and Sf9 suspension cells are collected first by centrifugation and washed with ice-cold PBS. Cells are then collected in Tris-HCl, PBS, or Hepes buffered solutions. To prevent or minimize receptor degradation by cellular proteases upon cell homogenization or lysis, EDTA/EGTA and protease inhibitors (combinations of benzamidine, leupeptin, phenylmethylsulfonyl fluoride, N-ethylmaleimide, pepstatin A, aprotinin, bacitracin, and soybean trypsin inhibitor) are routinely added in the buffer. In this laboratory, we use 10 μg/ml benzamidine, 5 μg/ml leupeptin, and 5 μg/ml soybean trypsin inhibitor. Polytron homogenization (Ng et al., 1994a; Kennedy and Limbird, 1993), sonication (O'Dowd et al., 1989; Moffett et al., 1993; Kennedy and Limbird, 1994), and aspiration through a syringe (Sadeghi et al., 1997; Kennedy and Limbird, 1994) have been utilized to fragment the cells, and membranes are usually pelleted by centrifugation and resuspended in appropriate buffer. The membrane preparations are then ready for receptor solubilization. Some researchers (van Koppen and Nathanson, 1991; Okamoto et al., 1997; Kawate and Menon, 1994; Karnik et al., 1993; Horstmeyer et al., 1996) proceed directly to the receptor solubilization step without prior mechanical fragmentation of the cells.

2.C.b. Receptor Protein Solubilization. Commonly used detergents such as digitonin (O'Dowd et al., 1989; Ng et al., 1994a; Moffett et al., 1993; Kennedy and Limbird, 1993), DM (*N*-dodecyl β-D-maltoside (Karnik et al., 1993), or Nonidet P-40 (Sadeghi et al., 1997; Kennedy and Limbird, 1994; Kawate and Menon, 1994; Horstmeyer et al., 1996) are required to solubilize receptors. A distinction should be made between detergents that maintain receptor binding capacity and those that do not, depending on the method of purification to be carried out (e.g., affinity chromatography requires that the detergent maintain the binding properties [such as digitonin], whereas immunoprecipitation does not).

The membranes are usually incubated in the lysis buffer with the appropriate detergent for a period of time on ice: 20 minutes (Kawate and Menon, 1994), 30 minutes (Kennedy and Limbird, 1993), 45 minutes (Kennedy and Limbird, 1994), 1 hour (Horstmeyer et al., 1996), 2 hours (Okamoto et al., 1997; Ng et al., 1994a; Moffett et al., 1993), and 3 hours (Karnik et al., 1993). The cell lysates are then centrifuged to remove cell debris. The solubilized receptors can be concentrated by centrifugation through Centriprep and Centricon membrane filters available from Amicon (Bedford, MA) (Ng et al., 1994a; Moffett et al., 1993). Subsequently, the solubilized and labeled receptors may be purified through immunoprecipitation or affinity purification/chromatography.

2.C.c. *Receptor Immunoprecipitation and Affinity Purification.* In many cases, researchers purify receptors by immunoprecipitation with antibodies that recognize the receptor or the epitope tags attached to receptors. Commonly used epitopes are hemagglutinin (HA) (Kennedy and Limbird, 1994) and c-myc (Ng et al., 1994a), and antibodies to these epitopes are commercially available. The solubilized receptors are incubated with the appropriate antibody for a period of time on ice: 1 hour (Horstmeyer et al., 1996), 2 hours (Ng et al., 1994a), and overnight (Sadeghi et al., 1997). The antibody-attached receptors are subsequently incubated with protein A–agarose/Sepharose beads on ice for 1 hour (Horstmeyer et al., 1996) or 2 hours (Sadeghi et al., 1997). Some of these antibodies do not recognize protein A, and therefore a secondary antibody needs to be used for the immunoprecipitation.

Some researchers take advantage of the high affinity and selectivity of ligands toward corresponding receptors to purify solubilized receptors. For instance, alprenolol-Sepharose affinity chromatography (O'Dowd et al., 1989; Moffett et al., 1993) was used to purify β_2AR. Yohimbine-agarose affinity chromatography (Kennedy and Limbird, 1993) and hCG-Affi-Gel 10 (Bio-Rad) (Kawate and Menon, 1994) were used to purify α_2-adrenergic receptor (α_2AR) and LH/hCG receptors, respectively. Receptors could be isolated to about 50% purity with alprenolol-Sepharose affinity chromatography (Moffett et al., 1993).

In the endothelin ET_B receptor study, cells were incubated with biotinylated ET1 (100 nM) for 60 minutes at 25°C (Okamoto et al., 1997), and the cells were then centrifuged and lysed for 2 hours at 4°C in lysis buffer containing PBS, EDTA, protease inhibitors, and digitonin. After removing the insoluble materials by centrifugation at 100,000*g* for 1 hour at 4°C, avidin-agarose was added to the supernatant and the reaction continued for 16 hours at 4°C. The avidin-agarose-biotin-ET1-receptor complex was recovered by centrifugation. After extensive washing with ice-cold lysis buffer, the receptors were eluted from the pellet with 2-mercaptoethanol at room temperature for 20 minutes. The recovered proteins were subsequently subjected to SDS-PAGE.

2.D. Gel Electrophoresis and X-Ray Film Exposure

The immunoprecipitated or affinity-purified receptors may be resolved by SDS-PAGE. Protein concentration is determined by the Bradford method using bovine serum albumin as the standard (Bradford, 1976). Because of the lability of the thioester bond between the ^{3}H-palmitate and the receptors, electrophoresis is usually run under nonreducing conditions (O'Dowd et al., 1989; Ng et al., 1994a; Moffett et al., 1993; Kennedy and Limbird, 1994; Kawate and Menon, 1994; Karnik et al., 1993). However, some groups have shown that reducing conditions could be used without disrupting the labile thioester bonds (Sadeghi et al., 1997; Okamoto et al., 1997; Horstmeyer et al., 1996).

After electrophoresis, the SDS-PAGE gels are fixed, treated for 30 minutes with Enlightning (Ng et al., 1994a,b; Moffett et al., 1993) or ENTENSIFY (Kennedy and Limbird, 1994) from DuPont/NEN, or Amplify from Amersham (Piscataway, NJ) (Sadeghi et al., 1997), dried at 60°–80°C for 2–3 hours under vacuum, and exposed to x-ray film at –70°C to –80°C. A less expensive alternative to Enlightning and Amplify is to incubate the gels in an aqueous solution of salicylic acid (1 M) following fixation of the gel. The fluorographic detection achieved with salicylic acid is identical to that obtained with the commercial compounds. Exposure times vary with different groups and receptors, from 1 week (Okamoto et al., 1997) to 6 weeks in this laboratory. A long exposure time is required even for the high expression baculovirus Sf9 system because the commercially available palmitic acid is labeled with a weak radioisotope, tritium. Other methods to enhance the detection of tritiated receptors include the treatment of the gel with 2,5-diphenyloxazole (PPO) (with the optimum concentration being 16%) before drying, and gel miniaturization (Magee et al., 1995), which involves soaking the PPO-impregnated gel in 50% polyethylene glycol 2000 or 4000 at 70°C for 15 minutes before drying, resulting in uniform gel shrinkage to one third to one half of the original size. The isotope density is therefore increased, causing shorter exposure time.

An alternative to the low specific activity of the tritiated palmitate is to use an iodinated analogue. Palmitic acid has been successfully labeled with ^{125}I to produce the [16-^{125}I]hexadecanoic acid (Peseckis et al., 1993; Berthiaume et al., 1995) . The use of this palmitate analogue (Alland et al., 1994) greatly reduced the exposure time to 4–10 days. However, this compound is not available commercially, and synthesis followed by isotopic labeling of the tracer is required.

2.E. How To Identify the Palmitoylated Receptors

Once palmitoylation of a protein has been documented by SDS-PAGE as described above, the identity of the acylated protein must be confirmed. For GPCRs, immunoblotting and photoaffinity labeling have been used.

2.E.a. Western Blot. The labeled proteins are transferred to nitrocellulose. The blots are blocked by skim milk to reduce background and incubated with the antibody that can recognize the receptor or the epitope. When the protein is purified by immunoprecipitation, the use of a different antibody from the one used for the immunoprecipitation is preferable. Blots are subsequently incubated with the secondary antibodies and conjugated with enzymes such as alkaline phosphatase that can generate color upon addition of development substrate.

Enhanced chemiluminescence (ECL) is a modified version of the above method. By taking advantage of enhanced light emission from secondary antibody-conjugated horseradish peroxidase (HRP) catalyzed oxidation of luminol in the presence of chemical enhancers such as phenols, ECL can be highly sensitive and rapid in detecting immobilized specific antigens, conjugated through primary and secondary antibodies with HRP. An ECL kit is available commercially from Amersham.

2.E.b. Photoaffinity Labeling. In the study of β_2AR, membrane receptors were photoaffinity labeled with ^{125}I-CYP diazirine in the absence and presence of 10 μM alprenolol before solubilization, and the photoaffinity ligand-treated membranes were processed in parallel with the palmitate-labeled membrane (O'Dowd et al., 1989). Photoaffinity labeling has also been carried out in α_2AR (Kennedy and Limbird, 1993) and D1 dopamine receptor (D1DR) (Ng et al., 1994a). In general, membrane proteins are resuspended in a buffer (e.g., PBS, 5 mM EDTA, pH 7.4, with protease inhibitors) and incubated in the dark with the iodinated photoaffinity agent for 3 hours at room temperature. The membrane is then exposed on ice to 360-nm ultraviolet light at 2 inches from the source for 10 minutes. The membrane is then solubilized in SDS sample buffer and subjected to SDS-PAGE, transferred onto nitrocellulose, and exposed to Kodak XAR film with an intensifying screen at −70°C. The photolabeled receptors were clearly visualized by autoradiography after 7 days.

To identify the palmitoylated receptors, several lines of evidence can be summoned. First, they are precipitated by receptor-specific antibody or adsorbed to the affinity agarose. Second, the ^{3}H-palmitate–labeled band(s) comigrate with the proteins labeled by the iodinated photolabeling agents. Third, the specific palmitate-labeled bands are not detected in preparations derived from wild-type cells that do not express receptors. The advantage of photoaffinity labeling is its specificity as revealed by the appearance of distinct bands for those receptors studied.

2.F. Identification of the Fatty Acid Attached to Receptor

2.F.a. Chromatographic Analysis. Radioactive fatty acids may be converted to other metabolites in cells. It was reported that in A431 cells (derived from epidermal carcinoma), as much as 50% of the protein-bound radioactive fatty acid was present as ^{3}H-palmitate after labeling for 4 hours with ^{3}H-myristate, presumably due to cellular acyl chain elongation (Olson et al., 1985). In addition, in 3T3 fibroblasts, two thirds of total protein-associated radioactivity obtained from ^{3}H-myristate labeling was present in amino acids, probably resulting from β-oxidation of ^{3}H-myristate with subsequent incorporation of

[3]H-intermediates into the amino acid carbon backbone (Olson et al., 1985). It has been shown (Peseckis et al., 1993) that approximately 3% 16-iodohexade-canoic acid (palmitate analogue) could be converted to 14- or 12-carbon derivatives. In view of the potential metabolic interconversion, it is necessary to prove that the radioactivity detected is really palmitate.

Fatty acyl chains can be identified following organic solvent extraction by thin-layer chromatography (TLC), reversed-phase high performance liquid chromatography (RP-HPLC), and gas chromatography. If the fatty acids extracted are shown to be palmitates by chromatography, it confirms the identity of the acyl group linked to the protein. Mass spectrometry was used for rhodopsin (Ovchinnikov et al., 1988).

Three reports in the field of GPCRs have analyzed the identity of released fatty acids by TLC. Hydrolysis with 100 mM hydroxylamine (pH 7.0) for 16 hours at 20°C (Karnik et al., 1993) or with 1 M KOH at 37°C for 12 hours (O'Dowd et al., 1989) and 24 hours (Zhu et al., 1995) were followed by hexane or hexane/methanol extraction.

The extracts were dried (under N_2) and developed on TLC silica gel with hexane/ethyl acetate/acetic acid (80:20:1) or chloroform/methanol/water (65:25:4). Zhu et al. (1995) used reversed-phase TLC with acetonitrile acetic acid (90:10) as the mobile phase. The solid support from the plates were then snipped into strips and counted for radioactivity.

Alternatively, RP-HPLC or gas chromatography can be utilized for palmitate identification, although no reports of the use of HPLC for the study of GPCRs have been published. However, RP-HPLC has been used for other palmitoylated proteins (Zeng et al., 1995; Stoffel et al., 1994; Olson et al., 1984; Linder et al., 1993). Gas chromatography has been used for rhodopsin (Ovchinnikov et al., 1988), myristoylated protein (Aitken and Cohen, 1984), and another palmitoylated protein (Gundersen et al., 1994).

2.F.b. The Nature of Palmitate-Receptor Bond. Covalent attachment of palmitate to receptors occurs via a thioester bond to cysteine whereas myristate is attached to proteins through an amide linkage (Casey, 1994).

Acid hydrolysis of acylproteins releases both ester-linked and amide-linked fatty acids, whereas the hydroxylamine at neutral and mild basic pH will only release ester-type linkages of fatty acids without cleaving fatty acyl amide bonds. Therefore, the release of radioactive fatty acid from metabolically labeled acylproteins with hydroxylamine or mild alkali had been used to identify ester-type bonds of acyl chains.

Zhu et al. (1995) resuspended the protein A–agarose beads containing the antibody–LH/CG complex in 1 M NH_2OH, pH 7.0 (or 1 M Tris, pH 7.0, as the control) and incubated with rotation for 4 hours at 20°C, and the treated receptors on the beads were eluted and analyzed by SDS-PAGE. Others (Ng et al., 1994a; Karnik et al., 1993) stripped the receptors of palmitates at a later stage. For opsin (Karnik et al., 1993), ^{3}H-palmitate–labeled receptor was treated with 100 mM hydroxylamine hydrochloride (pH 7.0) after elution from agarose (2 hours at 23°C). ^{3}H-Palmitoylated D1DR (Ng et al., 1994a) was incubated with 1.0 M hydroxylamine, pH 10.0, for 12 hours at 22°C after treating the fixed SDS-PAGE gel before it was dried and exposed. Despite the variations in timing of treatment, concentration and pH of hydroxylamine, and duration of treatment, all methods appeared successful.

2.G. Determination of the Site of Palmitoylation

At least three methods are available for determination of the palmitoylation site. First, isolated peptides derived from enzymatic or chemical cleavage of a palmitoylated protein can be sequenced to directly identify the site of palmitoylation. Such an approach was used to identify the palmitoylation sites of rhodopsin (Ovchinnikov et al., 1988).

A second method (Bizzozero et al., 1994) is to chemically deacylate the protein with such agents as iodoacetamide or 4-vinylpyridine and to alkylate simultaneously the newly formed sulfhydryl groups. The resultant peptides are more stable and can be procured in larger amounts than with the first method, leading to easier peptide sequencing. However, this method has never been used in GPCR palmitoylation studies.

The third method is currently most widely used and involves substitution of the candidate cysteine residue with another amino acid (glycine, serine, or alanine) using site-directed mutagenesis. Although there is some concern that the loss of acylation of the mutant receptors may be caused by induced conformational changes that render the acceptor amino acid unavailable for palmitoylation, site-directed mutagenesis has been the most common method employed in the investigations of GPCR palmitoylation. There are several methods of mutagenesis, and kits are available from commercial sources. All these methods fall into two categories: polymerase chain reaction (PCR) based and non-PCR based. Mutant D1DR (Jin et al., 1997), V2 vasopressin receptor (Sadeghi et al., 1997), ET_A receptor (Horstmeyer et al., 1996), LH/hCG receptor (Zhu et al., 1995), and $\alpha_{2A}AR$ (Kennedy and Limbird, 1994) were made based on PCR, using DNA polymerases (e.g., Taq, Vent, or pfu DNA polymerase) that work at high temperatures. TRH-R (Nussenzveig et al, 1993), $\alpha_{2A}AR$ (Eason et al., 1994), and M_2 AchR (van Koppen and Nathanson, 1991) mutants were engineered with a non-PCR method, namely, through DNA synthesis by DNA polymerase that works at 37°C. Both PCR-based and non-PCR–based methods require the use of oligonucleotides that are designed to incorporate the desired mutation, which can be truncation, deletion, or single or multiple point mutations.

3. FUNCTIONAL STUDIES OF RECEPTOR PALMITOYLATION

As summarized in several excellent reviews (Mumby, 1997; Casey, 1995; Bouvier et al., 1995; Bizzozero et al., 1994; Bizzozero, 1997), palmitoylation is a dynamic process and thus has the potential to be regulated, either enzymatically or nonenzymatically, for which there are reports supporting both.

3.A. Regulation of Protein Palmitoylation by Agonist

The palmitoylation of four proteins has been shown to be increased by agonist stimulation: β_2AR by isoproterenol (Mouillac et al., 1992), D1DR by dopamine

(Ng et al., 1994a), stimulatory GTP-binding protein (Gs) by β_2AR (Mumby et al., 1994; Degtyarev et al., 1993), and eNOS by bradykinin (Robinson et al., 1995).

The low level of receptor expression generally obtained in mammalian cells with the low specific activity of tritiated palmitate hinders the study of palmitoylation in these cells. Therefore, both β_2AR and D1DR were expressed in Sf9 cells to generate sufficient quantities of receptors for biochemical analysis of the palmitoylation phenomenon. Moulliac et al. (1992) treated the Sf9 cells expressing the c-myc epitope-tagged β_2AR with 1 μM isoproterenol for the final 15 minutes of metabolic labeling with [9,10-^{3}H]palmitate, purified the receptors with both alprenolol-Sepharose affinity chromatography and immunoprecipitation by 9E10 monoclonal antibody, and demonstrated with both methods that isoproterenol treatment led to a twofold increase in the incorporation of tritiated palmitate as measured by densitometric analysis of the fluorograph. The increase was agonist dependent because the effect could be blocked by the β-adrenergic antagonist alprenolol. In addition, the agonist-induced increase in palmitate incorporation into the β_2AR did not seem to result from a nonspecific alteration of palmitate uptake by the cells or other cellular metabolism, as the overall pattern of tritiated palmitate incorporation in the crude membrane preparation was not changed by this treatment. A similar phenomenon was observed in our D1DR study (Ng et al., 1994a). We treated Sf9 cells expressing the D1DR with 10 μM dopamine for the final 15 minutes during the palmitate labeling, immunoprecipitated the receptors, and observed a twofold increase in the palmitoylation signal.

However, a more complete study on β_2AR (Loisel et al., 1996) illustrated that the initial increased incorporation, which reflects an increased turnover rate of palmitate, is followed by decreased incorporation of tritiated palmitate, which suggests that sustained stimulation with the agonist probably favors the unpalmitoylated form of the receptor. This observation is corroborated with the V2 vasopressin receptor, in which no increase but rather a slight (26%) decrease in the extent of receptor palmitoylation was detected after 15 minutes exposure to vasopressin (Sadeghi et al., 1997).

3.B. Determination of the Half-Life of Receptor Palmitoylation

Half-lives of receptor palmitoylation have been investigated in both the α_2AR and β_2AR. To estimate the α_2AR palmitoylation half-life (Kennedy and Limbird, 1994), confluent dishes were incubated (pulsed) with ^{3}H-palmitate (1.5 mCi/ml) for 3 hours, and with 1% DMSO to facilitate the solubilization of the lipid. The cells were then washed extensively to remove radiolabeled fatty acid and chased with a 10-fold excess of unlabeled palmitate for 4 hours. At the specified time points, the labeled receptors were processed as described previously and analyzed to determine the half-life of palmitoylation. Pulse-chase labeling for the agonist effect on palmitoylation and receptor half-life were carried out similarly.

In the study of β_2AR (Loisel et al., 1996), the pulse-chase experiments with palmitate were conducted similarly as described above. To determine the receptor half-life, ^{35}S pulse-chase was carried out by adding 5 mCi of Trans^{35}S-label to methionine-free and cysteine-free Grace's supplemented medium following a 30-minute preincubation in this medium. The chase was initiated by adding complete Grace's medium containing 1 mM methionine and 1 mM cysteine. Data from tritiated palmitate and ^{35}S pulse-chase experiments were analyzed by nonlinear least-squares regression analysis, using the following equation: $q(t) = q(t \to \infty) + q(t = 0)e^{(-R)t}$, where t is the time of incubation (minutes), R the rate of decay, and q the level of labeling (percentage of control). The half-life was defined as t when $q(t) = 50\%$.

In β_2AR (Loisel et al., 1996), under basal conditions, the turnover of receptor-bound palmitate was rapid (half-life being 9.8 ± 1.8 minutes) compared with the turnover rate of the receptor protein itself (half-life being 109 ± 10 minutes), suggesting the existence of a dynamic equilibrium between the palmitoylated and nonpalmitoylated forms of receptor at basal state. Isoproterenol treatment of the receptor reduced the half-life of the β_2AR-bound palmitate by about twofold without affecting the half-life of the receptor itself. Furthermore, with sustained stimulation (over 30 minutes), the dynamic balance shifted toward the nonpalmitoylated form of receptors, suggesting that prolonged agonist stimulation may prevent receptor repalmitoylation in β_2AR. Similar half-life studies in other receptors are either lacking or less complete.

3.C. Functional Role of Receptor Palmitoylation

Primary sequence comparison among GPCRs reveals that at least one or two cysteine residues are conserved in the C terminus of the majority of these receptors (Jin et al., 1997). To determine the functional role of palmitoylation in GPCRs (see References in Table 5.2), several research groups have carried out a series of site-directed mutagenesis studies. The two main categories of receptor properties investigated were receptor–G protein coupling and receptor internalization/down-regulation.

Elimination of the palmitoylation site (C341G) of β_2AR resulted in a partial loss of receptor–G protein coupling (O'Dowd et al., 1989; Moffett et al., 1993). Likewise, ET$_B$ receptor (Okamoto et al., 1997) was shown to have impaired receptor–G protein coupling upon removal of the palmitoylation sites. In contrast, similar mutagenesis studies on other GPCRs have demonstrated that palmitoylation is not required for coupling. The elimination of the palmitoylation sites did not show any effect on rhodopsin-Gt coupling (Karnik et al., 1993), and, paradoxically, the chemical removal of palmitate from rhodopsin even increased the receptor–Gt coupling (Morrison et al., 1991).

A lack of effect on receptor–G protein coupling, assessed by the inhibition of adenylyl cyclase activity, has been demonstrated in both α_{2A}AR (Kennedy and Limbird, 1993) and the M_2 muscarinic receptors (van Koppen and Nathanson, 1991). Similarly, the substitution of glycine for ^{335}Cys and ^{337}Cys of the mouse TRH-R did not affect the receptor-mediated stimulation of IP$_3$ production (Nussenzveig et al., 1993). In LH/hCG receptors, the loss of palmitoylation did not affect the efficiency of coupling to Gs protein (Zhu et al., 1995; Kawate and Menon, 1994). In the single or double mutations in D1DR, which eliminated one or both of the palmitoylation sites, no difference was observed between the wild-type and mutant D1DRs (Jin et al., 1997). The lack of effect was also observed for ET_A and V2 receptors (Table 5.2). On the other hand, the effects of palmitoylation on receptor sequestration and down-regulation have been diverse (see Table 5.2), indicating that receptor palmitoylation may serve completely different roles among various GPCRs with respect to receptor sequestration and receptor down-regulation.

The membrane sublocalization of transmembrane proteins to a specific subdomain/microdomain of cell membrane, such as caveolae where caveolin and eNOS are aggregated (Mumby, 1997), may be regulated by palmitoylation. Caveolae (Anderson, 1993) are specialized parts of plasma membrane formed as flask-shaped invaginations, and evidence is accumulating that a variety of molecules involved directly or indirectly in signal transduction are enriched therein. Whether GPCRs locate within, or are internalized through, caveolae during desensitization is still unresolved due to conflicting data (Mumby, 1997). Another specialized membrane subdomain is clathrin-coated pits, which have been shown to mediate receptor internalization for some receptors (Lee et al., 1998; Goodman et al., 1996).

In summary, receptor palmitoylation is a common phenomenon. Further investigations will eventually clarify the role of palmitoylation in receptor function.

REFERENCES

Adamson P, Marshall CJ, Hall A, Tilbrook PA (1992): Post-translational modifications of p21rho proteins. J Biol Chem 267:20033–20038.

Agrawal HC, Sprinkle TJ, Agrawal D (1990): 2′,3′-cyclic nucleotide-3′-phosphodiesterase in the central nervous system is fatty acylated by thioester linkage. J Biol Chem 265:11849–11853.

Aitken A, Cohen P (1984): Identification of N-terminal myristyl blocking groups in proteins. Methods Enzymol 106:205–210.

Alaluf S, Mulvihill ER, McIlhinney RA (1995): Palmitoylation of metabotropic glutamate receptor subtype 4 but not 1 alpha expressed in permanently transfected BHK cells. Biochem Soc Trans 23:87S.

Alland L, Peseckis SM, Atherton RE, Berthiaume L, Resh MD (1994): Dual myristoylation and palmitoylation of Src family member p59fyn affects subcellular localization. J Biol Chem 269:16701–16705.

Alvarez E, Girones N, Davis RJ (1990): Inhibition of the receptor-mediated endocytosis of diferric transferrin is associated with the covalent modification of the transferrin receptor with palmitic acid. J Biol Chem 265:16644–16655.

Anderson RGW (1993): Caveolae: Where incoming and outgoing messengers meet. Proc Natl Acad Sci USA 90:10909–10913.

Barker PA, Barbee G, Misko TP, Shooter EM (1994): The low affinity neurotrophin receptor, p75LNTR, is palmitoylated by thioester formation through cysteine 279. J Biol Chem 269:30645–30650.

Beranger F, Tavitian A, de Gunzburg J (1991): Post-translational processing and subcellular localization of the Ras-related Rap2 protein. Oncogene 6:1835–1842.

Berthiaume L, Peseckis SM, Resh MD (1995): Synthesis and use of iodo-fatty acid analogs. Methods Enzymol 250:454–466.

Bizzozero OA (1997): The mechanism and functional roles of protein palmitoylation in the nervous system. Neuropediatrics 28:23–26.

Bizzozero OA, Good LK, Evans JE (1990): Cysteine-108 is an acylation site in myelin proteolipid protein. Biochem Biophys Res Commun 170:375–382.

Bizzozero OA, Tetzloff SU, Bharadwaj M (1994): Overview: Protein palmitoylation in the nervous system: Current views and unsolved problems. Neurochem Res 19:923–933.

Bonatti S, Migliaccio G, Simons K (1989): Palmitoylation of viral membrane glyco-proteins takes place after exit from the endoplasmic reticulum. J Biol Chem 264:12590–12595.

Boutin JA (1997): Myristoylation. Cell Signal 9:15–35.

Bouvier M, Moffett S, Loisel TP, Mouillac B, Hebert T, Chidiac P (1995): Palmitoyla-tion of G-protein–coupled receptors: A dynamic modification with functional con-sequences. Biochem Soc Trans 23:116–120.

Bradford MM (1976): A rapid and sensitive method for the quantitation of microgram quantities of protein using the principle of protein dye binding. Anal Biochem 72:248–254.

Butkerait P, Zheng Y, Hallak H, Graham TE, Miller HA, Burris KD, Molinoff PB, Man-ning DR (1995): Expression of the human 5-hydroxytryptamine1A receptor in Sf9 cells. J Biol Chem 270:18691–18699.

Caron JM (1997): Posttranslational modification of tubulin by palmitoylation: I. *In vivo* and cell-free studies. Mol Biol Cell 8:621–636.

Casey P (1995): Protein lipidation in cell signaling. Science 268:221–225.

Casey PJ (1994): Lipid modifications of G proteins. Curr Opin Cell Biol 6:219–225.

Chapman ER, Blasi J, An S, Brose N, Johnston PA, Sudhof TC, Jahn R (1996): Fatty acylation of synaptotagmin in PC12 cells and synaptosomes. Biochem Biophys Res Commun 225:326–332.

Chen C, Okayama H (1987): High-efficiency transformation of mammalian cells by plasmid DNA. Mol Cell Biol 7:2745–2752.

Chien AJ, Carr KM, Shirokov RE, Rios E, Hosey MM (1996): Identification of palmi-toylation sites within the L-type calcium channel β2a subunit and effects on channel function. J Biol Chem 271:26465–26468.

Christgau S, Aanstoot HJ, Schierbeck H, Begley K, Tullin S, Hejnaes K, Baekkeskov S (1992): Membrane anchoring of the autoantigen GAD65 to microvesicles in pan-creatic beta-cells by palmitoylation in the NH2-terminal domain. J Cell Biol 118:309–320.

Couve A, Protopopov V, Gerst JE (1995): Yeast synaptobrevin homologs are modi-fied posttranslationally by the addition of palmitate. Proc Natl Acad Sci USA 92:5987–5991.

Crise B, Rose JK (1992): Identification of palmitoylation sites on CD4, the human im-munodeficiency virus receptor. J Biol Chem 267:13593–13597.

Das AK, Kundu M, Chakrabarti P, Basu J (1992): Fatty acylation of a 55 kDa membrane protein of human erythrocytes. Biochim Biophys Acta 1108:128–132.

De Vries L, Elenko E, Hubler L, Jones TL, Farquhar MG (1996): GAIP is membrane-anchored by palmitoylation and interacts with the activated (GTP-bound) form of Gαi subunits. Proc Natl Acad Sci USA 93:15203–15208.

Degtyarev MY, Spiegel AM, Jones TL (1993): Increased palmitoylation of the Gs pro-tein α subunit after activation by the β-adrenergic receptor or cholera toxin. J Biol Chem 268:23769–23772.

Di Paolo G, Lutjens R, Pellier V, Stimpson SA, Beuchat MH, Catsicas S, Grenningloh G

(1997): Targeting of SCG10 to the area of the Golgi complex is mediated by its NH2-terminal region. J Biol Chem 272:5175–5182.

Dietzen DJ, Hastings WR, Lublin DM (1995): Caveolin is palmitoylated on multiple cysteine residues. Palmitoylation is not necessary for localization of caveolin to caveolae. J Biol Chem 270:6838–6842.

Eason MG, Jacinto MT, Theiss CT, Liggett SB (1994): The palmitoylated cysteine of the cytoplasmic tail of α2A-adrenergic receptors confers subtype-specific agonist-promoted downregulation. Proc Natl Acad Sci USA 91:11178–11182.

England PT (1993): The structure and biosynthesis of glycosyl phosphatidylinositol protein anchors. Annu Rev Biochem 62:121–138.

Evan GI, Lewis GK, Ramsay G, Biship JM (1985): Isolation of monoclonal antibodies specific for human c-myc proto-oncogene product. Mol Cell Biol 5:3610–3616.

Fujimoto T, Stroud E, Whatley RE, Prescott SM, Muszbek L, Laposata M, McEver RP (1993): P-selectin is acylated with palmitic acid and stearic acid at cysteine 766 through a thioester linkage. J Biol Chem 268:11394–11400.

Garcia-Cardena G, Oh P, Liu J, Schnitzer JE, Sessa WC (1996): Targeting of nitric oxide synthase to endothelial cell caveolae via palmitoylation: Implications for nitric oxide signaling. Proc Natl Acad Sci USA 93:6448–6453.

Gaudin Y, Tuffereau C, Benmansour A, Flamand A (1991): Fatty acylation of rabies virus proteins. Virology 184:441–444.

Goodman OB Jr, Krupnick JG, Santini F, Gurevich VV, Penn RB, Gagnon AW, Keen JH, Benovic JL (1996): Beta-arrestin acts as a clathrin adaptor in endocytosis of the β2-adrenergic receptor. Nature 383:447–450.

Grassie MA, McCallum JF, Parenti M, Magee AI, Milligan G (1993): Lack of N terminal palmitoylation of G protein α subunits reduces membrane association. Biochem Soc Trans 21:499S.

Gundersen CB, Mastrogiacomo A, Faull K, Umbach JA (1994): Extensive lipidation of a *Torpedo* cysteine string protein. J Biol Chem 269:19197–19199.

Guo YJ, Lin SC, Wang JH, Bigby M, Sy MS (1994): Palmitoylation of CD44 interferes with CD3-mediated signaling in human T lymphocytes. Int Immunol 6:213–221.

Habecker BA, Tietje KM, van Koppen CJ, Creason SA, Goldman PS, Migeon JC, Parenteau LA, Nathanson NM (1993): Regulation of expression and function of muscarinic receptors. Life Sci 52:429–432.

Hackett M, Walker CB, Guo L, Gray MC, Van Cuyk S, Ullmann A, Shabanowitz J, Hunt DF, Hewlett EL, Sebo P (1995): Hemolytic, but not cell-invasive activity, of adenylate cyclase toxin is selectively affected by differential fatty-acylation in *Escherichia coli*. J Biol Chem 270:20250–20253.

Hallak H, Muszbek L, Laposata M, Belmonte E, Brass LF, Manning DR (1994): Covalent binding of arachidonate to G protein α subunits of human platelets. J Biol Chem 269:4713–4716.

Hancock JF, Paterson H, Marshall CJ (1990): A polybasic domain or palmitoylation is required in addition to the CAAX motif to localize p21ras to the plasma membrane. Cell 63:133–139.

Hartel-Schenk S, Agre P (1992): Mammalian red cell membrane Rh polypeptides are selectively palmitoylated subunits of a macromolecular complex. J Biol Chem 267:5569–5574.

Hensel J, Hintz M, Karas M, Linder D, Stahl B, Geyer R (1995): Localization of the palmitoylation site in the transmembrane protein p12E of Friend murine leukaemia virus. Eur J Biochem 232:373–380.

Hess DT, Slater TM, Wilson MC, Skene JH (1992): The 25 kDa synaptosomal-associated protein SNAP-25 is the major methionine-rich polypeptide in rapid axonal transport and a major substrate for palmitoylation in adult CNS. J Neurosci 12:4634–4641.

Higgins JB, Casey PJ (1996): The role of prenylation in G-protein assembly and function. Cell Signal 8:433–437.

Horiuchi H, Kaibuchi K, Kawamura M, Matsuura Y, Suzuki N, Kuroda Y, Kataoka T, Takai Y (1992): The posttranslational processing of ras p21 is critical for its stimulation of yeast adenylate cyclase. Mol Cell Biol 12:4515–4520.

Horstmeyer A, Cramer H, Sauer T, Muller-Esterl W, Schroeder C (1996): Palmitoylation of endothelin receptor A. Differential modulation of signal transduction activity by post-translational modification. J Biol Chem 271:20811–20819.

Ivanova L, Schlesinger MJ (1993): Site-directed mutations in the Sindbis virus E2 glycoprotein identify palmitoylation sites and affect virus budding. J Virol 67:2546–2551.

Jackson JH, Cochrane CG, Bourne JR, Solski PA, Buss JE, Der CJ (1990): Farnesol modification of Kirsten-ras exon 4B protein is essential for transformation. Proc Natl Acad Sci USA 87:3042–3046.

Jensen AA, Pedersen UB, Kiemer A, Din N, Andersen PH (1995): Functional importance of the carboxyl tail cysteine residues in the human D1 dopamine receptor. J Neurochem 65:1325–1331.

Jin H, Zastawny R, George SR, O'Dowd BF (1997): Elimination of palmitoylation sites in the human dopamine D1 receptor does not affect receptor-G protein interaction. Eur J Pharmacol 324:109–116.

Jing S, Trowbridge IS (1990): Nonacylated human transferrin receptors are rapidly internalized and mediate iron uptake. J Biol Chem 265:11555–11559.

Jochen A, Hays J (1993): Purification of the major substrate for palmitoylation in rat adipocytes: N-terminal homology with CD36 and evidence for cell surface acylation. J Lipid Res 34:1783–1792.

Johnson DR, Bhatnagar RS, Knoll LJ, Gordon JI (1994): Genetic and biochemical studies of protein N-myristoylation. Annu Rev Biochem 63:869–914.

Kang D, Karbach D, Passow H (1994): Anion transport function of mouse erythroid band 3 protein (AE1) does not require acylation of cysteine residue 861. Biochim Biophys Acta 1194:341–344.

Karnik SS, Ridge KD, Bhattacharya S, Khorana HG (1993): Palmitoylation of bovine opsin and its cysteine mutants in COS cells. Proc Natl Acad Sci USA 90:40–44.

Kawate N, Menon KM (1994): Palmitoylation of luteinizing hormone/human choriogonadotropin receptors in transfected cells. Abolition of palmitoylation by mutation of Cys-621 and Cys-622 residues in the cytoplasmic tail increases ligand-induced internalization of the receptor. J Biol Chem 269:30651–30658.

Kawate N, Peegel H, Menon KM (1997): Role of palmitoylation of conserved cysteine residues of luteinizing hormone/human choriogonadotropin receptors in receptor down-regulation. Mol Cell Endocrinol 127:211–219.

Kennedy ME, Limbird LE (1993): Mutations of the α2A-adrenergic receptor that eliminate detectable palmitoylation do not perturb receptor-G–protein coupling. J Biol Chem 268:8003–8011.

Kennedy ME, Limbird LE (1994): Palmitoylation of the α2A-adrenergic receptor. Analysis of the sequence requirements for and the dynamic properties of α2A-adrenergic receptor palmitoylation. J Biol Chem 269:31915–31922.

Kloc M, Li XX, Etkin LD (1993): Two upstream cysteines and the CAAX motif but not the polybasic domain are required for membrane association of Xlcaax in *Xenopus* oocytes. Biochemistry 32:8207–8212.

Kloc M, Reddy B, Crawford S, Etkin LD (1991): A novel 110-kDa maternal CAAX box–containing protein from *Xenopus* is palmitoylated and isoprenylated when expressed in baculovirus. J Biol Chem 266:8206–8212.

Koegl M, Zlatkine P, Ley SC, Courtneidge SA, Magee AI (1994): Palmitoylation of multiple Src-family kinases at a homologous N-terminal motif. Biochem J 303:749–753.

Koeppe RE, Killian JA, Vogt TC, de Kruijff B, Taylor MJ, Mattice GL, Greathouse DV (1995): Palmitoylation-induced conformational changes of specific side chains in the gramicidin transmembrane channel. Biochemistry 34:9299–9306.

Kuroda Y, Suzuki N, Kataoka T (1993): The effect of posttranslational modifications on the interaction of Ras2 with adenylyl cyclase. Science 259:683–686.

Lee KB, Pals-Rylaarsdam R, Benovic JL, Hosey MM (1998): Arrestin-independent internalization of the m1, m3, and m4 subtypes of muscarinic cholinergic receptors. J Biol Chem 273:12967–12972.

Linder ME, Middleton P, Hepler JR, Taussig R, Gilman AG, Mumby SM (1993): Lipid modifications of G proteins: α subunits are palmitoylated. Proc Natl Acad Sci USA 90:3675–3679.

Liu Y, Fisher DA, Storm DR (1993): Analysis of the palmitoylation and membrane targeting domain of neuromodulin (GAP-43) by site-specific mutagenesis. Biochemistry 32:10714–10719.

Liu Y, Fisher DA, Storm DR (1994): Intracellular sorting of neuromodulin (GAP-43) mutants modified in the membrane targeting domain. J Neurosci 14:5807–5817.

Loisel TP, Adam L, Hebert TE, Bouvier M (1996): Agonist stimulation increases the turnover rate of beta 2AR-bound palmitate and promotes receptor depalmitoylation. Biochemistry 35:15923–15932.

Luthman H, Magnusson G (1983): High efficiency polyoma DNA transfection of chloroquine-treated cells. Nucleic Acids Res 11:1295–1308.

Magee AI, Wootton J, de Bony J (1995): Detecting radiolabeled lipid-modified proteins in polyacrylamide gels. Methods Enzymol 250:330–336.

Mattoo AK, Edelman M (1987): Intramembrane translocation and posttranslational palmitoylation of the chloroplast 32-kDa herbicide-binding protein. Proc Natl Acad Sci USA 84:1497–1501.

Moffett S, Adam L, Bonin H, Loisel TP, Bouvier M, Mouillac B (1996): Palmitoylated cysteine 341 modulates phosphorylation of the β2-adrenergic receptor by the cAMP-dependent protein kinase. J Biol Chem 271:21490–21497.

Moffett S, Mouillac B, Bonin H, Bouvier M (1993): Altered phosphorylation and desensitization patterns of a human β2-adrenergic receptor lacking the palmitoylated Cys341. EMBO J 12:349–356.

Mollner S, Becl K, Pfeuffer T (1995): Acylation of adenylyl cyclase catalyst is important for enzymic activity. FEBS Lett 371:241–244.

Monier S, Dietzen DJ, Hastings WR, Lublin DM, Kurzchalia TV (1996): Oligomerization of VIP21-caveolin *in vitro* is stabilized by long chain fatty acylation or cholesterol. FEBS Lett 388:143–149.

Morrison DF, O'Brien PJ, Pepperberg DR (1991): Depalmitoylation with hydroxylamine alters the functional properties of the rhodopsin. J Biol Chem 266:20118–20123.

Mouillac B, Caron M, Bonin H, Dennis M, Bouvier M (1992): Agonist-modulated palmitoylation of β2-adrenergic receptor in Sf9 cells. J Biol Chem 267:21733–21737.

Mumby SM (1997): Reversible palmitoylation of signaling proteins. Curr Opin Cell Biol 9:148–154.

Mumby SM, Kleuss C, Gilman AG (1994): Receptor regulation of G-protein palmitoylation. Proc Natl Acad Sci USA 91:2800–2804.

Mundy DI, Warren G (1992): Mitosis and inhibition of intracellular transport stimulate palmitoylation of a 62-kD protein. J Cell Biol 116:135–146.

Murray BA, Hoffman S, Cunningham BS (1987): Molecular features of cell-adhesion molecules. Prog Brain Res 71:35–45.

Newman CM, Giannakouros T, Hancock JF, Fawell EH, Armstrong J, Magee AI (1992): Post-translational processing of Schizosaccharomyces pombe YPT proteins. J Biol Chem 267:11329–11336.

Ng GY, George SR, Zastawny RL, Caron M, Bouvier M, Dennis M, O'Dowd BF (1993): Human serotonin1B receptor expression in Sf9 cells: Phosphorylation, palmitoylation, and adenylyl cyclase inhibition. Biochemistry 32:11727–11733.

Ng GY, Mouillac B, George SR, Caron M, Dennis M, Bouvier M, O'Dowd BF (1994a): Desensitization, phosphorylation and palmitoylation of the human dopamine D1 receptor. Eur J Pharmacol 267:7–19.

Ng GY, O'Dowd BF, Caron M, Dennis M, Brann MR, George SR (1994b): Phosphorylation and palmitoylation of the human D2L dopamine receptor in Sf9 cells. J Neurochem 63:1589–1595.

Nussenzveig DR, Heinflink M, Gershengorn MC (1993): Agonist-stimulated internalization of the thyrotropin-releasing hormone receptor is dependent on two domains in the receptor carboxyl terminus. J Biol Chem 268:2389–2392.

O'Dowd BF, Hnatowich M, Caron MG, Lefkowitz RJ, Bouvier M (1989): Palmitoylation of the human beta 2-adrenergic receptor. Mutation of Cys341 in the carboxyl tail leads to an uncoupled nonpalmitoylated form of the receptor. J Biol Chem 264:7564–7569.

Okamoto Y, Ninomiya H, Tanioka M, Sakamoto A, Miwa S, Masaki T (1997): Palmitoylation of human endothelinB. Its critical role in G protein coupling and a differential requirement for the cytoplasmic tail by G protein subtypes. J Biol Chem 272:21589–21596.

Okubo K, Hamasaki N, Hara K, Kageura M (1991): Palmitoylation of cysteine 69 from the COOH-terminal of band 3 protein in the human erythrocyte membrane. Acylation occurs in the middle of the consensus sequence of F—I-IICLAVL found in band 3 protein and G2 protein of Rift Valley fever virus. J Biol Chem 266:16420–16424.

Olson EN, Glaser L, Merlie JP (1984): α and β subunits of the nicotinic acetylcholine receptor contain covalently bound lipid. J Biol Chem 259:5364–5367.

Olson EN, Towler DA, Glaser L (1985): Specificity of fatty acid acylation of cellular proteins. J Biol Chem 260:3784–3790.

Ovchinnikov YA, Abdulaev NG, Bgachuk AS (1988): Two adjecent cysteine residues in the C-terminal cytoplasmic fragment of bovine rhodopsin are palmitoylated. FEBS Lett 230:1–5.

Ozols J, Caron JM (1997): Posttranslational modification of tubulin by palmitoylation: II. Identification of sites of palmitoylation. Mol Biol Cell 8:637–645.

Paige LA, Nadler MJ, Harrison ML, Cassady JM, Geahlen RL (1993): Reversible palmitoylation of the protein-tyrosine kinase p56lck. J Biol Chem 268:8669–8674.

Parenti M, Vigano MA, Newman CM, Milligan G, Magee AI (1993): A novel N-terminal motif for palmitoylation of G-protein α subunits. Biochem J 291:349–353.

Pedraza L, Owens GC, Green LA, Salzer JL (1990): The myelin-associated glyco-proteins: Membrane disposition, evidence of a novel disulfide linkage between immunoglobin-like domains, and post-translational palmitoylation. J Cell Biol 111:2651–2661.

Peseckis SM, Deichaite I, Resh MD (1993): Iodinated fatty acids as probes for myristate processing and function. Incorporation into pp60v-src. J Biol Chem 268:5107–5114.

Philipp HC, Schroth B, Veit M, Krumbiegel M, Herrmann A, Schmidt MF (1995): As-sessment of fusogenic properties of influenza virus hemagglutinin deacylated by site-directed mutagenesis and hydroxylamine treatment. Virology 210:20–28.

Pickering DS, Taverna FA, Salter MW, Hampson DR (1995): Palmitoylation of the GluR6 kainate receptor. Proc Natl Acad Sci USA 92:12090–12094.

Portincasa P, Conti G, Chezzi C (1992): Role of acylation of viral haemagglutinin dur-ing the influenza virus infectious cycle. Res Virol 143:401–406.

Pouliot JF, Beliveau R (1995): Palmitoylation of the glucose transporter in blood–brain barrier capillaries. Biochim Biophys Acta 1234:191–196.

Premont RT, Macrae AD, Stoffel RH, Chung N, Pitcher JA, Ambrose C, Inglese J, Mac-Donald ME, Lefkowitz RJ (1996): Characterization of the G protein–coupled recep-tor kinase GRK4. Identification of four splice variants. J Biol Chem 271:6403–6410.

Randall WR (1994): Cellular expression of a cloned, hydrophilic, murine acetyl-cholinesterase. J Biol Chem 269:12367–12374.

Reddy BA, Kloc M, Etkin LD (1991): Identification of the cDNA for xlcaax-1, a mem-brane associated *Xenopus* maternal protein. Biochem Biophys Res Commun 179:1635–1641.

Resh MD (1994): Myristylation and palmitoylation of Src family members: The fats of the matter. Cell 76:411–413.

Resh MD (1996): Regulation of cellular signalling by fatty acid acylation and prenyla-tion of signal transduction proteins. Cell Signal 8:403–412.

Robbins SM, Quintrell NA, Bishop JM (1995): Myristoylation and differential palmi-toylation of the HCK protein-tyrosine kinases govern their attachment to membranes and association with caveolae. Mol Cell Biol 15:3507–3515.

Robinson LJ, Busconi L, Michel T (1995): Agonist-modulated palmitoylation of en-dothelial nitric oxide synthase. J Biol Chem 270:995–998.

Rodgers W, Crise B, Rose JK (1994): Signals determining protein tyrosine kinase and glycosyl-phosphatidylinositol-anchored protein targeting to a glycolipid-enriched membrane fraction. Mol Cell Biol 14:5384–5391.

Rose JK, Adams GA, Gallione CJ (1984): The presence of cysteine in the cytoplasmic domain of the vesicular stomatitis virus glycoprotein is required for palmitate addi-tion. Proc Natl Acad Sci USA 81:2050–2054.

Sadeghi HM, Innamorati G, Dagarag M, Birnbaumer M (1997): Palmitoylation of the V2 vasopressin receptor. Mol Pharmacol 52:21–29.

Scharer CG, Naim HY, Koblet H (1993): Palmitoylation of Semliki Forest virus glyco-proteins in insect cells (C6/36) occurs in an early compartment and is coupled to the cleavage of the precursor p62. Arch Virol 132:237–254.

Schick PK, Walker J (1996): The acylation of megakaryocyte proteins: Glycoprotein IX is primarily myristoylated while glycoprotein Ib is palmitoylated. Blood 87:1377–1384.

Schmidt JW, Catterall WA (1987): Palmitoylation, sulfation, and glycosylation of the al-pha-subunit of the sodium channel: Role of post-translational modifications in chan-nel assembly. J Biol Chem 262:13713–13723.

Schmidt MF, Burns GR (1989): Hydrophobic modifications of membrane proteins by palmitoylation in vitro. Biochem Soc Trans 17:625–626.

Schweizer A, Kornfeld S, Rohrer J (1996): Cysteine34 of the cytoplasmic tail of the cation-dependent mannose 6-phosphate receptor is reversibly palmitoylated and required for normal trafficking and lysosomal enzyme sorting. J Cell Biol 132:577–584.

Schweizer A, Rohrer J, Kornfeld S (1995): Determination of the structural requirements for palmitoylation of p63. J Biol Chem 270:9638–9644.

Shenoy-Scaria AM, Dietzen DJ, Kwong J, Link DC, Lublin DM (1994): Cysteine3 of Src family protein tyrosine kinase determines palmitoylation and localization in caveolae. J Cell Biol 126:353–363.

Shum L, Turck CW, Derynck R (1996): Cysteines 153 and 154 of transmembrane transforming growth factor-alpha are palmitoylated and mediate cytoplasmic protein association. J Biol Chem 271:28502–28508.

Solimena M, Dirkx R Jr., Radzynski M, Mundigl O, De Camilli P (1994): A signal located within amino acids 1–27 of GAD65 is required for its targeting to the Golgi complex region. J Cell Biol 126:331–341.

Staufenbiel M (1987): Ankyrin-bound fatty acid turns over rapidly at the erythrocyte plasma membrane. Mol Cell Biol 7:2981–2984.

Steinhauer DA, Wharton SA, Wiley DC, Skehel JJ (1991): Deacylation of the hemagglutinin of influenza A/Aichi/2/68 has no effect on membrane fusion properties. Virology 184:445–448.

Stevens VL (1995): Biosynthesis of glycosylphosphatidylinositol membrane anchors. Biochem J 310:361–370.

Stoffel RH, Randall RR, Premont RT, Lefkowitz RJ, Inglese J (1994): Palmitoylation of G protein–coupled receptor kinase, GRK6. Lipid modification diversity in the GRK family. J Biol Chem 269:27791–27794.

Stormann TM, Adula DD, Weiner DM, Brann MR (1990): Molecular cloning and expression of a dopamine D2 receptor from human retina. Mol Pharmacol 37:1–6.

Sudo Y, Valenzuela D, Beck-Sickinger AG, Fishman MC, Strittmatter SM (1992): Palmitoylation alters protein activity: Blockade of G(o) stimulation by GAP-43. EMBO J 11:2095–2102.

Sugrue RJ, Belshe RB, Hay, AJ (1990): Palmitoylation of the influenza A virus M2 protein. Virology 179:51–56.

Summers MD, Smith GE (1987): A manual of methods for baculovirus vectors and insect cell culture procedures. Tex Agric Exp Stn Bull 1555:1–56.

Tao N, Wagner SJ, Lublin DM (1996): CD36 is palmitoylated on both N- and C-terminal cytoplasmic tails. J Biol Chem 271:22315–22320.

Towler DA, Gordon JI, Adams SP, Glaser L (1988): The biology and enzymology of eukaryotic protein acylation. Annu Rev Biochem 57:69–99.

Udenfriend S, Kodukula K (1995): How glycosylphosphatidylinositol-anchored membrane proteins are made. Annu Rev Biochem 64:563–591.

van Koppen CJ, Nathanson NM (1991): The cysteine residue in the carboxyl-terminal domain of the m2 acetylcholine receptor is not required for receptor-mediated inhibition of adenylate cyclase. J Neurochem 57:1873–1877.

Veit M, Klenk HD, Kendal A, Rott R (1991): The M2 protein of influenza A virus is acylated. J Gen Virol 72:1461–1465.

Veit M, Nurnberg B, Spicher K, Harteneck C, Ponimaskin E, Schultz G, Schmidt MF (1994): The alpha-subunits of G-proteins G12 and G13 are palmitoylated, but not amidically myristoylated. FEBS Lett 339:160–164.

Veit M, Schmidt MF (1993): Timing of palmitoylation of influenza virus hemagglutinin. FEBS Lett 336:243–247.

Veit M, Sollner TH, Rothman JE (1996): Multiple palmitoylation of synaptotagmin and the t-SNARE SNAP-25. FEBS Lett 385:119–123.

Vialard J, Lalumiere M, Vernet T, Briedis D, Alkhatib G, Henning D, Levin D, Rechardson C (1990): Synthesis of the membrane fusion and hemagglutinin proteins of measles virus, using a novel baculovirus vector containing the β-galactosidase gene. J Virol 64:37–50.

Wedegaertner PB, Chu DH, Wilson PT, Levis MJ, Bourne HR (1993): Palmitoylation is required for signaling functions and membrane attachment of Gq α and Gs α. J Biol Chem 268:25001–25008.

Wolven A, Okamura H, Rosenblatt Y, Resh MD (1997): Palmitoylation of p59fyn is reversible and sufficient for plasma membrane association. Mol Biol Cell 8:1159–1173.

Yang C, Compans RW (1996): Palmitoylation of the murine leukemia virus envelope glycoprotein transmembrane subunits. Virol 221:87–97.

Yang C, Spies CP, Compans RW (1995): The human and simian immunodeficiency virus envelope glycoprotein transmembrane subunits are palmitoylated. Proc Natl Acad Sci USA 92:9871–9875.

Zeng FY, Bhupendra SK, Ansari GAS, Weigel PH (1995): Fatty acylation of the rat asialoglycoprotein receptor. J Biol Chem 270:21382–21387.

Zhang FL, Casey PJ (1996): Protein prenylation: Molecular mechanisms and functional consequences. Annu Rev Biochem 65:241–269.

Zhu H, Wang H, Ascoli M (1995): The lutropin/choriogonadotropin receptor is palmitoylated at intracellular cysteine residues. Mol Endocrinol 9:141–150.

CHAPTER 6

ASSAYS FOR MEASURING RECEPTOR–G PROTEIN COUPLING

QIUBO LI and RICHARD A. CERIONE

Regulation of G Protein–Coupled Receptor Function and Expression,
Edited by Jeffrey L. Benovic.
ISBN 0-471-25277-8   Copyright © 2000 Wiley-Liss, Inc.

1. INTRODUCTION

The G protein–coupled receptor signaling pathways typically comprise three components, a heptahelical receptor, a heterotrimeric guanine nucleotide–binding protein (G protein), and an effector that generates a cellular response, often by changing the levels of a second messenger (reviewed by Gilman, 1987; Lefkowitz and Caron, 1988; Kaziro et al., 1991; Simon et al., 1991; Neer, 1995). The transduction of an extracellular signal into an intracellular response relies on two critical coupling events. The first involves the binding of the receptor to the G protein and results in its activation. The second involves the interactions of the activated G protein with its target/effector molecule. Thus, in a sense, the G protein mediates the signal transduction process like a computer transistor that is able to both turn on and turn off the signaling circuit.

This function of the G protein is dependent on its ability to undergo receptor-mediated GDP/GTP exchange (signal-on state) and hydrolyze its bound GTP (signal-off state). The coupling between the receptor and the G protein, which essentially triggers the on-switch by stimulating G protein activation, is the key first step in the signaling pathway. Understanding the molecular mechanisms underlying receptor–G protein coupling has necessitated the development of sensitive assays that can monitor and measure the functional outcomes of these protein–protein interactions. In this chapter, we first provide a general consideration of the approaches that have been commonly used in assaying receptor–G protein coupling. We then use the rhodopsin/transducin-mediated phototransduction system, which operates in vertebrate vision, to provide specific examples of assays that measure receptor–G protein interactions.

2. ASSAYS FOR MEASURING RECEPTOR–G PROTEIN COUPLING

Functional coupling between the receptor and the G protein is a highly regulated event, which only occurs when the receptor is activated by an extracellular stimuli (e.g., hormone, growth factor, light, odorant). Following stimulation, the receptor has an increased affinity for the heterotrimeric G protein, and, upon the formation of the receptor–G protein complex, the rapid exchange of GDP for GTP on the α subunit of the G protein (Gα) is catalyzed. Due to conformational changes induced by GTP binding, the Gα subunit loses its affinity for the G protein β and γ subunits but now has an enhanced affinity for its target/effector. In some cases, the G protein $\beta\gamma$ complex is also able to bind to a target/effector molecule and influences its activity (for reviews, see Clapham and Neer, 1993; Sternweis, 1994). Thus, the activation of the heterotrimeric G protein provides for a possible bifurcation of signaling events (i.e., Gα and G$\beta\gamma$ interact with distinct targets), as well as for synergistic effector regulation (Gα and G$\beta\gamma$ interact with the same target). In all cases, the resultant signaling events that are triggered by these G protein–target interactions are halted through the GTP hydrolytic activity of the Gα subunit, which returns the activated G protein to its GDP-bound (basal) state and enables the Gα subunit to reassociate with G$\beta\gamma$.

Two types of strategies are possible for assaying receptor–G protein coupling. One strategy involves monitoring the activation event that accompanies GDP–GTP exchange and includes direct measurements of GTP binding (e.g., ^{35}S-GTPγS–binding assays) or assays that monitor different aspects of the activating conformational change within the Gα subunit (changes in tryptophan fluorescence that accompany alterations in the Switch II domain of the G protein or the dissociation of Gα from Gβγ). The second commonly used strategy involves monitoring the increased GTP hydrolysis that occurs as an outcome of the receptor-stimulated GTP binding to the G protein.

2.A. Assays Monitoring Receptor-Mediated G Protein Activation: General Strategies

2.A.a. Receptor-Stimulated GTP Binding to Gα Subunits. Receptor-stimulated guanine nucleotide binding to a Gα subunit, which occurs as an outcome of nucleotide exchange, can be directly measured by assaying the incorporation of radiolabeled guanine nucleotide (^{35}S-GTPγS or $α^{32}$P-GTP is commonly used). Briefly, purified G protein subunits (Gα and Gβγ) are incubated with the receptor (either in a cell-free membrane preparation or incorporated into artificial lipid membranes; see below) in a solution containing ^{35}S-GTPγS. Upon exposure to a signal that activates the receptor for a period of time, the ^{35}S-GTPγS bound to the G protein is separated from the unbound nucleotide with a filtration devise and then counted. The extent of G protein activation is directly reflected by the amount of ^{35}S-GTPγS incorporation. This method allows for the quantitation of the stoichiometry of nucleotide binding and the kinetics of G protein activation.

2.A.b. Changes in the Intrinsic Tryptophan Fluorescence of Gα Subunits.
It is well documented that the intrinsic tryptophan fluorescence of many Gα subunits is highly sensitive to G protein activation (Higashijima et al., 1987a,b; Phillips and Cerione, 1988; Guy et al., 1990). For example, the α subunits of G_O ($α_O$) and transducin ($α_T$) each contains two tryptophan residues and undergoes an ~2-fold enhancement in total fluorescence upon the exchange of GDP for GTP. Structural studies now provide a mechanistic basis for these types of fluorescence changes.

Specifically, virtually all of the heterotrimeric Gα subunits contain a conserved tryptophan residue within their Switch II domain (i.e., one of the regions on the Gα subunit that undergoes a conformational change during the transition from the GDP-bound state to the GTP-bound state). X-ray crystallographic studies of $Gα_{i1}$ and $Gα_T$ (Noel et al., 1993; Coleman et al., 1994; Lambright et al., 1994) indicate that the GTP-induced alteration of Switch II results in a significant movement of the conserved tryptophan residue (e.g., ^{207}Trp in $α_T$), reducing its exposure to solvent and thereby accounting for its enhanced fluorescence. Thus, this change in tryptophan fluorescence provides for a convenient real-time spectroscopic read-out for G protein activation.

Nonetheless, there are some important factors that need to be considered when using this method. One concerns the ratio of signal to noise in detecting the fluorescence change in the Gα subunits (signal). Because Gα subunit activation requires the presence of an activated receptor and the G protein $\beta\gamma$ subunit complex (which increases the affinity of $G\alpha$ for the receptor [Phillips et al., 1992]), there is the potential for significant "background tryptophan fluorescence" (i.e., noise) contributed by the presence of the other proteins in the assay system. One way to circumvent this problem is to use minimal (substoichiometric) amounts of the other required proteins relative to an excess of the $G\alpha$ subunits. This is feasible because it has already been shown that both receptors and $G\beta\gamma$ subunit complexes can in effect act catalytically to promote the activation of the $G\alpha$ subunits (Fung, 1983). The catalytic nature of receptor–G protein–coupled signaling stems from the fact that upon G protein activation (GTP binding), both the receptor and the $\beta\gamma$ subunit complex dissociate from the GTP-bound α subunit. Because we and others have found that a single receptor molecule, and a single $\beta\gamma$ subunit complex, can activate 10–40 $G\alpha$ subunits within several seconds in a reconstituted system (Guy et al., 1990), it is possible to establish fluorescence assays in which the intrinsic tryptophan fluorescence contributed by receptors and $\beta\gamma$ complexes is minimal.

A second issue that must be taken into account when using tryptophan fluorescence to monitor G protein activation is the need to establish detergent-free systems to monitor receptor–G protein coupling. The detergents that have been used to solubilize and purify seven-membrane–spanning receptors and heterotrimeric G proteins interfere with receptor–G protein interactions. Thus, methods have been established to insert purified receptors and heterotrimeric G proteins into reconstituted phospholipid vesicles (Cerione and Ross, 1991). These reconstituted systems provide the necessary phospholipid milieu for efficient receptor–G protein coupling. As long as the molar amount of G protein exceeds the molar amount of lipid vesicles (such conditions are feasible and have been successfully established [Cerione and Ross, 1991]), the background scattering contributions from the vesicles will not obscure the tryptophan fluorescence signal from the $G\alpha$ subunit. The vertebrate vision system (see below) provides a particular advantage along these lines because the G protein transducin is soluble and stable in the absence of detergent. Thus, it is possible to establish phospholipid vesicles that have been reconstituted with rhodopsin and then measure (rhodopsin-stimulated) GTP binding to purified α_T, which together with purified $\beta\gamma_T$ subunits, was added directly to these vesicles. The reconstitution of the rhodopsin/transducin system has worked extremely well for a variety of spectroscopic read-outs of signal transduction (see below).

2.A.c. Assays for Monitoring the Dissociation of Activated $G\alpha$ Subunits From $G\beta\gamma$ Subunit Complexes.

An additional approach that can be used to monitor receptor-stimulated G protein activation involves spectroscopic read-outs of the dissociation of an activated $G\alpha$ subunit from the $G\beta\gamma$ subunit complex. There are two approaches that have been taken. One involves the use of resonance energy transfer in which fluorescent donor and acceptor chromophores are introduced into the $G\alpha$ subunit and $G\beta\gamma$ complex (Mittal et al., 1994; Cerione, 1994). The idea is that, because the emission spectrum of the donor fluorophore overlaps the absorption spectrum of the acceptor chromophore, the association of the donor-labeled $G\alpha$ with the acceptor-labeled $G\beta\gamma$ should result in resonance energy transfer between the chromophores that can be read out by a quenching of the donor fluorescence. Although in principle this read-out provides a direct monitor for the association of GDP-bound Gα subunits with Gβγ subunit complexes, it has not been possible to use it to directly follow the G protein subunit dissociation that accompanies G protein activation. This is because the labeled Gα subunits are not able to undergo a receptor-stimulated activation (GDP–GTP exchange) event (Mittal et al., 1994).

An alternative approach that has been used successfully to monitor the G protein subunit dissociation that accompanies receptor-stimulated activation of Gα subunits involves the introduction of fluorescent reporter groups into reactive cysteine residues within Gβγ subunit complexes (see below). In this case, the idea is to use an environmentally sensitive reporter group attached to βγ cysteine residues that undergoes a change in its fluorescence emission upon the binding of the GDP-bound Gα subunit. Receptor-stimulated activation and accompanying subunit dissociation should then be accompanied by a reversal of the change in the reporter group fluorescence.

An important consideration for both of these approaches is the determination of the stoichiometry and specificity of G protein subunit labeling. A potential problem is that the labeling of a G protein subunit with a chromophore may interfere with function (e.g., by preventing receptor–G protein coupling or G protein activation). This has been a difficulty associated with the fluorescence resonance energy transfer approach, and so it becomes important to carefully ascertain that the chemical modification procedures yield a fully functional, labeled protein. A potential alternative to the chemical modification of reactive lysine or cysteine residues within the wild-type protein is to use site-directed mutagenesis to introduce a single specific reactive residue at a site that will not interfere with protein function.

2.B. Assays for Monitoring Receptor-Stimulated GTPase Activity: General Strategies

The hydrolysis of GTP by Gα is an ultimate outcome of GTP binding (i.e., due to receptor-stimulated GDP–GTP exchange). Although the receptor-stimulated guanine nucleotide exchange event is often extremely rapid (within seconds), the GTP hydrolytic event occurs on a time scale of 0.5 to 1 minute (with turnover numbers usually on the order of $1-4$ min^{-1}). If sufficient GTP is provided to the assay solution, a single receptor can stimulate multiple GTP-binding/GTPase cycles within several G proteins. Thus, direct measurements of the receptor-stimulated GTPase activity, using $\gamma^{32}P$-GTP and as quantitated by ^{32}Pi release, represents a read-out for a receptor-catalyzed GTP-binding/GTP hydrolysis cycle. Therefore, measurements of the hydrolysis of GTP catalyzed by Gα provide an indirect way of monitoring receptor–G protein coupling.

3. THE VERTEBRATE VISION SYSTEM AS A MODEL FOR DEVELOPING ASSAYS FOR RECEPTOR–G PROTEIN COUPLING

The vertebrate vision system offers an especially attractive model for studying receptor–G protein coupling or the ensuing coupling between the G protein and its effector because it is relatively easy to obtain milligram quantities of each of the component proteins from rod outer segments (ROS) and because the relevant signaling interactions can be fully reconstituted *in vitro*. Thus, as described below, we have used this signaling system as a model for developing fluorescence spectroscopic read-outs for different aspects of receptor–G protein coupling.

The primary components of the vision system include the photoreceptor rhodopsin, the heterotrimeric G protein transducin, and the effector enzyme cGMP phosphodiesterase (PDE). The signaling cascade is initiated by a photon-induced conformational change in rhodopsin, which maximizes its affinity for transducin and results in conformational changes within the α_T subunit that allow GDP–GTP exchange. The activation of the α_T subunit is accompanied by its dissociation from the $\beta\gamma_T$ complex and its subsequent binding and activation of the cGMP PDE. The stimulation of PDE activity, which causes a hyperpolarization of the retinal rod membranes and triggers the signal sent to the optic nerve, continues until the GTP bound to α_T is hydrolyzed to GDP. This GTP hydrolytic event effectively returns the signaling system to its starting point.

4. EXPERIMENTAL APPROACHES

4.A. Preparing Functional Rhodopsin

A necessary first step in assaying the functional interactions between rhodopsin and transducin requires the functional reconstitution of rhodopsin with α_T and the $\beta\gamma_T$ subunit complex. The isolation of G protein–coupled receptors in a highly purified and functional form is a very challenging task. This is due to the unique molecular architecture of this family of receptors whose members are seven-membrane–spanning proteins. To optimally couple to the G protein, the receptor must be associated with a lipid membrane to maintain its membrane-spanning conformation. This situation is usually achieved in two ways; one is to prepare native membrane fractions from the cells in which the receptor is present, and the other is to incorporate the purified receptor into artificial lipid vesicles.

4.A.a. Preparation of Rhodopsin-Containing Rod Outer Segment (ROS) Membranes. A number of methods can be used to obtain cell-free membrane preparations that are essentially devoid of peripheral membrane proteins and still retain functional rhodopsin. For the purpose of assaying receptor–G protein coupling, we have found the following method to be the most reliable and convenient.

The ROS are isolated from frozen, dark-adapted bovine retinas following the procedure originally described by Fung (1983). First, the ROS are subjected to a series of extensive washes, using an isotonic buffer (10 mM Hepes, pH 7.5, 100 mM NaCl, 5 mM $MgCl_2$, 1 mM DTT) and a hypotonic buffer (10 mM Hepes, pH 7.5, 1 mM DTT). The ROS then are extracted sequentially with solutions containing decreasing concentrations of urea (4 M, 2 M, and 1 M urea in 50 mM Hepes, pH 7.5, 2 mM EDTA, 1 mM DTT). This step removes most of the peripheral membrane proteins and some of the integral membrane proteins from the ROS membranes. Finally, the ROS are washed twice and resuspended in a solution of 20 mM Hepes, pH 7.5, and 1 mM DTT. During the washes, the ROS membranes are collected by centrifugation at 100,000g for 30 minutes at 4°C. All steps are carried out in the dark or dim red light to prevent the photobleaching of rhodopsin. The rhodopsin concentration in the final ROS membrane preparation (typically 10 μM) can be determined from the absorbance at 498 nm, using an extinction coefficient of 42,700 M^{-1} cm^{-1}.

The membrane preparation described above contains rhodopsin molecules that retain their ability to couple to transducin. Although these preparations are mostly free of endogenous transducin and cGMP PDE, a number of other membrane proteins are still present. This then becomes a consideration when using assays in which low background of contaminating proteins is required (e.g., when assaying rhodopsin-stimulated tryptophan fluorescence changes in transducin). In these cases, it is advantageous to use pure rhodopsin in reconstituted phospholipid vesicle systems.

4.A.b. Incorporation of Rhodopsin Into Lipid Vesicles. Procedures for the purification of rhodopsin from the ROS of bovine retina have been well established and basically follow the original methods outlined by Litman (1982). The first descriptions of the insertion of purified rhodopsin into phospholipid vesicles involved extensive dialysis to remove the detergent used in its solubilization and purification. However, subsequently, a more flexible and rapid approach was designed taking advantage of the detergent-binding resin Extracti-gel (Pierce Chemical Co., Rockford, IL). A typical protocol is as follows.

The purified rhodopsin is added in the dark to a lipid-detergent mixture, which is prepared by incubating 50 μl of a 17 mg/ml solution of phosphatidylcholine (Sigma, St. Louis, MO, type II-S; sonicated to clarity in a bath sonicator) with 25 μl of a 17% (w/v) solution of octyl glucoside (Calbiochem, La Jolla, CA) and 225–325 μl of 10 mM Hepes (pH 7.5), 100 mM NaCl, 1 mM dithiothreitol (DTT), and 5 mM MgCl$_2$ for 30–45 minutes on ice. Usually, 100–200 μl of 30 μM purified rhodopsin (as determined from the absorbance of the retinal mixture at 498 nm, using an extinction coefficient of 42,700 M^{-1} cm^{-1}) is added to the phospholipid/detergent incubation to make a final volume of 500 μl. The entire mixture is then added to a 1-ml column of Extracti-gel resin. The idea is that the resin will bind detergent and thereby allow the reformation of the phosphatidylcholine-containing vesicles (which had been solubilized by their initial incubation with octyl glucoside). The removal of detergent allows the rhodopsin molecules to be incorporated into the re-formed lipid vesicles.

Before loading the rhodopsin-lipid-detergent mixture, the Extracti-gel column is pre-equilibrated with 2 ml of 10 mM Hepes (pH 7.5), 100 mM NaCl, and 1 mg/ml bovine serum albumin (to prevent the nonspecific adsorption of rhodopsin to the Extract-gel) and washed with 2 ml of elution buffer (10 mM Hepes, pH 7.5, 100 mM NaCl, 1 mM DTT, and 5 mM MgCl$_2$). The rhodopsin-containing lipid vesicles are then collected by washing the column with 2 ml of elution buffer. Using this procedure, we have typically obtained efficiencies of incorporation of approximately 50%. Half of the incorporated rhodopsin will have the cytoplasmic domains facing the outside of the vesicles and then are able to couple to transducin.

A slight modification of this procedure has been used to prepare rhodopsin-containing phospholipid vesicles in smaller volumes (Guy et al., 1990). In these cases, 16 μl of soybean phosphatidylcholine is incubated with 9 μl of octyl glucoside and 75 μl of 10 mM Hepes (pH 7.5)/100 mM NaCl for 30 minutes on ice. Purified rhodopsin (1–5 μl from a 5-μM stock solution) is then added to the lipid/octyl glucoside mixture. This mixture (150 ml) is added to 0.5 ml Extracti-gel resin in a 1-ml syringe. The Extracti-gel is pretreated as outlined above, and then the rhodopsin-containing vesicles are eluted from the resin with 250 μl of vesicle elution buffer (this yields ~400 μl of rhodopsin/vesicles).

4.B.   Specific Assays for Monitoring Rhodopsin–G Protein Coupling

4.B.a. Rhodopsin-Stimulated ^{35}S-GTPγS Binding to Transducin.   Assays of the rhodopsin-stimulated binding of ^{35}S-GTPγS to α_T are performed by first inserting rhodopsin into phosphatidylcholine vesicles (as outlined in the preceding section) and then adding purified α_T and $\beta\gamma_T$ subunits to the rhodopsin-containing lipid vesicles. The α_T and $\beta\gamma_T$ subunit complexes are purified from ROS as follows. The ROS isolated from bovine retina are washed extensively in the isotonic and hypotonic buffer solutions as described above. The holotransducin molecules are extracted from the ROS in hypotonic buffer containing 100 μM GTP, following incubation on ice for 30 minutes under room light. The α_T subunits are then resolved from the $\beta\gamma_T$ subunit complexes by Blue Sepharose chromatography as outlined in detail by Phillips et al. (1989). To measure rhodopsin-stimulated GDP/GTPγS exchange on α_T (i.e., as measured by the binding of ^{35}S-GTPγS), 100–250 nM of the purified α_T subunit and 50 nM of the $\beta\gamma_T$ complex are mixed with 25 nM of rhodopsin (either in a cell-free membrane preparation or in phospholipid vesicles; see preceding section) in a final volume of 200 μl of 20 mM Hepes (pH 7.5), 5 mM MgCl$_2$, 1 mM DTT, and 100 mM NaCl (GTP-binding buffer). This solution is incubated in room light for 2–3 minutes at room temperature to allow photon excitation of rhodopsin to occur. The GDP/GTPγS exchange is initiated by the addition of a mixture of GTPγS and ^{35}S-GTPγS to a final concentration of 1 μM. The exchange reaction is allowed to proceed for 20 minutes at room temperature and is then stopped by the addition of 2 ml of ice-cold GTP-binding buffer. The assay solution is then subjected to filtration through a 0.45-μm nitrocellulose membrane with a suction manifold devise. The filter membranes are further washed twice with 3 ml of ice-cold GTP-binding buffer, air-dried, and counted for retention of ^{35}S-GTPγS in 3 ml of Liquiscint (National Diagnotics). The intrinsic ability of α_T to undergo nucleotide exchange is assayed under the same conditions as above except that rhodopsin is omitted; nonspecific binding of nucleotide to the filters is assayed in the absence of α_T.

Because GTPγS is poorly hydrolyzable, it will remain tightly bound to α_T such that this assay represents a single-turnover measurement for receptor-mediated G protein activation. The number of α_T molecules that are activated within a certain time period by a given number of rhodopsin molecules can be calculated based on the ratio of the bound ^{35}S-GTPγS versus total α_T protein. Figure 6.1 shows an example of a time course for rhodopsin-stimulated ^{35}S-GTPγS binding obtained at different levels of rhodopsin.

4.B.b. Rhodopsin-Stimulated Intrinsic Fluorescence Changes.   The intrinsic tryptophan fluorescence of α_T is enhanced in response to GTP binding (ac-

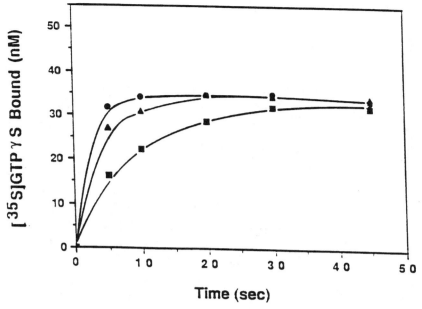

Figure 6.1. Rhodopsin-stimulated GTPγS binding to α_T. Rhodopsin-containing phosphatidylcholine vesicles were prepared as described in the text and added to the assay solution to yield different final concentrations of rhodopsin: 13 nm (■), 25 nM (▲), or 40 nm (●). These vesicles were incubated with $\beta\gamma_T$ (67 nM), α_T (100 nM), and 60 nM ^{35}S-GTPγS at room temperature. At the indicated time, an aliquot (40 μl) was removed from the binding assay and applied to nitrocellulose filters to measure the amount of bound nucleotide. (Data from Guy et al., 1990.)

tivation), and this fluorescence change is essentially fully reversed upon GTP hydrolysis (de-activation) (Fig. 6.2). These fluorescence changes appear to directly reflect the movement of the Switch II domain of α_T during its activation–deactivation cycle (see above) and thus provide a real-time spectroscopic read-out for G protein activation.

Typically, a fluorescence experiment is performed by incubating 500 nM of the purified α_T subunit with 50 nM of $\beta\gamma_T$ and 50 nM of rhodopsin-containing phospholipid vesicles. This incubation is performed in 150 μl of 10 mM Hepes (pH 7.5), 5 mM MgCl$_2$, 1 mM DTT, 100 mM NaCl, together with 0.5–1 μM of GTP or GTPγS. It is important to use conditions in which there is a significant excess of α_T relative to rhodopsin and $\beta\gamma_T$ (and thus take advantage of the abilities of rhodopsin and $\beta\gamma_T$ to act catalytically in activating α_T); otherwise, the background tryptophan fluorescence from rhodopsin and $\beta\gamma_T$ will mask the changes in α_T fluorescence. This mixture is then added to a stirred quartz cuvette in the dark (i.e., to prevent premature activation of α_T by light-activated rhodopsin). Following exposure to room light, the tryptophan fluorescence emission from α_T is monitored at 1-second intervals by an SLM 8000c spectrofluorometer (SLM, Urbana, IL) operated in the ratio mode (with emission at 335 nm and excitation at 280 nm). The fluorescence signals can be corrected for background scattering and for contributions from rhodopsin and $\beta\gamma_T$ by making the same measurements in the absence of added α_T.

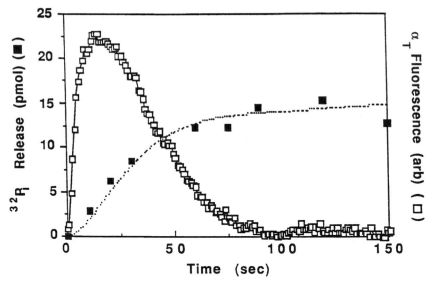

Figure 6.2. Comparisons of the rates of the rhodopsin- and GTP-induced fluorescence decay with the rate for GTP hydrolysis. Rhodopsin-containing phosphatidylcholine vesicles (final concentration of rhodopsin in the assay = 20 nM) were prepared as described in the text and incubated with $\beta\gamma_T$ (250 nM) and α_T (250 nM) for 10 minutes in the light at room temperature. GTP or γ^{32}P-GTP (final concentration = 300 nM) was added to the incubation. In the former case, the fluorescence emission at 335 nm was continuously monitored (excitation = 280 nM), while in the latter case ^{32}P-P$_i$ release, due to GTP hydrolysis, was measured (see text). The dotted line represents the integration of the fluorescence signal (described by Guy et al., 1990). (Data from Guy et al., 1990.)

When GTP is used in the assay, the expected fluorescence changes are an immediate increase in tryptophan emission followed by a slower reversal of the fluorescence change (with kinetics indistinguishable from those for GTP hydrolysis as measured by the release of ^{32}P$_i$; see below). However, when GTPγS is used in the assay solution, deactivation is essentially prevented because of the extremely slow hydrolysis of this nucleotide analogue. Thus, a persistence of the enhanced fluorescence state is observed.

4.B.c. Additional Fluorescence Read-Outs. The use of labeled G protein subunits provides an alternative approach for monitoring G protein–α subunit activation in real time. In particular, we have used $\beta\gamma_T$ subunits labeled with fluorescent cysteine reagents for this purpose (Phillips and Cerione, 1991). The first step (the chemical modification of $\beta\gamma_T$) is performed by dialyzing the purified retinal $\beta\gamma_T$ complex versus 10 mM Hepes (pH 7.5), 5 mM MgCl$_2$, 0.15 M NaCl, and 20% glycerol to remove the DTT that is present from the final stages of the purification of $\beta\gamma_T$. The dialyzed $\beta\gamma_T$ (usually 100–500 μg) is then incubated with a fluorescent cysteine reagent such as iodoacetamidofluorescein or 2-(4′-maleimidylanilino) naphthalene-6-sulfonic acid (MIANS).

We have had especially good success with MIANS; the reaction is usually performed by incubating the $\beta\gamma_T$ with 1 mM MIANS (in a volume of 100–500 μl) for 1 hour at room temperature. The MIANS-labeled $\beta\gamma_T$ can then be removed from free MIANS by hydroxylapatite chromatography. A 0.5-ml bed-volume column can be used, with about 0.5 ml of the reaction mixture being applied. The column is then washed with 4–6 volumes of 20 mM sodium phosphate, and the MIANS-labeled $\beta\gamma_T$ is eluted with 0.1 M phosphate. The stoichiometry of incorporation of MIANS into the $\beta\gamma_T$ complex (which is typically 3–5 labels per mole complex) can be determined by using the molar extinction coefficient of MIANS (17,000 at 325 nM) and the known amount of protein (determined using the Cu^+/bicinchoninic acid method; Pierce).

We have previously shown that the MIANS-labeled $\beta\gamma_T$ serves as an excellent extrinsic reporter complex for the binding of this subunit complex to α_TGDP (Phillips and Cerione, 1991). The addition of purified α_TGDP (0.7 μM) to a mixture of rhodopsin-containing lipid vesicles (~0.04 μM rhodopsin) and MIANS-$\beta\gamma_T$ (0.75 μM) results in an immediate enhancement of the MIANS fluorescence (excitation = 320 nM, emission = 420 nm) due to the interactions of the MIANS-labeled $\beta\gamma_T$ with the α_T subunit. The addition of GTP then elicits a decrease in the MIANS fluorescence (reflecting some of the α_TGTP species dissociating from the MIANS–$\beta\gamma_T$ complex); however, the change is small and levels off (presumably due to GTP hydrolysis, which allows reassociation of MIANS–$\beta\gamma_T$ with α_T-GDP) (Fig. 6.3, bottom). After GTP hydrolysis is complete, the addition of AlF_4^-

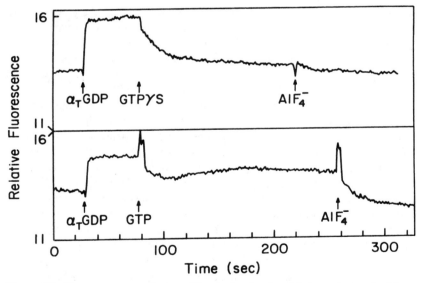

Figure 6.3. Fluorescence monitoring of the rhodopsin-stimulated guanine nucleotide exchange activity of α_T using MIANS-labeled $\beta\gamma_T$ subunit complexes. **Top:** MIANS fluorescence (excitation = 322 nm; emission = 420 nm) of an incubation that included rhodopsin-containing phosphatidylcholine vesicles (0.034 μM rhodopsin) and MIANS-labeled $\beta\gamma_T$ (0.37 μM). The fluorescence was continually monitored at one determination per second, and then α_T (0.68 μM) was added at the time indicated and the contents of the cuvette were mixed. GTPγS (3.3 μM) and AlF_4^- (5 mM NaF/20 μM $AlCl_3$, final concentrations) were added and mixed at the indicated times. **Bottom:** An identical experiment was performed except that GTP (0.67 μM) was added instead of GTPγS. All incubations were performed at room temperature. (Data from Phillips and Cerione, 1991.)

will fully reverse the fluorescence enhancement, presumably due to the complete dissociation of the α_TGDP/AlF$_4^-$ species from the MIANS–$\beta\gamma_T$ complex. The traces at the top of Figure 6.3 show that, when GTPγS is used instead of GTP, the enhancement of the MIANS fluorescence is fully reversed due to the activation and dissociation of (GTPγS-bound) α_T from MIANS–$\beta\gamma_T$. This then represents a true real-time read-out for the rhodopsin-promoted exchange of GTPγS for GDP on α_T and its accompanying dissociation from the $\beta\gamma_T$ complex.

4.B.d. Rhodopsin-Stimulated GTP Hydrolysis by Transducin.

A relatively convenient assay for effective receptor–G protein coupling involves measurements of the GTP hydrolytic activity that follows the activation of the G protein and reflects its deactivation and return to the basal state. This can be most conveniently assayed with γ^{32}P-GTP.

The assay is set up in much the same way as that for measuring ^{35}S-GTPγS binding (see above), starting with rhodopsin-containing lipid vesicles to which are added pure α_T and the $\beta\gamma_T$ complex. The rhodopsin-stimulated binding of γ^{32}P-GTP to α_T is accompanied by the hydrolysis of the bound, radiolabeled GTP and the release of ^{32}P$_i$. The turnover number of the hydrolytic reaction is on the order of four per minute. An example of a time course for rhodopsin-stimulated GTP hydrolysis obtained under essentially single-turnover conditions is shown in Figure 6.2. However, at concentrations of GTP >> [α_T], the ability of rhodopsin to continue to catalyze the exchange of GDP (i.e., the product of the GTPase reaction) for fresh substrate (γ^{32}P-GTP) allows a linear measurement of product (^{32}P$_i$) release over a period of 30 minutes to 1 hour. This then can provide a relatively sensitive read-out of rhodopsin–G protein coupling because the free α_T subunit (in the absence of rhodopsin) can only slowly exchange GTP for GDP and thus does not contribute much background GTP hydrolysis.

The assay is typically started by adding γ^{32}P-GTP to a final concentration of 1 μM to a reaction mixture containing rhodopsin-containing lipid vesicles, α_T, and $\beta\gamma_T$ under similar conditions to those used to measure ^{35}S-GTPγS–binding activity. At the end of a 1–30 minute incubation, the reaction is quenched by the addition of 5 ml of 10% ammonium molybdate solution (1 M HCl). The released [^{32}P$_i$] is extracted in 5 ml of a 1:1 mixture of isobutanol and benzene and quantitated by liquid scintillation counting. Nonspecific or intrinsic GTP hydrolysis is measured in the absence of α_T or rhodopsin, respectively. Various investigators have used charcoal to adsorb γ^{32}P-GTP and separate the nonhydrolyzed guanine nucleotide from the free inorganic phosphate (Higashijima et al., 1987a). However, in our experience, we have found that the reproducibility of the assay is better with the isobutanol/benzene extraction procedure.

5. SUMMARY

In this chapter we describe a number of reconstitution assays that can be used to obtain quantitative information regarding receptor–G protein coupling. The advantage of using these *in vitro* reconstituted systems is that one can modify or vary the amount of each of the signaling components in a well-defined manner. However, a key consideration is that these assay systems should faithfully reproduce the characteristics of the signaling molecules *in vivo*.

This obviously requires the development of purification procedures for isolating each of the signaling components in a functional form. In the case of the G protein–coupled receptors, it is essential that they be associated with a lipid membrane in order to be functional. Incorporation of purified receptors into artificial lipid bilayers with the aid of an appropriate detergent has proven to be an effective way to reconstitute receptor function, as read out by receptor-stimulated activation of its appropriate G protein. With the wide use of recombinant protein expression systems, it is now possible to obtain receptors and/or G protein subunits in relatively high quantity and in modified forms by site-directed mutagenesis (Li and Cerione, 1997). Thus, the types of read-outs described above for rhodopsin and transducin are becoming increasingly feasible for a wide range of receptor–G protein-coupled signaling systems. This then should provide for exciting new insights into the detailed mechanisms by which receptors activate G proteins.

REFERENCES

Cerione RA (1994): Fluorescence assays for G-protein interactions. Methods Enzymol 237:409–423.

Cerione RA, Ross EM (1991): Reconstitution and quantitation of regulated adenylate cyclase activity. Methods Enzymol 195:329–342.

Clapham DE, Neer EJ (1993): New roles for G-protein βγ-dimers in transmembrane signaling. Nature 365:403–406.

Coleman DE, Berghuis AM, Lee E, Linder ME, Gilman AG, Sprang SR (1994): Structures of active conformations of Gi alpha 1 and the mechanism of GTP hydrolysis. Science 265:1405–1412.

Fung BKK (1983): Characterization of transducin from bovine retinal rod outer segments. I. Separation and reconstitution of the subunits. J Biol Chem 25:10495–10502.

Gilman AG (1987): G Proteins: Transducers of receptor-generated signals. Annu Rev Biochem 56:615–649.

Guy PM, Koland JG, Cerione RA (1990): Rhodopsin-stimulated activation–deactivation cycle of transducin. Biochemistry 29:6954–6963.

Higashijima T, Ferguson KM, Smigel MD, Gilman AG (1987a): The effect of GTP and Mg^{2+} on the GTPase activity and the fluorescent properties of G_o. J Biol Chem 262:757–761.

Higashijima T, Ferguson KM, Sternweis PC, Ross EM, Smigel MD, Gilman AG (1987b): The effect of Mg^{2+} and the βγ subunit complex on the interactions of guanine nucleotides with G proteins. J Biol Chem 262:762–766.

Kaziro Y, Itoh H, Kozasa T, Nakafuku M, Satoh T (1991): Structure and function of signal-transducing GTP-binding proteins. Annu Rev Biochem 60:349–400.

Lambright DG, Noel JP, Hamm HE, Sigler PB (1994): Structural determinants for activation of the alpha-subunit of a heterotrimeric G protein. Nature 369:621–628.

Lefkowitz RJ, Caron MG (1988) Adrenergic receptors: Models for the study of receptors coupled to guanine nucleotide regulatory proteins. J Biol Chem 263:4993–4996.

Li Q, Cerione RA (1997): Communication between Switch II and Switch III of the transducin α subunit is essential for target activation. J Biol Chem 272:21673–21676.

Litman BJ (1982): Ultraviolet circular dichroism of rhodopsin in disk membranes and
detergent solution. Methods Enzymol 81:150–153.

Mittal R, Cerione RA, Erickson JW (1994): Aluminum fluoride activation of bovine
transducin induces two distinct conformational changes in the α subunit. Biochemistry 33:10178–10184.

Neer EJ (1995): Heterotrimeric G proteins: Organizers of transmembrane signals. Cell
80:249–257.

Noel JP, Hamm HE, Sigler PB (1993): The 2.2 Å crystal structure of transducin-alpha
complexed with GTPγS. Nature 366:654–663.

Phillips WJ, Cerione RA (1988): The intrinsic fluorescence of the alpha subunit of transducin. Measurement of receptor-dependent guanine nucleotide exchange. J Biol
Chem 263:15498–15505.

Phillips WJ, Cerione RA (1991): Labeling of the βγ subunit complex of transducin with
an environmentally sensitive cysteine reagent. Use of fluorescence spectroscopy to
monitor transducin subunit interactions. J Biol Chem 266:11017–11024.

Phillips WJ, Trukawinski S, Cerione RA (1989): An antibody-induced enhancement of
the transducin-stimulated enhancement of the cyclic GMP phosphodiesterase. J Biol
Chem 264:16679–16688.

Phillips WJ, Wong SC, Cerione RA (1992): Rhodopsin/transducin interactions: II, Influence of the transducin–βγ subunit complex on the coupling of the transducin-α
subunit to rhodopsin. J Biol Chem 267:17040–17046.

Simon MI, Strathmann MP, Gautam N (1991): Diversity of G proteins in signal transduction. Science 252:802–808.

Sternweis PC (1994): The active role of βγ in signal transduction. Curr Opin Cell Biol
6:198–203.

PART 2

PROTEINS THAT REGULATE RECEPTOR FUNCTION

CHAPTER 7

EXPRESSION AND ACTIVITY OF G PROTEIN–COUPLED RECEPTOR KINASES

ROBERT P. LOUDON, ALEXEY N. PRONIN, and
JEFFREY L. BENOVIC

Regulation of G Protein–Coupled Receptor Function and Expression,
Edited by Jeffrey L. Benovic.
ISBN 0-471-25277-8   Copyright © 2000 Wiley-Liss, Inc.

I. INTRODUCTION

Many transmembrane signaling systems consist of specific G protein–coupled receptors (GPCRs) that transduce the binding of a diverse array of extracellular stimuli into intracellular signaling events. GPCRs modulate the activity of numerous effector molecules, including various adenylyl cyclases, phosphoinositol $(PI)_3$ kinases, nonreceptor tyrosine kinases, small G proteins, phospholipases, phosphodiesterases, and ion channels (Gutkind, 1998). To ensure that extracellular signals are translated into intracellular signals of appropriate magnitude and specificity, these signaling cascades need to be tightly controlled. It is well recognized that GPCRs are subject to three principle modes of regulation: (1) desensitization, the process by which a receptor becomes refractory to continued stimuli; (2) internalization, whereby receptors are physically removed from the cell surface by endocytosis; and (3) down-regulation, where total cellular receptor levels are decreased (Penn and Benovic, 1998).

GPCR desensitization is primarily mediated by second-messenger responsive kinases, such as protein kinase A (PKA) and protein kinase C (PKC), and by G protein–coupled receptor kinases (GRKs). GRKs specifically phosphorylate activated GPCRs and initiate the recruitment of additional proteins, termed *arrestins,* that further receptor desensitization and internalization. The six mammalian GRKs that have been identified can be divided into three subfamilies based on their overall structural organization and homology: GRK1 (rhodopsin kinase); GRK2 (βARK1) and GRK3 (βARK2); and GRK4, GRK5, and GRK6. Common features shared by the GRKs include a centrally localized catalytic domain of ~270 amino acids, an N-terminal domain of ~184 amino acids that has been implicated in GPCR interaction, and a variable length C-terminal domain of 105–233 amino acids that is involved in phospholipid association (Fig. 7.1). The ability of GRKs to phosphorylate and regulate the activity of GPCRs has been extensively studied and has been the subject of several recent reviews (Pitcher et al., 1998; Krupnick and Benovic, 1998; Carman and Benovic, 1998; Iacovelli et al., 1999).

In this chapter, we describe procedures for analyzing endogenous and overexpressed GRKs using immunoblotting techniques as well as analysis of GPCR phosphorylation. In addition, procedures for expressing GRKs in mammalian cells and expressing and purifying GRKs from Sf9 insect cells are also described.

2. GRK EXPRESSION IN MAMMALIAN CELLS

Two basic strategies have been successfully used to analyze GRK expression in mammalian cells. The first utilizes GRK-specific antibodies to analyze cell or

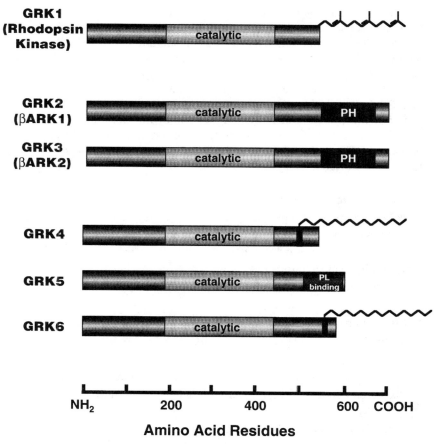

Figure 7.1. Domain architecture of GRKs. The sequences of the six known mammalian GRKs are represented schematically. The putative sites of farnesylation of GRK1 and palmitoylation of GRK4 and 6 are shown. PH, pleckstrin homology domain; PL, phospholipid.

tissue extracts for GRK levels by immunoblotting analysis. This method often provides sufficient specificity and sensitivity to enable detection of multiple GRKs in crude cellular extracts. Another method of measuring GRK expression involves analysis of GRK activity using a receptor substrate such as rhodopsin. This is discussed in Section 4; although it has proven useful, it is generally less sensitive than immunoblotting and also does not enable discrimination among the various GRK subtypes.

2.A. Immunoblotting of GRKs

There are three major considerations when attempting to analyze GRK levels by immunoblotting. The first is that GRKs contain phospholipid-binding domains that often result in the kinases being present in both cytosolic and membrane-associated fractions following cell lysis. Although such fractions can be individually generated and analyzed, it is often advantageous to homogenize cells or tissues using a buffer that results in the GRKs being primarily in the soluble fraction. A second consideration is the availability and quality of GRK antibodies. Several useful GRK-specific antibodies have been generated by various laboratories and GRK antibodies are also commercially available (Table 7.1). A third consideration is the level of a particular GRK in a given cell or tissue. Although endogenous GRK levels are generally low in most cells, GRK2 appears to be the most abundant and can be readily detected by immunoblotting in most mammalian cells. By comparison, GRK3, 4, 5, and 6 are typically found at lower levels and are thus less readily detected.

2.A.a. Materials Required

Polyacrylamide gel apparatus and Western blot transfer set (Bio-Rad miniprotean II)

Power supply (500 V minimum)

1.5-ml Eppendorf tubes

Pipettes capable of dispensing 1–20 μl and 20–200 μl

Disposable pipette tips

Microcentrifuge

Vortex

Disposable gloves

Nitrocellulose transfer membrane (NitroME from Micron Separations, Inc.)

Plastic containers for immunoblotting and washing nitrocellulose membranes

X-ray film (Kodak X-OMAT AR)

Enhanced chemiluminescence reagents (Amersham ECL)

Various chemicals as outlined below (Sigma)

2.A.b. Preparation of Cellular Lysate.
The general lysis procedure that we use helps to extract both cytosolic and membrane-associated proteins because most GRKs can bind to phospholipids. However, although these extracts are optimal for assessing endogenous (or overexpressed) GRK levels, these preparations should not be used for directly assessing activity because GRKs are potently inhibited by the detergent present in the lysis buffer. We describe a procedure in this section for partially purifying GRKs from such cell homogenates. This results in preparations that can be readily assayed.

1. Prepare 5 ml of lysis buffer as follows:

Component	Volume	Final Concentration
0.5 M Na Hepes, pH 7.5 (in H$_2$O)	200 μl	20 mM
10% Triton X-100 (in H$_2$O)	500 μl	1%
1 M NaCl (in H$_2$O)	750 μl	150 mM
0.5 M Na EDTA, pH 7 (in H$_2$O)	100 μl	10 mM

TABLE 7.1. GRK Antibodies

Antibody Name	Specificity	Type	Epitope	Reference
G8	GRK1 specific	Mouse mAb	Human GRK1 C-terminal epitope	Zhou et al. (1998)
GRK1 (G-8)	GRK1 specific	Mouse mAb	Human GRK1 C-terminal epitope	Santa Cruz Biotechnology
D11	GRK1 specific	Mouse mAb	Human GRK1 N-terminal epitope	Zhao et al. (1998)
3A10	GRK2 specific	Mouse mAb	Bovine GRK2 (aa 525–530)	Loudon et al. (1996)
C5/1	GRK2 ~GRK3	Mouse mAb	Rat GRK3 (aa 466–510)	Opperman et al. (1996)
C5/1				Upstate Biotechnology
A16/17	GRK4 ~GRK5 ~GRK6	Mouse mAb	Bovine GRK5 (aa 463–521)	Opperman et al. (1996)
Ab17–34	GRK1	Rabbit polyclonal	Bovine GRK1 (aa 17–34)	Palczewski et al. (1993)
Ab216–237	GRK1	Rabbit polyclonal	Bovine GRK1 (aa 216–237)	Palczewski et al. (1993)
Ab483–497	GRK1	Rabbit polyclonal	Bovine GRK1 (aa 483–497)	Palczewski et al. (1993)
Ab539–556	GRK1	Rabbit polyclonal	Bovine GRK1 (aa 539–556)	Palczewski et al. (1993)
Anti-RK	GRK1	Rabbit polyclonal	Bovine GRK1	Palczewski et al. (1993)
Anti-GST-RKCT	GRK1	Rabbit polyclonal	Bovine GRK1 (aa 502–561)	Inglese et al. (1992)
UW54	GRK1b specific	Rabbit polyclonal	Human GRK1 (aa 463–598)	Zhao et al. (1998)
Anti-βARK1	GRK2 specific	Rabbit polyclonal	Rat GRK2 (aa 468–689)	Arriza et al. (1992)
Ab1	GRK2	Rabbit polyclonal	Bovine GRK2 (aa 648–655)	Penela et al. (1998)
Ab9	GRK2	Rabbit polyclonal	Bovine GRK2	Penela et al. (1998)
AbFP1	GRK2	Rabbit polyclonal	Bovine GRK2 (aa 50–145)	Penela et al. (1998)
AbFP2	GRK2	Rabbit polyclonal	Bovine GRK2 (aa 436–689)	Penela et al. (1998)
βARK-1	GRK2	Rabbit polyclonal	Bovine GRK2	Winstel et al. (1996)

(continued)

TABLE 7.1. GRK Antibodies (*Continued*)

Antibody Name	Specificity	Type	Epitope	Reference
Anti-βARK2	GRK3 specific	Rabbit polyclonal	Rat GRK3 (aa 468–688)	Arriza et al. (1992)
GRK4-CT	GRK4	Rabbit polyclonal	Human GRK4α (aa 464–578)	Premont et al. (1996)
N-Ab	GRK4	Rabbit polyclonal	Human GRK4δ (aa 52–114)	Sallese et al. (1997)
Anti-GRK5	GRK5 ~GRK6	Rabbit polyclonal	Bovine GRK5 (aa 463–590)	Premont et al. (1994)
TJ54	GRK5 specific	Rabbit polyclonal	Human GRK5 (aa 98–136)	Pronin and Benovic (1997)
PK2	GRK5 specific	Rabbit polyclonal	Human GRK5 (aa 571–590)	Pronin and Benovic (1997)
PF2	GRK5 ~GRK6	Rabbit polyclonal	Human GRK5 (aa 489–590)	Pronin and Benovic (1997)
TJ56	GRK6 specific	Rabbit polyclonal	Human GRK6 (aa 98–136)	Loudon et al. (1996)
GRK1 (C-20)	GRK1 specific	Rabbit polyclonal	C-terminal epitope	Santa Cruz Biotechnology
GRK2 (C-15)	GRK2 specific	Rabbit polyclonal	C-terminal epitope	Santa Cruz Biotechnology
GRK3 (C-14)	GRK3 specific	Rabbit polyclonal	C-terminal epitope	Santa Cruz Biotechnology
GRK4 (1-20)	GRK4(γ,δ) specific	Rabbit polyclonal	C-terminal epitope	Santa Cruz Biotechnology
GRK4 (K-20)	GRK4(α,β) specific	Rabbit polyclonal	N-terminal epitope	Santa Cruz Biotechnology
GRK5 (C-20)	GRK5 specific	Rabbit polyclonal	C-terminal epitope	Santa Cruz Biotechnology
GRK6 (C-20)	GRK6 specific	Rabbit polyclonal	C-terminal epitope	Santa Cruz Biotechnology

aa, amino acids.

0.1 M PMSF (in isopropanol)	25 μl	0.5 mM
10 mg/ml leupeptin (in H_2O)	10 μl	20 μg/ml
10 mg/ml benzamidine (in H_2O)	50 μl	100 μg/ml
Deionized H_2O	3.36 ml	

Note: Stock solutions can be made ahead of time and stored either at room temperature (Hepes, Triton X-100, NaCl, EDTA) or at −20°C (leupeptin, benzamidine). Phenylmethylsulfonyl fluoride (PMSF) should be made fresh weekly and stored at room temperature. The volume of lysis buffer can be scaled up or down as needed but should be made fresh and chilled on ice before use.

2. Cells are grown in complete media to 75%–90% confluence and then rinsed several times with phosphate-buffered saline (PBS). Ice-cold lysis buffer is then added (0.5–1 ml of buffer per 100-mm dish of cells); the cells are scraped off using a rubber policeman, transferred to a centrifuge tube, and then vortexed several times. Tissues can be homogenized with an ~10:1 (vol:wet weight) ratio of buffer to tissue (i.e., 1 ml of buffer per 100 mg of minced tissue) using a polytron homogenizer (two bursts for 20 seconds at maximum speed).

3. Lysates are centrifuged (40,000g for 10 minutes in an SS34 rotor or equivalent) to remove particulate matter, and the supernatant is transferred to an Eppendorf tube and assayed for protein (Bio-Rad Protein Assay using bovine serum albumin as a standard). Cell/tissue supernatants can either be further analyzed immediately or aliquoted, frozen in liquid nitrogen, and stored at −80°C for later analysis. It is important to keep everything on ice and to perform the procedures as rapidly as possible to prevent proteolysis.

2.A.c. Electrophoresis and Immunoblotting Analysis. While GRKs are found at relatively low concentrations in most tissues, there are a number of GRK antibodies that have proven useful for immunoblotting of mammalian cell lysates (see Table 7.1). Here we describe a general procedure for analyzing GRK levels in crude homogenates.

1. Incubate 20–40 μg of protein from the cell supernatant prepared as described in Section 2.A.b. with SDS sample buffer (Laemmli, 1970) for 10 minutes at room temperature. It is advantageous to also include samples containing 1–5 ng of the purified GRK being assessed (see Section 3.A.) as well as a prestained protein standard.

2. Electrophorese the samples on a 10% SDS-PAGE at a constant voltage of 120–150 V until the dye front is within a few millimeters of the bottom (~1 hour).

3. Carefully remove the stacking gel and then set up the running gel for transfer to nitrocellulose. This is accomplished by layering a transfer sponge, one piece of Whatman 3 MM paper, the running gel, one piece of nitrocellulose (with one corner cut on the protein standard side), one piece of Whatman 3 MM paper, and another sponge. Everything

should initially be wetted in transfer buffer (25 mM Tris base, 192 mM glycine, 20% methanol; *do not adjust pH*) before layering. It is also important to make sure that all air bubbles are removed as each layer is added (a 13 × 100 mm plastic test tube can be rolled over the layer at each step to accomplish this). Transfer the protein samples to a nitrocellulose membrane for 1 hour at 100 V (constant voltage) using cold transfer buffer.

4. After transfer, sample loading and transfer efficiency can be checked by staining the nitrocellulose with 0.2% Ponceau S (prepared in deionized water) for ~1 minute with the transfer side up. Transferred proteins can then be visualized by rinsing the nitrocellulose with deionized water to remove excess stain.

5. Rinse the nitrocellulose several times in Tris-buffered saline (20 mM Tris-Cl, pH 7.5, 150 mM NaCl) containing 0.05% Tween 20 (TTBS). The membrane is then blocked for 1 hour in 10 ml of TTBS containing 5% (wt/vol) nonfat dry milk.

6. The nitrocellulose is next incubated with 5 ml of a diluted GRK-specific antibody (see Table 7.1) for ≥1 hour at room temperature. The antibody dilution will vary depending on the particular antibody being used and the tissue being analyzed.

7. Wash the nitrocellulose three to five times for 10 minutes each in TTBS.

8. Incubate the nitrocellulose for 1 hour with an ~1:3,000 dilution of affinity-purified goat antimouse or antirabbit IgG conjugated to horseradish peroxidase (Bio-Rad) in TTBS-5% milk.

9. Wash the membrane three to fives times for 10 minutes each in TTBS.

10. Overlay the nitrocellulose with 1–2 ml of enhanced chemiluminescence (ECL) reagent for ~1 minute, allow the blot to drip dry, wrap in Saran wrap, and visualize on x-ray film. An example of a GRK2 immunoblot analysis of various mammalian cell lines using the GRK2–specific monoclonal antibody 3A10 is shown in Figure 7.2.

2.B. Methods for GRK Isolation

Although immunoblotting of crude homogenates is the preferred method for determining the GRKs present in a particular cell, such crude preparations are often not very useful for analysis of GRK activity. Here we outline a procedure for partially purifying either endogenous or overexpressed GRKs to enable more detailed characterization. This procedure not only provides a GRK preparation that can be readily assayed, but it also effectively separates several of the GRK isoforms. Such fractions can then be analyzed by activity analysis and/or immunoblotting.

2.B.a. Partial Purification of GRKs by SP-Sepharose Chromatography.

To perform biochemical studies of overexpressed or endogenous GRKs, we have found it useful to partially purify these proteins because various factors within crude lysates significantly inhibit GRK activity. Stepwise fractionation on the cation exchange resin SP-Sepharose (Pharmacia) permits adequate res-

Figure 7.2. GRK2 expression in various mammalian cell lines. Twenty-five micrograms of protein from eight different mammalian cell lysates were electrophoresed on a 10% SDS polyacrylamide gel, transferred to nitocellulose, and probed with the GRK2–specific monoclonal antibody 3A10. The blot was then incubated with an HRP-conjugated goat antimouse secondary antibody (1:2,000 dilution) and visualized by ECL.

olution of GRKs from the majority of lysate proteins and, in addition, results in partial separation of specific GRK isoforms. For example, while GRK2, 3, 5, 6 all bind to SP-Sepharose at low ionic strength (e.g., 50 mM NaCl), GRK2 and 3 can be eluted with 150 mM NaCl, GRK6 with 400 mM NaCl, and GRK5 with 600 mM NaCl. Below we describe a procedure for the batchwise purification of GRKs from mammalian cell lysates.

1. Prepare a cell or tissue lysate as outlined in section 2.A.2. Dilute the lysate 1:4 with buffer A (20 mM Hepes, pH 7.5, 10 mM EDTA, 0.5 mM PMSF, 20 µg/ml leupeptin, 200 µg/ml benzamidine) to reduce the concentration of salt (this is particularly important for the purification of GRK2/3).
2. Add the diluted lysate to SP-Sepharose resin (~1 mg total protein/30 µl resin) prewashed in buffer A containing 50 mM NaCl and incubate for 1 hour on a rotator at 4°C. Pellet the resin by centrifugation (~10 seconds in a microcentrifuge is adequate), and discard the supernatant.
3. Wash the resin two to three times with ~1 ml of cold buffer A containing 50 mM NaCl to remove unbound proteins.
4. GRK2 (and GRK3) are eluted three times with 100 µl buffer A containing 150–200 mM NaCl by incubating the resin for 10 minutes with elution buffer on a rotator at 4°C. Briefly centrifuge (~10 seconds) the resin between elutions. The eluants can be pooled or assayed separately for peak activity.

5. The resin can then be eluted with cold buffer A containing 400 mM NaCl to obtain a GRK6-enriched fraction, following the same procedure as used in step 4 above. GRK5 can subsequently be eluted with buffer A containing 600 mM NaCl. It is important to note, however, that some GRK5 will likely elute in the 400 mM NaCl fraction. Thus, if complete separation of GRK5 and 6 is needed, it is recommended that the sample be fractionated on a Mono S FPLC column using a linear NaCl gradient.

It is best to aliquot and store the samples at $-80°C$ and use the preparations within several weeks of purification. Expression and recovery of GRKs can be evaluated by immunoblotting and rhodopsin phosphorylation assays (see Sections 2.A. and 4.A.). If the GRKs are overexpressed, typical yields are 1–10 µg of partially purified GRK per T75 flask of COS-1 cells. Recovery of endogenous GRKs using this procedure will vary depending on the cell type or tissue being analyzed.

2.B.b. Immunoprecipitation of GRKs.

2.B.b. Immunoprecipitation of GRKs. When detection of low levels of endogenous GRK activity is desired or when ion exchange chromatography does not sufficiently resolve the activities of two or more GRKs present in the same cell type, it may be useful to isolate a particular GRK isoform using immunoaffinity-based techniques. Such preparations can be analyzed by immunoblotting or activity analysis. For example, immune complexes containing GRK6 from hematopoietic cells and cell lines have been used to assay kinase activity *in vitro* (Loudon et al., 1996).

The number of cells to be used for such analysis will depend on the level of GRK expression in a particular cell line. The following procedure describes the immunoprecipitation of GRK6 from HL-60 cells.

1. Incubate 3 µg of affinity-purified polyclonal rabbit anti-GRK6 antibody (Santa Cruz Biotechnology, Inc.) with 20 µl of a 50% suspension of protein A–agarose beads (Boehringer Mannheim). This is done in the presence of 300 µl of buffer B (0.2% Triton X-100, 150 mM NaCl, 20 mM Hepes, pH 7.5, 10 mM EDTA, 0.5 mM PMSF, 20 µg/ml leupeptin, 200 µg/ml benzamidine) in an Eppendorf tube for 1 hour at 4°C.

2. Wash the beads three times with 300 µl of ice-cold buffer B. Completely remove the buffer from the beads after the third wash, and keep the beads on ice.

3. Wash $5–10 × 10^6$ cells several times in ice-cold PBS and lyse them in 100 µl of 1% Triton X-100 lysis buffer as described in Section 2.A.b.

4. Add the lysate to the beads, and rotate the mixture for 1 hour at 4°C.

5. Wash the beads three times with 300 µl buffer B followed by a single wash with 300 µl of 20 mM Hepes, pH 7.5, and 10 mM EDTA to remove the detergent and salt from the kinase immunocomplex.

6. The kinase can then be eluted by boiling the beads for 10 minutes with 25 µl SDS sample buffer followed by SDS-PAGE and immunoblotting.

Alternatively, the beads can be directly assayed for GRK activity using a nonreceptor substrate such as casein or phosvitin (Loudon et al., 1996). Receptor substrates can also be used, although the ability of such preparations to phosphorylate GPCRs is often inhibited (possibly due to the antibody blocking GPCR interaction sites on the GRK).

2.C. Methods for GRK Expression

It is often advantageous to manipulate cellular GRK levels and/or activity to enable an assessment of the effect of a GRK on a particular receptor. In this section, we briefly describe the strategies for transiently expressing wild-type and dominant-negative mutant GRKs in mammalian cells. Subsequent assessment of GRK levels and activity using *in vitro* methods can be achieved as outlined in sections 2.A. and 4.A. In addition, intact cell assessment of the effect of wild-type or dominant-negative GRK on the function of a particular GPCR can also be analyzed by assessing such processes as GPCR phosphorylation (see Chapter 2), GPCR/G protein coupling (see Chapter 6), and receptor internalization (see Chapters 8, 10, and 11).

2.C.a. Materials Required

Laminar flow hood dedicated to tissue culture use

Tissue culture incubator capable of maintaining 37°C, 5% CO_2

Dulbecco's modified Eagle's medium (DMEM), containing 10% fetal calf serum and 100 µg/ml penicillin and streptomycin (Gibco/BRL)

Dulbecco's PBS (Gibco/BRL)

Trypsin/EDTA (Gibco/BRL)

FuGENE Transfection Reagent (Boehringer Mannheim)

Disposable, sterile serological pipettes and tissue culture dishes

Plasmid DNA containing the gene of interest (column or cesium chloride purified)

Disposable, sterile pipette tips

Disposable gloves

1.5-ml sterile Eppendorf tubes

15-ml and 50-ml sterile, conical polypropylene tubes (Falcon)

Light microscope

2.C.b. Transient Overexpression of GRKs.
Overexpression of GRKs in mammalian cells enables functional assessment of GRK–GPCR interaction as well as the recovery of sufficient quantities of GRKs for biochemical analysis. Although a number of transfection strategies exist, transient transfection of GRK cDNAs using polycationic lipid reagents, such as LipofectAMINE (Life Technologies, Inc.) and FuGENE (Boehringer Mannheim), is a relatively easy and reproducible method of overexpressing GRKs. It should be noted that careful optimization of the transfection protocol should be done to ensure efficient

transfection of the cell line of interest. Attention should be paid to factors such as cell type, cell density (confluence), DNA and transfection reagent concentrations, and duration of transfection.

Typically, we use an adherent cell line such as the African green monkey kidney COS-1 or the human embryonic kidney cell line HEK 293 to perform our transfections. The cells should be seeded 24 hours before transfection such that they are 50%–90% confluent at the time of transfection. The procedure provided is for the use of FuGENE as the transfection reagent, although LipofectAMINE also works well.

1. Aliquot the plasmid DNA (~3–5 μg per 60-mm dish) into a 1.5-ml Eppendorf tube. In a separate Eppendorf tube, add 0.3 ml of warmed DMEM followed by 10 μl of FuGENE added dropwise with swirling. Do not allow the FuGENE to directly contact the sides of the tube. Allow the FuGENE:DMEM mixture to incubate for 5 minutes at room temperature. We generally maintain a 3:1 ratio of FuGENE:DNA (e.g., 12 μl FuGENE to 4 μg DNA). Additional guidelines for optimizing transfection efficiency depending on the cell type and plasmid are provided in the manufacturer's directions.

2. Add the DMEM/FuGENE mixture to the DNA dropwise, mix by gentle inversion of the tube, and leave at room temperature for 15 minutes. The FuGENE:DNA mixture is then added evenly to the cell monolayer dropwise while swirling gently. There is no need to change the existing medium.

3. The transfection mixture can be left on the cells until the time of harvest. The cells are typically harvested 24–48 hours post-transfection for immunoblot analysis. Cells may also be split 5 hours after transfection, or on the following day, for subsequent functional analysis (see Section 4).

3. EXPRESSION AND PURIFICATION OF GRKs USING Sf9 INSECT CELLS

Overexpression and purification of GRKs has provided a useful tool for detailed analysis of enzyme kinetics, assessment of *in vitro* substrate specificity, and use as a protein standard for quantitative immunoblotting. While all of the GRKs have been successfully expressed in mammalian cells, expression in insect cells results in a high level of expression and enables purification of milligram quantities of protein at a low cost. Attempts at expressing functional GRKs in bacteria have thus far proved unsuccessful. Although most of the GRKs have been successfully expressed in insect cells, we mainly describe the procedures for overexpressing and purifying GRK2 (Kim et al., 1993). The reader should refer to previous publications for specific details of GRK1 (Cha et al., 1997), GRK3 (Kim et al., 1993), GRK5 (Kunapuli et al., 1994), and GRK6 (Loudon and Benovic, 1994) overexpression and purification.

3.A. Sf9 Cell Culture and Baculovirus Preparation

3.A.a. Materials Required

Wild-type *Spodoptera frugiperda* (Sf9) cells from American Type Culture
   Collection

TNM-FH medium (Sigma)

Fetal bovine serum (Sigma or Life Technologies, Inc.)

Fungizone

Streptomycin (Sigma)

Penicillin (Sigma)

Pluronic F-68 (Life Technologies, Inc.)

3.A.b. Sf9 Cell Culture. Sf9 cells can be cultured in a 27°C incubator either
on a monolayer or in suspension (spinner flask, 70 rpm) using TNM-FH
medium containing 10% fetal bovine serum and antibiotics (0.25 μg/ml Fungizone, 50 μg/ml streptomycin, 50 μg/ml penicillin). For large scale cultures
(≥1 liter), Sf9 cells are grown in a shaking incubator (120 rpm) at 27°C in complete medium supplemented with 0.1% Pluronic F-68. Cells in monolayer are
typically used during the early stages of baculovirus isolation and amplification, whereas suspension cells are used for larger scale baculovirus amplification and protein expression. In general, although Sf9 cells are easy to grow,
they need to be split every 2–3 days and kept at a cell density between 0.5 and
2×10^6 cells/ml.

3.A.c. Recombinant Baculovirus Preparation. In this section we briefly describe the preparation of a recombinant baculovirus containing the bovine
GRK2 cDNA. Preparation of a GRK2–containing baculovirus was accomplished by initially excising the open reading frame of the GRK2 cDNA by
cleavage of the clone pβARK3A (Benovic et al., 1989) with the restriction enzymes *Nhe*I and *Eco*RI. The resulting 2.1 kb insert, containing 22 bp of 5' and
11 bp of 3' untranslated sequence, was isolated by electrophoresis on a 1%
agarose gel, purified, and then treated with Klenow in the presence of dNTPs
to produce blunt ends. The isolated GRK2 cDNA was ligated into the baculovirus transfer vector pJVP10 (cut with *Nhe*I and blunted) and used to trans-
form DH5α cells. Plasmid DNA from the transformed colonies was restriction
mapped to determine the insert orientation and the resulting transfer vector was
termed pJV-GRK2 (Kim et al., 1993).

Sf9 insect cells grown on a monolayer were co-transfected with 2 μg of pJV-
GRK2 and 1 μg of wild-type AcNPV DNA by the calcium-phosphate precipitation technique (Summers and Smith, 1987). The transfected cells were
allowed to recover and produce phage particles in culture media for 6 days. The
virus laden medium was then centrifuged and used to infect a fresh monolayer
of Sf9 cells. The infected cells were overlaid with low melting agarose in
medium containing the β-galactosidase indicator X-Gal (150 μg/ml). After a 4-
day incubation, recombinant viruses were selected by picking blue plaques produced in the agarose overlaid cell culture. The virus was eluted into TNM-FH
culture medium and used to reinfect Sf9 cells. This procedure was repeated

three times to isolate a recombinant virus free of the wild-type baculovirus. The isolated virus was then amplified and used to infect Sf9 cells in suspension. The presence of the full-length GRK2 cDNA within the genome of the recombinant baculovirus were confirmed by Southern blotting. The recombinant baculovirus was identified as AcNPV-GRK2.

3.B. Expression and Purification of GRK2

3.B.a. Materials Required

GRK2–containing recombinant baculovirus

Sf9 insect cells

Reagents for culturing Sf9 cells as described in Section 3.A.a.

Chromatography resins such as S-Sepharose and Heparin-Sepharose (Pharmacia)

3.B.b. Expression of GRK2 in Sf9 Cells

1. Culture 1 liter of Sf9 cells at 27°C in a 2.8-liter Fernbach flask to a density of ~2.5 × 10^6 cells/ml. This is best accomplished in a shaking incubator, but the cells can also be cultured in a spinner flask to a density of ~2 × 10^6 cells/ml. If cultured in a shaking incubator, the medium should be supplemented with 0.1% Pluronic F-68.

2. The cells are removed from the incubator and then infected with the recombinant baculovirus (we typically use ~25 ml of amplified virus per liter of cells, resulting in a multiplicity of infection of 5–10). The cells are incubated for 1 hour at room temperature and then returned to the incubator.

3.B.c. Purification of GRK2 From Sf9 Cells

1. Baculovirus-infected cells are normally harvested at 40–48 hours postinfection. The cells should be carefully monitored during this period because longer infection times often result in cell lysis and loss of cytosolic proteins. The cells are centrifuged at 1,000g for 15 minutes in a Beckman J-6B Centrifuge or equivalant. The supernatant is discarded, and the pellet is resuspended in ice-cold phosphate buffered saline (PBS) and centrifuged again at 1,000g for 15 minutes.

2. The washed pellet is resuspended in 150 ml of ice-cold homogenization buffer (20 mM Na-Hepes, pH 7.2, 250 mM NaCl, 5 mM EDTA, 1 mM PMSF, 3 mM benzamidine) and homogenized using a Brinkman Polytron (2 × 30 sec at 25,000 rpm). The homogenization and all subsequent steps are performed at 4°C. The homogenate is centrifuged at 45,000g to remove unbroken cells and nuclei (SS34 rotor or equivalent), and the supernatant is then recentrifuged at 300,000g for 60 minutes (TI60 rotor or equivalent).

3. Dilute the high-speed supernatant to ~600 ml with ice cold buffer C (20 mM Na-Hepes, pH 7.2, 5 mM EDTA, 0.02 % Triton X-100) and load the sample on an 18-ml S-Sepharose column (1.5 × 10 cm) equi-

librated with buffer C. The column is washed with 50 mM NaCl in buffer C and then eluted with a 120-ml linear gradient from 50 to 300 mM NaCl in buffer C at a flow rate of 1 ml/min (GRK2 elutes at ~150 mM NaCl). Fractions from the S-Sepharose column are monitored for GRK2 activity by rhodopsin phosphorylation as described in Section 4. In addition, protein purity is monitored by SDS-PAGE and Coomassie blue staining.

4. The peak fractions are pooled (~ 50 ml), diluted threefold with buffer C, and loaded on a 7-ml Heparin-Sepharose column (1.5 × 4 cm). The column is washed with 100 mM NaCl in buffer C, and GRK2 activity is then eluted with a 100-ml linear gradient from 100 to 600 mM NaCl in buffer C at a flow rate of 1 ml/min. The fractions are analyzed for GRK2 purity, activity, and protein concentration. The appropriate fractions are then pooled, concentrated (typically to 1–1.5 mg/ml), stabilized by addition of ~25% glycerol, and stored at –20°C.

4. ANALYSIS OF GPCR PHOSPHORYLATION BY GRKs

4.A. Assay of GRK Activity Using Rhodopsin

Many protein kinases can be assayed using readily available substrates such as histones, phosvitin, casein, or specific synthetic peptides. Unfortunately, these general kinase substrates are relatively poor substrates for GRKs and thus cannot be readily used to assay GRK activity in crude cell or tissue homogenates. However, GRKs do have the ability to specifically phosphorylate activated GPCRs. Thus, one very useful assay utilizes the ability of GRKs to phosphorylate rhodopsin in a light-dependent fashion. This assay uses urea-treated bovine rod outer segments (ROS) as the substrate. Enough ROS for a few thousand assays can be readily prepared from bovine retinas in about 12 hours. The procedure described below details the preparation of urea-treated ROS and a phosphorylation procedure using either crude cellular homogenates or purified GRK preparations.

4.A.a. Materials Required

Polyacrylamide gel apparatus (Bio-Rad mini-protean II)

Power supply (500 V minimum)

1.5-ml Eppendorf tubes

Pipettes capable of dispensing 1–20 µl and 20–200 µl

Disposable pipette tips

High-speed centrifuge (RC5 or equilvalent) and rotors (SS34 and SW27 or equilvalent)

Vortex

Disposable gloves

40-ml Wheaton Dounce homogenizer with an A pestle

50-ml syringe with a 20.5-gauge needle

X-ray film (Kodak X-OMAT AR)
Various chemicals as outlined below (Sigma)
Frozen bovine retinas (dark adapted if available)

4.A.b. Preparation of Urea-Treated Rod Outer Segments.

The procedure for preparing urea-treated ROS is described for 50 bovine retinas but can be readily scaled up or down as needed. All steps in this procedure should be carried out in a darkroom using a safelight. All procedures should be performed on ice except where noted.

1. Thaw 50 frozen bovine retinas in 50 ml of ice-cold homogenization buffer (prepared by dissolving 34 g sucrose in 66 ml 65 mM NaCl, 2 mM MgCl$_2$, 10 mM Tris-acetate, pH 7.4) in a 200-ml Erlenmeyer flask (keep the flask on ice). Shake vigorously for 4 minutes and then centrifuge at 4,000 rpm for 4 minutes in an SS34 rotor. Carefully remove the supernatant with a sucrose needle and transfer to a clean 200-ml Erlenmeyer flask.

2. Add 100 ml of 10 mM Tris-acetate, pH 7.4, buffer to the supernatant and centrifuge at 19,000 rpm for 20 minutes in an SS34 rotor. Check a small amount of the supernatant and if it is strongly orange, dilute with additional buffer and recentrifuge. Resuspend the pellets in a minimal volume of 0.77 M sucrose buffer (prepared in 10 mM Tris-acetate, pH 7.4, 1 mM MgCl$_2$). Pool the tubes and adjust the volume to ~30 ml with the 0.77 M sucrose buffer.

3. Homogenize the sample with 15 vigorous strokes using a 40-ml Wheaton Dounce homogenizer equipped with an A pestle. Draw up the homogenate through a 20.5-gauge needle several times followed by an additional Dounce homogenization with 5–10 strokes. The sample can then be layered onto a stepwise sucrose gradient composed of 5 ml of 1.14 M sucrose, 5 ml of 1.00 M sucrose, and 5 ml of 0.84 M sucrose (all dissolved in 10 mM Tris-acetate, pH 7.4, 1 mM MgCl$_2$).

4. Centrifuge the tubes at 26,000 rpm for 30 minutes in an SW27 rotor (36-ml swinging bucket). Carefully remove the band at the 0.84/1.00 sucrose interface with a sucrose needle (the total volume should be ~25 ml). Dilute the sample with an equal volume of 0.77 M sucrose buffer, vortex briefly, and then centrifuge at 18,000 rpm for 20 minutes in an SS34 rotor. Discard the supernatant, and resuspend the pellets as described below. Alternatively, this is a convenient stopping point, so the tubes can also be wrapped in aluminum foil and stored at −80°C.

5. Resuspend the pellets in a total volume of 25 ml of 50 mM Tris-HCl, pH 7.4, 5 mM EDTA, 5 M urea. Sonicate the sample for 4 minutes in a tube immersed in ice. This step denatures the endogenous rhodopsin kinase activity without harm to the rhodopsin.

6. Dilute the sample with 100 ml of 50 mM Tris-HCl, pH 7.4, and spin in an ultracentifuge at 45,000 rpm for 45 minutes (Ti60 rotor or equivalent). Resuspend the pellets in 50 mM Tris-HCl, pH 7.4, and recentrifuge a total of three times. The final pellet should be resuspended in a total vol-

ume of 5–10 ml of 50 mM Tris-HCl, pH 7.4 (using a polytron on low speed helps with the resuspension). The urea-treated ROS can be aliquoted (20–30 μl is convenient), wrapped in aluminum foil, frozen in liquid nitrogen, and stored at –80°C until needed. The typical yield is 5–10 mg of rhodopsin from a 50 retina preparation.

4.A.c. *Phosphorylation of Rhodopsin by GRKs.* The phosphorylation of urea-treated ROS by crude kinase preparations requires the separation of the phosphorylated rhodopsin from other endogenous phosphorylated proteins. The separation of rhodopsin is readily accomplished by quenching the reaction with sodium dodecyl sulfate (SDS) sample buffer followed by electrophoresis on a 10% polyacrylamide gel and autoradiography. The resultant bands can then be cut and counted if quantitative results are needed. It is important to note that GRKs are very sensitive to inhibition by salt (e.g., 0.1 M NaCl inhibits GRK2 activity by ~90%). Thus, the ionic strength in the reaction should be kept as low as possible. The procedure for assaying GRK activity using urea-treated ROS is detailed below.

1. Prepare an assay mixture containing (per 10 assays)

    100 μl 20/2 buffer (20 mM Tris-HCl, pH 7.5, 2 mM EDTA)

    1 μl 0.5 M MgCl$_2$

    1 μl 10 mM ATP

    ~10 μCi [γ^{32}P]ATP (NEG-002A from NEN Dupont)

    10 μl urea-treated ROS (~1 mg/ml rhodopsin) (add just before use)

2. Assays are set up containing 10 μl of assay mixture (from step 1) and ~2 μl of the kinase preparation. The samples are incubated for 5–20 minutes at 30°C in room light (control incubations should be kept in the dark). It is important that the ROS be kept in the dark until just before use or else the amount of phosphate incorporated into rhodopsin will be significantly reduced.

3. Following the incubation period, reactions can be quenched several different ways. If the kinase preparation is a crude homogenate, the reactions can be stopped by the addition of 0.2 ml of cold 100 mM sodium phosphate, pH 7, and 5 mM EDTA buffer followed by centrifugation in a tabletop ultracentrifuge. The pellets can be resuspended in 20 μl of SDS sample buffer (sonication and/or vortexing helps with the resuspension) followed by electrophoresis on a homogeneous 10% polyacrylamide gel. The gel is dried and exposed to x-ray film. The centrifugation step is not absolutely necessary; however, it does significantly improve the signal-to-noise ratio on the autoradiogram. With more purified kinase preparations, the reactions can be directly quenched with SDS sample buffer followed by electrophoresis. For both methods, it is important that the samples be incubated at room temperature for ~30 minutes following the addition of SDS sample buffer. This is needed to obtain adequate denaturation of the rhodopsin. It is also important to note that the sample should not be boiled because this will cause aggregation of the receptor.

4.B. GPCR Phosphorylation by GRKs

Although the ability of GRKs to phosphorylate rhodopsin has proved useful for assessing GRK activity and regulation, it is also useful to be able to characterize the phosphorylation of other GPCRs with GRKs. In this section, we describe some of the *in vitro* strategies for characterizing GRK phosphorylation of additional GPCRs. Successful strategies for studying GPCR phosphorylation by GRKs have utilized purified GPCRs that are present in either phospholipid vesicles or mixed micelles. In addition, strategies for phosphorylating membrane preparations containing overexpressed GPCRs have also been developed. In both cases, it is advantageous to use purified GRK preparations for such analysis (see Section 3). Endogenous and/or overexpressed GRKs, however, can also be utilized for many of these studies.

4.B.a. Phosphorylation of Purified GPCRs. Although a relatively small number of GPCRs have been purified, GRK phosphorylation of several purified receptors (e.g., β_2AR, M_2 AChR) has provided useful kinetic information (Kim et al., 1993; Kunapuli et al., 1994; Loudon and Benovic, 1994). Such studies require a relatively highly purified receptor that is then reinserted into a phospholipid bilayer (GPCR purification and reconstitution techniques are discussed in Chapter 3). Alternatively, the β_2AR has also been successfully used as a GRK substrate in mixed micelle preparations (Onorato et al., 1995). While such purified preparations can be phosphorylated with either highly purified or even relatively crude GRK preparations, a key feature of these assays is the lipid requirement. Reconstitution of the receptor into phospholipid vesicles is often necessary to generate a functionally active receptor that can interact with GRKs in an agonist-dependent manner. In addition, the GRKs appear to be phospholipid-dependent enzymes whose activity is largely controlled by association with acidic lipids (DebBurman et al., 1995a; Onorato et al., 1995; Pitcher et al., 1998). Once a purified reconstituted receptor preparation is available, GRK phosphorylation can be studied using a procedure similar to that described for the phosphorylation of rhodopsin in Section 4.A (Fig. 7.3). Such analysis can provide information about the ability of a receptor to serve as a GRK substrate, the specificity of the phosphorylation (e.g., different GRKs, PKA, PKC), the stoichiometry and potential sites of phosphorylation, and the kinetics of phosphorylation (K_m, V_{max}). In addition, the ability to stoichiometrically phosphorylate a purified receptor preparation enables additional analysis to be performed such as for the effects of phosphorylation on G protein coupling (see Chapter 6). Overall, such results provide important information concerning potential mechanisms of GPCR regulation.

4.B.b. Phosphorylation of GPCRs in Membrane Preparations. Because it is not always possible to generate a highly purified receptor preparation, strategies for phosphorylating GPCRs enriched in membrane preparations have also been developed. An important feature of such strategies is an adequate overexpression system for preparing an enriched-receptor preparation. The most successful assays for analyzing GRK phosphorylation in membrane preparations have involved the use of GPCRs expressed in Sf9 insect cells. This

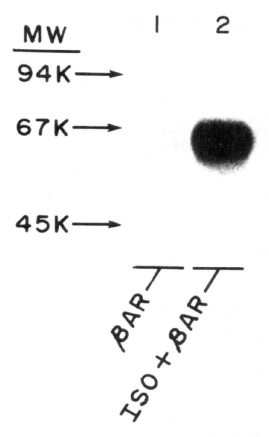

Figure 7.3. Analysis of β_2AR phosphorylation by GRK2. Reconstituted β_2AR was incubated with partially purified GRK2 for 30 minutes at 30°C in the absence (lane 1) or presence (lane 2) of 10 μM (–)isoproterenol. The phosphorylated receptor was electrophoresed on a 10% SDS polyacrylamide gel and then analyzed by autoradiography.

is partly due to the somewhat higher level of expression that one can achieve in insect cells than in mammalian cells as well as to the more homogenous glycosylation that occurs in insect cells (the heterogeneity of receptor glycosylation observed in mammalian cells often reduces the signal-to-noise ratio of phosphorylation reactions). Once cells that express a high level of a particular GPCR are generated (ideally ≥10 pmol/mg membrane protein), the membranes need to be treated to generate a usable substrate for GRK phosphorylation.

One way of treating such preparations is essentially identical to the preparation of urea-treated ROS (i.e., the membranes are treated with urea). This treatment appears to effectively dissociate/denature many of the peripherally associated proteins and protein kinases without harming the GPCR. Although such a strategy has been successfully used with the M_2 AChR (DebBurman et al., 1995b) (see Chapter 3), one needs to empirically determine that such a treatment is not harmful to the receptor of interest. If the receptor appears to be denatured by this treatment, antagonist binding to the receptor usually helps to stabilize the receptor during urea treatment. Alternatively, other protein denaturants can also be tested.

A second strategy that has been used to generate a GPCR-containing membrane preparation that could serve as an effective GRK substrate involved differential sucrose gradient fractionation (Pei et al., 1994). This procedure makes use of the finding that some GPCRs occur in a light vesicle fraction that can be readily separated from plasma membranes. Such preparations yield a receptor that is highly enriched (100–300 pmol/mg) and can serve as an effective GRK substrate (see Chapter 9 for a detailed procedure). As discussed above, the use of a purified GRK has proved most effective for looking at GPCR phosphorylation in membranes; however, overexpressed and/or partially purified GRKs should also work in such assays.

REFERENCES

Arriza JL, Dawson TM, Simerly RB, Martin LJ, Caron MG, Snyder SH, Lefkowitz RJ (1992): The G-protein–coupled receptor kinases βARK1 and βARK2 are widely distributed at synapses in rat brain. J Neurosci 12:4045–4055.

Benovic JL, DeBlasi A, Stone WC, Caron MG, Lefkowitz RJ (1989): Primary structure of the β-adrenergic receptor kinase delineates a potential multigene family of receptor specific kinases. Science 246:235–240.

Carman CV, Benovic JL (1998): G protein–coupled receptors: Turn ons and turn offs. Curr Opin Neurobiol 8:335–344.

Cha K, Bruel C, Inglese J, Khorana HG (1997): Rhodopsin kinase: Expression in baculovirus-infected insect cells and characterization of post-translational modifications. Proc Natl Acad Sci USA 94:10577–10582.

DebBurman SK, Ptasienski J, Benovic JL, Hosey MM (1995a): Lipid-mediated regulation of G protein–coupled receptor kinases. J Biol Chem 270:5742–5747.

DebBurman SK, Kunapuli P, Benovic JL, Hosey MM (1995b): Agonist-dependent phosphorylation of human muscarinic receptors in insect Sf9 cell membranes by G protein–coupled receptor kinases. Mol Pharmacol 47:224–233.

Gutkind JS (1998): The pathways connecting G protein–coupled receptors to the nucleus through divergent mitogen-activated protein kinase cascades. J Biol Chem 273:1839–1842.

Iacovelli L, Sallese M, Mariggio S, de Blasi A (1999): Regulation of G-protein–coupled receptor kinases by calcium sensor proteins. FASEB J 13:1–8.

Inglese J, Glickman JF, Lorenz W, Caron MG, Lefkowitz RJ (1992): Isoprenylation of a protein kinase: Requirement of farnesylation/α-carboxyl methylation for full enzymatic activity of rhodopsin kinase. J Biol Chem 267:1422–1425.

Kim CM, Dion SB, Onorato JJ, Benovic JL (1993): Expression and characterization of two β-adrenergic receptor kinase isoforms using the baculovirus expression system. Receptor 3:39–55.

Krupnick JG, Benovic JL (1998): The role of receptor kinases and arrestins in G protein–coupled receptor regulation. Annu Rev Pharmacol Toxicol 38:289–319.

Kunapuli P, Onorato JJ, Hosey MM, Benovic JL (1994): Expression purification and characterization of the G protein–coupled receptor kinase GRK5. J Biol Chem 269:1099–1105.

Laemmli UK (1970): Cleavage of structural proteins during the assembly of the head of bacteriophage T4. Nature 227:680–685.

Loudon RP, Benovic JL (1994): Expression purification and characterization of the G protein–coupled receptor kinase GRK6. J Biol Chem 269:22691–22697.

Loudon RP, Perussia B, Benovic JL (1996): Differentially regulated expression of the G-protein–coupled receptor kinases βARK and GRK6 during myelomonocytic cell development in vitro. Blood 88:4547–4557.

Onorato JJ, Gillis ME, Liu Y, Benovic JL, Ruoho AE (1995): Activation of the β-adrenergic receptor kinase by phospholipids. J Biol Chem 270:21346–21353.

Opperman M, Diverse-Pierluissi M, Drazner MH, Dyer SL, Freedman NJ, Peppel K.C, Lefkowitz RJ (1996): Monoclonal antibodies reveal receptor specificity among G-protein–coupled receptor kinases. Proc Natl Acad Sci USA 93:7649–7654.

Palczewski K, Buczylko J, Lebioda L, Crabb JW, Polans AS (1993): Identification of the N-terminal region in rhodopsin kinase involved in its interaction with rhodopsin. J Biol Chem 268:6004–6013.

Pei G, Tiberi M, Caron MG, Lefkowitz RJ (1994): An approach to the study of G-protein–coupled receptor kinases: An *in vitro*–purified membrane assay reveals differential receptor specificity and regulation by G beta gamma subunits. Proc Natl Acad Sci USA 91:3633–3636.

Penela P, Ruiz-Gomez A, Castano JG, Mayor F Jr (1998): Degradation of the G protein–coupled receptor kinase 2 by the proteasome pathway. J Biol Chem 273:35238–35244.

Penn RB, Benovic JL (1998): Mechanisms of receptor regulation In Conn PM (ed): Handbook of Physiology, Vol 1, New York: Oxford University Press, pp 125–164.

Pitcher JA, Freedman NJ, Lefkowitz RJ (1998): G protein–coupled receptor kinases. Annu Rev Biochem 67:653–692.

Premont RT, Koch WJ, Inglese J, Lefkowitz RJ (1994): Identification purification and characterization of GRK5 a member of the family of G protein–coupled receptor kinases. J Biol Chem 269:6832–6841.

Premont RT, Macrae AD, Stoffel RH, Chung N, Pitcher JA, Ambrose C, Inglese J, MacDonald ME, Lefkowitz RJ (1996): Characterization of the G protein–coupled receptor kinase GRK4: Identification of four splice variants. J Biol Chem 271:6403–6410.

Pronin AN, Benovic JL (1997): Regulation of the G protein–coupled receptor kinase GRK5 by protein kinase C. J Biol Chem 272:3806–3812.

Sallese M, Mariggio S, Collodel G, Moretti E, Piomboni P, Baccetti B, de Blasi A (1997): G protein–coupled receptor kinase GRK4: Molecular analysis of the four isoforms and ultrastructural localization in spermatozoa and germinal cells. J Biol Chem 272:10188–10195.

Summers MD, Smith GE (1987): A manual of methods for baculovirus vectors and insect cell culture procedures. Texas Agricultural Experiment Station Bulletin No. 1555:1–56.

Winstel R, Freund S, Krasel C, Hoppe E, Lohse M (1996): Protein kinase cross-talk: Membrane targeting of the β-adrenergic receptor kinase by protein kinase C. Proc Natl Acad Sci USA 93:2105–2109.

Zhao X, Huang J, Khani SC, Palczewski K (1998): Molecular forms of human rhodopsin kinase (GRK1). J Biol Chem 273:5124–5131.

CHAPTER 8

CHARACTERIZATION OF ARRESTIN EXPRESSION AND FUNCTION

VSEVOLOD V. GUREVICH, MICHAEL J. ORSINI, and
JEFFREY L. BENOVIC

Regulation of G Protein–Coupled Receptor Function and Expression,
Edited by Jeffrey L. Benovic.
ISBN 0-471-25277-8   Copyright © 2000 Wiley-Liss, Inc.

I. INTRODUCTION

Arrestins are thought to play a general role in mediating G protein–coupled receptor (GPCR) desensitization. The role of arrestins in quenching or desensitizing receptor signaling has been most extensively studied in the visual system. Visual arrestin (termed *arrestin-1* in this chapter) was initially identified as a major protein that redistributed from the cytoplasm to the disc membrane following light activation of rod outer segments (Kuhn, 1978). Kuhn and coworkers (1984) further demonstrated that the binding of arrestin-1 to photoreceptor membranes was significantly enhanced by the phosphorylation of rhodopsin. The cloning of a bovine arrestin-1 cDNA revealed that it encodes a protein of 404 amino acids (Yamaki et al., 1987; Shinohara et al., 1987). Several studies have demonstrated that while rhodopsin phosphorylation is important to the desensitization process, it is the arrestin-1 binding that effectively blocks further rhodopsin–G protein coupling (Kuhn et al., 1984; Krupnick et al., 1997b).

Evidence for the involvement of an arrestin in nonvisual systems was initially suggested by studies using the β_2-adrenergic receptor (β_2AR) as a model (Benovic et al., 1987). A cDNA encoding an arrestin homologue, termed β-arrestin (or *arrestin-2*), was subsequently cloned from bovine brain (Lohse et al., 1990). This cDNA encodes a protein of 418 amino acids that has 58% identity with arrestin-1 and appears to uncouple G protein–coupled receptor kinase (GRK)-phosphorylated β_2AR from G proteins in a reconstituted system (Lohse et al., 1990, 1992). Utilizing various cloning strategies, a third arrestin cDNA was subsequently cloned (Rapaport et al., 1992; Attramadal et al., 1992; Sterne-Marr et al., 1993). This arrestin, which is termed β-arrestin-2 (or *arrestin-3*), has 79% amino acid identity with arrestin-2 and 56% with arrestin-1. A fourth homologue, termed X-arrestin or arrestin-C (*arrestin-4*), has also been cloned and likely plays a role in regulating cone phototransduction because it is primarily localized to cone cells (Murakami et al., 1993; Craft et al., 1994). A comparison of the overall amino acid identities and known structural domains of the four mammalian arrestins is depicted in Figure 8.1.

Arrestin-1, arrestin-2, and arrestin-3 are each expressed as polypeptide variants that arise as a result of alternative splicing. Arrestin-1 is predominately expressed as a 404 amino acid form, although a 370 amino acid form (Smith et al., 1994), in which the last 35 residues are replaced by a single alanine, and a 396 residue form (Parruti et al., 1993) that lacks the eight amino acids (residues 338–345) encoded by exon 13 of the human gene (Yamaki et al., 1990) have also been identified. Arrestin-2 exists as two polypeptide variants, the 418 amino acid form and a 410 residue form that lacks the comparable eight amino acid stretch (residues 334–341) that is spliced out in arrestin-1 (Sterne-Marr et al., 1993; Parruti et al., 1993). Arrestin-3 is predominantly expressed as a 409 amino acid protein, although in some tissues a 420 residue form that contains an 11 amino acid insert following residue 361 is observed (Sterne-Marr et al., 1993). Northern analysis demonstrates that both arrestin-1 (Lohse et al., 1990) and arrestin-4 (Murakami et al., 1993) are expressed most abundantly in the retina and pineal gland. However, reverse transcription

Figure 8.1. Domain architecture of arrestins. The sequences of the four known mammalian arrestins are represented schematically. The solid areas are regions of invariant amino acid sequence. The bifurcation near the C terminus represents divergence in sequence between the visual (arrestin-1 and arrestin-4) and nonvisual (arrestin-2 and arrestin-3) arrestins. A represents the activation-recognition domain, P the phosphorylation-recognition domain, S the secondary hydrophobic interaction domain, C the clathrin-binding domain, + the basic N terminus, and (−) the acidic C terminus.

polymerase chain reaction (RT-PCR) has also identified low levels of arrestin-1 in heart, kidney, lung, and brain (Smith et al., 1994). Arrestin-2 and arrestin-3 are widely expressed, with the highest levels in the brain, spleen, and prostate (Sterne-Marr et al., 1993; Parruti et al., 1993). In most tissues examined, arrestin-2 appears to be expressed at higher levels than arrestin-3. A notable exception is olfactory epithelium, where arrestin-3 is the major isoform and appears to play a role in the desensitization of odorant receptors (Dawson et al., 1993).

In this chapter, we outline procedures for characterizing the expression and function of mammalian arrestins in cell lines and *in vitro*. These procedures should provide the framework for enabling one to establish whether an arrestin can bind, desensitize, and/or promote internalization of a given GPCR.

2. EXPRESSION OF ARRESTINS

2.A. Detection of Endogenous Arrestins by Immunoblotting

The detection of endogenous arrestins in mammalian cell lines or tissues is best accomplished by immunoblotting with arrestin-specific antibodies. There are a number of antibodies that have been successfully used for such purposes (Table 8.1). Perhaps the best antibody for detecting mammalian arrestins is one that was initially generated against bovine arrestin-1 (Donoso et al., 1990). This antibody (termed *F4C1*) is a mouse monoclonal that detects the epitope DGVVLVD, a sequence that is present in the four known mammalian arrestins. This antibody appears to detect arrestin-1, arrestin-2, and arrestin-3 equally, and, thus, it can be used to compare the levels of various arrestins in mammalian cells. Additional antibodies that are specific for arrestin-1 (Knospe et al., 1988; Donoso et al., 1990; Razaghi et al., 1997), arrestin-2 (Attramadal et al., 1992; Sterne-Marr et al., 1993), and arrestin-3 (Attramadal et. al. 1992; Orsini and Benovic, 1998) have also been reported. These antibodies are obviously advantageous for studies that focus on a single arrestin species. Below, we outline an immunoblotting protocol that we have used to detect arrestins in numerous cell lines.

TABLE 8.1. Arrestin Antibodies

Antibody Name	Specificity	Type	Epitope	Reference
mAbF4C1	arr1~arr2~arr3	Mouse mAb	DGVVLVD	Donoso et al. (1990)
mAbA2G5	arr1 specific	Mouse mAb	PEDPDTAKE	Donoso et al. (1990)
mAbC10C10	?	Mouse mAb	RERRGIALD	Donoso et al. (1990)
mAbH11A2	arr1>arr3>arr2	Mouse mAb	NLASSTIIKE	Donoso et al. (1990)
mAbB11A11	?	Mouse mAb	VFEEFARHNLK	Donoso et al. (1990)
S2D2	?	mAb	N-terminal epitope	Razaghi et al. (1997)
S6D8	?	Mouse mAb	PVDGVVLVDPE	Razaghi et al. (1997)
S8D8	?	mAb	N-terminal epitope	Razaghi et al. (1997)
M3A9	?	Rat mAb	EAQEKVPPNSSLTKTLVPL	Razaghi et al. (1997)
M4H10	?	Rat mAb	RGIALDGKIKH	Razaghi et al. (1997)
S10H9	?	Mouse mAb	EKIDQEAAMD	Razaghi et al. (1997)
S1A3	?	Mouse mAb	VFEEFARQNLKD	Razaghi et al. (1997)
S6H8	arr1~arr2~arr3	Mouse mAb	PVDGVVLVDPE	Faure et al. (1984)
S85-63	arr1 specific	Mouse mAb	Bovine arr1 (aa 383–394)	Smith et al. (1994)
RDI-BARREST1abm	arr2	Mouse mAb	Rat arr2 (aa 262–409)	Research Diagnostics, Inc.
RDI-BARREST2abm	arr3	Mouse mAb	Rat arr3 (aa 112–131)	Research Diagnostics, Inc.
β-Arrestin-1	arr2	Mouse mAb	Rat arr2 (aa 262–409)	Transduction Laboratories
β-Arrestin-2	arr3	Mouse mAb	Rat arr3 (aa 112–131)	Transduction Laboratories

β-Arrestin-1	arr2 specific	Rabbit polyclonal	Rat arr2 (aa 331–418)	Attramadal et al. (1992)
β-Arrestin-2	arr3 specific	Rabbit polyclonal	Rat arr3 (aa 333–410)	Attramadal et al. (1992)
HDH	arr3 specific	Rabbit polyclonal	HDHIALPRPQSA	Sterne-Marr et al. (1993)
KEE	arr2 specific	Rabbit polyclonal	KEEEEDGTGSPRLNDR	Sterne-Marr et al. (1993)
Val170-Arg182	?	Rabbit polyclonal	VRLLIRKVQHAPR	Kieselbach et al. (1994)
C-p44	arr1 (p44 variant)	Rabbit polyclonal	CPEDPDTAKESA	Smith et al. (1994)
Arrestin-2	arr2>arr3	Rabbit polyclonal	Bovine arr2 (aa 357–418)	Benovic (unpublished)
Arrestin-3	arr3>arr2	Rabbit polyclonal	Bovine arr3 (aa 350–409)	Orsini and Benovic (1998)
β-Arrestin-1 (N-19)	arr2 specific	Goat polyclonal	N-terminal epitope	Santa Cruz Biotechnology
β-Arrestin-2 (N-16)	arr3 specific	Goat polyclonal	N-terminal epitope	Santa Cruz Biotechnology
β-Arrestin-2 (C-18)	arr3 specific	Goat polyclonal	C-terminal epitope	Santa Cruz Biotechnology

aa, amino acids.

2.A.a. Materials Required

Polyacrylamide gel apparatus and Western blot transfer set (Bio-Rad mini-protean II)

Power supply (500 V minimum)

1.5-ml Eppendorf tubes

Pipettes capable of dispensing 1–20 µl and 20–200 µl

Disposable pipette tips

Microcentrifuge

Vortex

Disposable gloves

Nitrocellulose transfer membrane (NitroME from Micron Separations, Inc.)

Plastic containers for immunoblotting and washing nitrocellulose membranes

X-ray film (Kodak X-OMAT AR)

Enhanced chemiluminescence reagents (Amersham ECL)

Various chemicals as outlined below (Sigma)

2.A.b. Preparation of Cellular Lysate.

Although arrestins are primarily cytosolic proteins, we generally use a lysis procedure that extracts both soluble and membrane-associated proteins. While these extracts are optimal for assessing endogenous (or overexpressed) arrestin levels, however, these preparations should not be used for assessing arrestin function because functional effects may be disrupted by the detergent present in the lysis buffer.

1. Prepare 5 ml of lysis buffer as follows:

Component	Volume	Final Concentration
0.5 M Na Hepes, pH 7.5 (in H_2O)	200 µl	20 mM
10% Triton X-100 (in H_2O)	500 µl	1%
1 M NaCl (in H_2O)	750 µl	150 mM
0.5 M Na EDTA, pH 7 (in H_2O)	100 µl	10 mM
0.1 M PMSF (in isopropanol)	25 µl	0.5 mM
10 mg/ml leupeptin (in H_2O)	10 µl	20 µg/ml
10 mg/ml benzamidine (in H_2O)	50 µl	100 µg/ml
Deionized H_2O	3.36 ml	

Note: Stock solutions can be made ahead of time and stored either at room temperature (Hepes, Triton X-100, NaCl, EDTA) or at –20°C (leupeptin, benzamidine). Phenylmethylsulfonyl fluoride (PMSF) should be made fresh weekly and stored at room temperature. The volume of lysis buffer can be scaled up or down as needed but should be made fresh and chilled on ice before use.

2. Cells are grown in complete media to 75%–90% confluence and then rinsed several times with phosphate-buffered saline (PBS). Ice-cold lysis buffer is then added (~1 ml of buffer per 100 mm dish of cells), and the

cells are scraped off with a rubber policeman and transferred to a centrifuge tube.

3. The cells are lysed by freezing in liquid nitrogen or a dry ice/ethanol bath, thawing in a room temperature water bath, and vortexing several times. The cell lysate is centrifuged (40,000g for 10 minutes in an SS34 rotor or equivalent) to remove particulate matter, and the supernatant is transferred to a microfuge tube and assayed for protein (Bio-Rad Protein Assay with bovine serum albumin as a standard). The cell supernatant can either be analyzed immediately or aliquoted, frozen in liquid nitrogen, and stored at $-80°C$ for later analysis.

2.A.c. Electrophoresis and Western Blot Analysis. Arrestins are found at relatively low concentrations in most nonvisual tissues (typically 0.1–2 pmol/mg protein). However, because several of the arrestin antibodies can detect low nanogram quantities, endogenous arrestins in most mammalian cells are readily detectable by immunoblotting.

1. Incubate 20–40 μg of protein from the cell supernatant prepared in Section 2.A.b. with SDS sample buffer (Laemmli, 1970) for 10 minutes at room temperature. It is advantageous to also include samples containing ~1 ng of purified arrestin-2 or arrestin-3 (Goodman et al., 1996) as controls as well as a prestained protein standard.

2. Electrophorese the samples on a 10% SDS-polyacrylamide gel at a constant voltage of 120–150 V until the dye front is within a few millimeters of the bottom (~1 hour).

3. Carefully remove the stacking gel and then set up the running gel for transfer to nitrocellulose. This is accomplished by layering a transfer sponge, one piece of Whatman 3 MM paper, the running gel, one piece of nitrocellulose (with one corner cut on the protein standard side), one piece of Whatman 3 MM paper, and another sponge. Everything should initially be wetted in transfer buffer (25 mM Tris base, 192 mM glycine, 20% methanol; *do not adjust pH*) before layering. It is also important to make sure that all air bubbles are removed as each layer is added (a 13 × 100 mm plastic test tube can be rolled over the layer at each step to accomplish this). Transfer the protein samples to a nitrocellulose membrane for 1 hour at 100 V (constant voltage) with cold transfer buffer.

4. After transfer, sample loading and transfer efficiency can be checked by staining the nitrocellulose with 0.2% Ponceau S (prepared in deionized water) for ~1 minute with the transfer side up. Transferred proteins can then be visualized by rinsing the nitrocellulose with deionized water to remove excess stain.

5. Rinse the nitrocellulose several times in Tris-buffered saline (20 mM Tris-HCl, pH 7.5, 150 mM NaCl) containing 0.05% Tween 20 (TTBS). The membrane is then blocked for 1 hour in 10 ml of TTBS containing 5% (wt/vol) nonfat dry milk.

6. The nitrocellulose is next incubated with 5 ml of either the mouse mono-clonal antibody F4C1 (~1:2,000 dilution) or an arrestin-specific antibody for ≥1 hour at room temperature. As discussed above, F4C1 recognizes an epitope that is common to all mammalian arrestins, and thus it can detect arrestin-2 and arrestin-3 as well as arrestin-1 (see Table 8.1).

7. Wash the nitrocellulose three to five times for 10 minutes each in TTBS.

8. Incubate the nitrocellulose for 1 hour with a 1:3,000 dilution of affinity-purified goat antimouse IgG conjugated to horseradish peroxidase (Bio-Rad) in TTBS–5% milk.

9. Wash the membrane three to five times for 10 minutes each in TTBS.

10. Overlay the nitrocellulose with 1–2 ml of enhanced chemiluminescence (ECL) reagent for ~1 minute, allow the blot to drip dry, wrap in Saran wrap, and visualize on x-ray film. An example of an arrestin Western blot analysis of various mammalian cell lines with F4C1 is shown in Figure 8.2.

2.B. Expression of Arrestins in Mammalian Cells

The overexpression of wild-type or mutant arrestins in mammalian cells can provide information regarding the role of arrestins in the desensitization and in-ternalization of a given GPCR. In general, HEK 293 and COS-7 cells have been used extensively for such studies because they are widely available, easy to grow, and permit the expression of high levels of GPCRs and arrestins with either transient or stable (HEK 293) transfection. The mammalian expression vector pcDNA3, available from Invitrogen, is suitable for such studies. This vector features a convenient multiple cloning site, the strong cytomegalovirus (CMV) promoter, the neomycin resistence gene, and a bovine growth hormone polyadenylation signal. Moreover, the presence of SP6 and T7 promoters flank-ing the multiple cloning site can also be used to express the gene of interest by

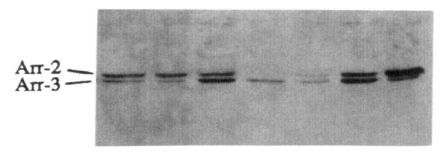

Figure 8.2. Arrestin expression in various mammalian cell lines. Twenty-five micro-grams of protein from seven different mammalian cell lysates were electrophoresed on a 10% SDS-PAGE, transferred to nitocellulose, and probed with the general arrestin mon-oclonal antibody F4C1 (1:2,000 dilution). The blot was then incubated with an HRP-con-jugated goat antimouse secondary antibody (1:2,000 dilution) and visualized by ECL.

in vitro transcription and translation (see Section 2.C.). In COS cells, pcDNA3 will replicate episomally by virtue of the SV40 origin of replication present in the vector and SV40 large T antigen that is constitutively expressed in COS cells. The level of arrestin expression in HEK 293 cells is comparable with the levels achieved in COS cells, probably because of the cell's inherent capacity to overexpress the protein of interest and the strength of the CMV promoter. Using 1–5 μg of plasmid DNA, we routinely obtain 25–50-fold overexpression of transfected wild-type and mutant arrestins in both cell lines. Both Fugene (Boehringer Mannheim) and Lipofectamine (Gibco/BRL) cationic lipid-based transfection reagents have yielded comparable results. However, due to its ease of use and apparent lower toxicity, we routinely use Fugene, which is the protocol provided below.

2.B.a. Materials Required

Laminar flow hood dedicated to tissue culture use

Tissue culture incubator, capable of maintaining 37°C, 5% CO_2

Dulbecco's modified Eagle's medium (DMEM), containing 10% fetal calf serum and 100 μg/ml penicillin and streptomycin (Gibco/BRL)

Dulbecco's PBS (Gibco/BRL)

Trypsin/EDTA (Gibco/BRL)

Fugene Transfection Reagent (Boehringer Mannheim)

Disposable, sterile serological pipettes and tissue culture dishes

Plasmid DNA containing the gene of interest (column or cesium chloride purified)

Disposable, sterile pipette tips

Disposable gloves

1.5-ml sterile Eppendorf tubes

15-ml and 50-ml sterile, conical polypropylene tubes (Falcon)

Light microscope

2.B.b. Cell Culture

1. COS and HEK 293 cells are maintained in DMEM, supplemented with 10% fetal calf serum and 100 μg/ml penicillin and streptomycin. In general, cells are split 1:10 twice per week for maintenance. Cells are maintained using ~3 ml of complete medium per 60-mm dish.

2. To split COS cells, aspirate the medium and rinse the monolayer twice with sterile PBS. Incubate with 0.05% trypsin, 0.5 mM EDTA (Gibco/BRL) for 1–2 minutes at room temperature. Aspirate trypsin, tap the plate gently to dislodge cells, and then add ~3 ml of complete medium to quench the remaining trypsin. Occasionally, cells may need a short incubation at 37°C for complete trypsinization. HEK 293 cells are less adherent than COS and can be split by aspirating the medium, gently washing with PBS lacking Ca^{2+} or Mg^{2+}, and gently tapping the dish or pipetting. Split COS or HEK 293 cells the day before transfection such that they are 50%–80% confluent the following day (generally a 1:3 to 1:5 split

of a previously confluent dish). The degree of confluency required also depends on the toxicity of the transfection reagent used (e.g., in our experience Fugene works effectively over a broad confluence range whereas Lipofectamine works optimally when the cells are more confluent).

2.B.c. Transfection of Cells

1. Aliquot the plasmid DNA (~3–5 μg per 60-mm dish) into a 1.5-ml Eppendorf tube. In a separate 15-ml conical tube, add 0.3 ml of warmed DMEM followed by 10 μl of Fugene added dropwise with swirling. Do not allow the Fugene to directly contact the sides of the tube. Allow the Fugene/DMEM mixture to incubate for 5 minutes at room temperature. We generally maintain a 3:1 ratio of Fugene:DNA; successful transfections in 60-mm dishes have been obtained with a maximum of 5–7 μg of DNA. Additional guidelines for optimizing transfection efficiency depending on the cell type and plasmid are provided in the manufacturer's directions.

2. Add the Fugene/DMEM mixture to the DNA dropwise, mix by gentle inversion of the tube, and leave at room temperature for 15 minutes. The Fugene/DNA mixture is then added evenly to the cell monolayer dropwise while swirling gently. There is no need to change the existing medium.

3. The transfection mixture can be left on the cells until the time of harvest. The cells are typically harvested 24 to 48 hours post-transfection for Western blot analysis (see Section 2.A.). Cells may also be split 5 hours after transfection, or the following day, for subsequent functional analysis (see Section 3.B.).

2.C. Expression of Arrestins in a Cell-Free Translation System

A systematic analysis of the functional domains within a protein requires the construction, expression, and analysis of a considerable number of mutants. A simple, rapid, and cost-effective expression system is a prerequisite for such studies. Because a major function of arrestins is to bind to phosphorylated activated receptors, the ability to produce radiolabeled arrestins is advantageous for such functional analysis. Expression by *in vitro* translation satisfies many of these requirements. Commercially available ^{3}H-leucine (140–150 Ci/mmol) makes it possible to translate arrestin-1 (which has 34 leucines) with a specific activity of up to ~5,000 Ci/mmol. Moreover, varying the ratio of cold and hot leucine in the translation mix enables proteins with specific activities from 0–5,000 Ci/mmol to be translated. In our own studies, we typically translate arrestins with specific activities of 400–600 Ci/mmol (i.e., 900–1,400 dpm/fmol) and obtain yields of 30–200 pmol per milliliter of translation mix. Because the recombinant arrestin is the only radiolabeled protein in the translation mix (Fig. 8.3), it can be used without further purification (Gurevich and Benovic, 1992; Gurevich et al., 1993). A simple and sensitive direct binding assay suitable for quantitative analysis of wild-type or mutant *in vitro*–translated arrestin binding to purified GPCRs is described below.

Figure 8.3. Analysis of arrestin expression and function by *in vitro* translation. Arrestin-1, arrestin-2, and arrestin-3 cDNAs were *in vitro* transcribed using SP6 polymerase and then translated using rabbit reticulocyte lysate. Binding of the *in vitro*–translated arrestins to different functional forms of rhodopsin was then measured, and bound arrestins were electrophoresed on a 10% SDS-PAGE and the gel was fixed, dried, and exposed to x-ray film.

2.C.a. Materials Required

Rabbit reticulocyte lysate (RRL) (numerous suppliers)

Amino acid mixture (20×) (prepare by mixing 20 mM solutions of all individual amino acids except for leucine as described by Jackson and Hunt (1983)

1 M creatine phosphate in water

Protease inhibitor mixture (100×) (contains 10 μg/ml each of pepstatin and leupeptin in water)

0.2 M cAMP in water

2 M potassium acetate, pH 7.4, in water

^{14}C-leucine (NEC-279 from NEN Dupont, concentrated in a Speed-Vac to ~10 mCi/ml)

^{3}H-leucine (NET-460A from NEN Dupont, 5 mCi/ml in water)

20 mg/ml creatine phosphokinase in 50% glycerol (CPK type I from Sigma) (dissolve at 40 mg/ml in 100 mM Tris-HCl, pH 7.4, 1 mM DTT, and then mix with an equal volume of glycerol, stable for 3–6 months at –20°C)

RNasin, 40 U/ml (Promega or Boehringer Mannheim)

DEPC-treated water (use throughout)

Note that all reagents can be aliquoted and stored at –80°C (stable for several years) except for creatine phosphokinase and RNasin, which are kept at –20°C. Except for RRL, all solutions can be thawed and refrozen several times without detrimental effect.

2.C.b. In Vitro Transcription.

The methods employed for *in vitro* transcription have been described in detail elsewhere (Gurevich, 1996).

1. A purified plasmid containing an arrestin cDNA under control of an SP6 (or alternative) promoter is linearized with a restriction enzyme that uniquely cuts downstream of the stop codon.

2. Linearized DNA (30 μg/ml) is incubated in 120 mM Hepes-K, pH 7.5; 2 mM spermidine; 16 mM MgCl$_2$; 40 mM DTT; 3 mM each of ATP, GTP, CTP, and UTP; 2.5 U/ml inorganic pyrophosphatase; 200 U/ml RNasin; and 1,500 U/ml SP6 RNA polymerase at 38°C for 2 hours.

3. The mRNA is precipitated by addition of 0.4 volume of 9 M LiCl, 10 minutes incubation on ice, and centrifugation at 8,000g for 10 minutes at 4°C.

4. Rinse the pellet with 1 ml of cold 2.5 M LiCl and 1 ml of room temperature 70% ethanol, dissolve in 1 volume of DEPC-treated water, and then reprecipitate by addition of 0.1 volume of 3 M sodium acetate and 3.3 volumes of ethanol (mRNAs in this suspension are stable for several years at −80°C).

5. For translation, pellet the mRNA in a microcentrifuge for 5–10 minutes; wash the pellet with 1 ml of 70% ethanol by vortexing and recentrifugation, allow to dry, and then dissolve in DEPC-treated water (all steps at room temperature).

2.C.c. In Vitro Translation. Translation with RRL has been described in detail elsewhere (Jackson and Hunt, 1983). While arrestin yields with either RRL or wheat germ extract are similar, 85%–90% of the synthesized protein is soluble and functionally active with RRL, while ~90% of the product appears to be denatured and aggregated when wheat germ extract is used. The resulting supernatant from the RRL translation can be used for receptor-binding assays either directly or after gel-filtration on a Sephadex G-25 column equilibrated with 50 mM Tris-HCl, pH 7.5, 50 mM potassium acetate, 2 mM EDTA. Protein synthesis can be determined from the amount of ^{3}H- or ^{14}C-leucine incorporated into a hot trichloroacetic acid–insoluble fraction or into the respective band after SDS-PAGE and autoradiography (Gurevich and Benovic, 1992, 1993). The radioactivity in the arrestin band along with the specific activity of the leucine is then used to determine yields. The procedure for a 200-μl translation reaction is provided below but can be readily scaled up or down as needed. Arrestins prepared in this manner are stable for several months, and a few freeze–thaw cycles do not seem to appreciably reduce their functional activity.

1. Add the following to an Eppendorf tube on ice: 140 μl RRL, 10 μl amino acid mix, 6 μl 1 M creatine phosphate, 2 μl protease inhibitor mix, 5 μl 0.2 M cAMP, 11 μl 2 M potassium acetate, pH 7.4, 0.8 μl concentrated ^{14}C-leucine (supplies 30–35 μM leucine), and 2.2 μl ^{3}H-leucine (supplies 0.3–0.35 μM leucine), vortex briefly, then add 1 μl RNasin and 2 μl creatine kinase, and vortex again.

2. Dissolve dry mRNA (10–30 μg) in 20 μl DEPC-treated water, immediately add to the translation mix, and vortex thoroughly. Incubate 2 hours at 22°–23°C.

3. To complete the translation, add 4 μl of a solution containing 50 mM ATP and GTP, incubate 7 mintues at 37°C, cool on ice, and centrifuge in a TLA 100.1 rotor (Beckman) at 100,000 rpm for 60 minutes at 4°C.

4. Remove the supernatant (carefully avoiding the dark pellet), aliquot, and freeze at −80°C.

Yield Calculation. To accurately determine the yield, the specific activity of leucine in the translation mix needs to be determined. While most batches of RRL contain ~5–7 μM leucine, the easiest way to precisely determine the endogenous leucine concentration is by isotope dilution. This involves performing three identical analytical-scale (20–50 μl) translations in the presence of 30, 60, or 90 μM ^{14}C-leucine. In addition, a control translation using water in place of the mRNA is also performed. A comparison of the amounts of trichloroacetic acid (TCA)–insoluble radioactivity (see below) from the various translations then enables calculation of the endogenous leucine concentration.

To determine the concentration of ^{14}C- and ^{3}H-leucine in the translation mix, we dilute a 2-μl aliquot 10-fold with water and count 5-μl aliquots in triplicate in a dual-label dpm mode. To determine the amount of protein-incorporated radioactivity, we dilute a 2-μl aliquot of translation mix 10-fold after the incubation (both before and after high-speed centrifugation to enable the percentage of soluble product to be determined) and apply 5 μl of diluted mix to 1×1 cm squares of Whatman 3 MM paper in duplicate. Let the paper dry for 3–5 minutes, immerse it in ice-cold 10% TCA for 10–15 minutes, boil 5% TCA for 10 minutes, rinse briefly with ethanol (to remove water) and diethyl ether (to remove TCA), let the paper dry, and then put each square into a scintillation vial containing 0.6 ml of 1% SDS/50 mM NaOH. After the protein dissolves (30–50 minutes at room temperature), add 5 ml of water-miscible scintillation fluid, mix, and count.

The total leucine concentration in the translation mix (endogenous + ^{14}C-leucine + ^{3}H-leucine) should be in the 30–50 μM range (i.e., 30,000–50,000 fmol/μl). From the leucine concentration (in fmol/μl) and the radioactivity (in dpm/μl) in the translation mix, the specific activity of leucine (in dpm/fmol) can be calculated. The specific activities of individual arrestins can then be determined by multiplying the leucine specific activity by 34, 42, or 41 (the number of leucine residues in bovine arrestin-1, arrestin-2, and arrestin-3, respectively). Dividing the total protein-incorporated (TCA-insoluble) radioactivity (dpm per μl of translation mix after subtracting the control value) by the arrestin specific activity (dpm/fmol) gives the yield in fmol/μl. We typically get 70–150 fmol/μl for arrestin-1 and 25–70 fmol/μl for arrestin-2 and arrestin-3.

Electrophoresis and Fluorography. It is advisable to analyze all samples by SDS-PAGE before functional analysis. *In vitro* translated arrestins usually yield a single band, but this is important to ascertain for every translation reaction. We add 2 μl of each translation mix to 18 μl of SDS sample buffer, mix, and run on a standard 10% SDS-PAGE. Gels are stained with Coomassie Blue G-250, destained for 30 minutes, and then soaked in 20% 2,5,-diphenyloxazole in glacial acetic acid for 10 minutes. The fluorochrome is then precipitated by washing the gel with two to three changes of water; the gels are dried and then exposed to x-ray film at −80°C overnight. The labeled protein bands also can be excised and counted in a liquid scintillation counter to confirm yields.

Troubleshooting. In general, because so many components go into the translation reaction, it is always easier to throw away all the "old" aliquots and to use new ones as problems arise. Nevertheless, it is always a good idea to keep a control mRNA that has produced high yields in the past that can be used to test the various components when troubleshooting. Several specific problems that can occur are outlined below. If the yields decrease by about twofold, a fresh CPK solution should be used (yields are two to threefold lower in the absence of CPK). If the yield drops sharply (to <10 fmol/µl), most likely the RRL (each freeze–thaw cycle reduces the RRL activity by ~20%–30%) or mRNA needs replaced. If a smear or more than one protein band is consistently observed, the mRNA is most likely degraded. The ^{3}H-leucine might also be the problem if it is over 1 year old. This can be determined by translating with ^{14}C-leucine alone (^{14}C-leucine is stable at –80°C for at least 5 years). Several additional problems that may arise with ^{35}S-methionine are discussed in detail elsewhere (Jackson and Hunt, 1983).

3. FUNCTION OF ARRESTINS

Arrestins function by specifically binding to phosphorylated GPCRs, thereby inhibiting GPCR–G protein coupling. In addition, nonvisual arrestins (arrestin-2 and arrestin-3) can augment agonist-promoted internalization of numerous GPCRs. In this section, we describe an *in vitro* assay for directly measuring arrestin binding to GPCRs. We also describe methods for measuring arrestin stimulated GPCR internalization with either ligand binding or ELISA assays.

3.A. Direct Receptor-Binding Assay

To characterize arrestin interaction with GPCRs, it is often informative to compare the ability of an arrestin to bind to different functional forms of a receptor (e.g., ± agonist, ± GPCR phosphorylation). Studies assessing arrestin binding to rhodopsin, the β_2AR, and the M_2 muscarinic cholinergic receptor (m2AChR) revealed that arrestin-1 is very selective because it preferentially binds to activated phosphorylated receptors, while arrestin-2 and arrestin-3 are primarily selective for the phosphorylation state of a receptor (Gurevich and Benovic, 1993; Gurevich et al., 1993, 1995) (see Fig. 8.3). Arrestin-1 is also the most specific because it binds well to rhodopsin but poorly to other GPCRs (e.g., β_2AR and m2AChR). In contrast, arrestin-2 and arrestin-3 are less specific, although they interact better with the β_2AR and m2AChR than the rhodopsin.

We have developed a simple and sensitive direct binding assay using radiolabeled *in vitro*–translated arrestins. This assay takes advantage of the high specific activity of *in vitro*–translated arrestins (up to 1,600 dpm/fmol), which provides sufficient sensitivity to study both high-affinity arrestin binding to phosphorylated activated receptors as well as low-affinity interactions with other receptor forms (Gurevich and Benovic, 1993; Gurevich et al., 1993, 1995). Moreover, while these procedures work well with purified receptor preparations, *in vitro* translated arrestins can also be used to study arrestin binding to receptor domains (Wu et al., 1997) as well as additional proteins such as clathrin (Goodman et al., 1996, 1997; Krupnick et al., 1997a).

3.A.a. Arrestin Binding to Rhodopsin

1. Prepare *in vitro*–translated tritiated arrestin as outlined in Section 2.C. This preparation should be analyzed by SDS-PAGE and normally can be used without further purification. The arrestin is diluted to a concentration of 2–4 fmol/μl with 50 mM Tris-HCl, pH 7.5, 50 mM potassium acetate, 0.5 mM MgCl$_2$ (RB buffer) containing 0.5 mM DTT. Aliquot 25 μl of this solution per Eppendorf tube (50–100 fmol per assay, 1–2 nM final concentration).

2. Prepare urea-treated rod outer segments (ROS) as described in Chapter 7. The rhodopsin can be phosphorylated with GRK1 or GRK2 and regenerated as described in Chapter 7 (a stoichiometry of 1 mol/mol with GRK1 or 2 mol/mol with GRK2 is adequate for measuring phosphorylation-dependent arrestin binding). Under dim red light, dilute the rhodopsin (Rh) and phosphorylated rhodopsin (P-Rh) to 12 μg/ml with RB buffer containing 0.5 mM DTT and aliquot 25 μl per Eppendorf tube (~7.5 pmol rhodopsin per tube). Nonspecific binding is determined with 25 μl of 12 μg/ml phospholipid vesicles (prepared with either crude soybean phosphatidylcholine or ROS-extracted lipids).

3. Incubate the arrestin and rhodopsin in the dark or with illumination at 37°C for 5 minutes. The timing of this incubation is critical because light-activated rhodopsin has a relatively short half-life.

4. Cool the samples on ice and then separate bound and free arrestins by gel filtration at 4°C (under dim red light for dark rhodopsin) on 2-ml Sepharose CL-2B columns equilibrated with 20 mM Tris-HCl, pH 7.5, 2 mM EDTA (20/2 buffer). The samples are allowed to enter the column completely, and the column is then washed successively with 100, 400, and 600 μl of 20/2 buffer. The 600 μl wash is collected directly into a scintillation vial and counted after addition of scintillation fluid. Columns can be reused but should be washed with 4 × 3 ml of 20/2 buffer and stored capped at 4°C.

3.A.b. Binding of Arrestins to Purified Reconstituted Receptors

1. Prepare *in vitro*–translated tritiated arrestins as outlined in Section 2.C. These preparations should be analyzed by SDS-PAGE but normally can be used without further purification. The arrestin is diluted to a concentration of ~2 fmol/μl with 50 mM Tris-HCl, pH 7.5, 50 mM potassium acetate, 0.5 mM MgCl$_2$ (RB buffer) containing 0.5 mM DTT. Aliquot 25 μl of this solution per Eppendorf tube (50 fmol per assay, 1 nM final concentration).

2. Prepare purified reconstituted receptor with or without phosphorylation (e.g., see Chapter 3 for m2AchR). Incubate 50–200 fmol of receptor per 50 μl assay in RB buffer (supplemented with 0.2 M DTT for m2AChR) with 50 fmol of arrestin at 30°C for 30–40 minutes. We use saturating concentrations of agonist or antagonist to keep receptors in an active or inactive state (e.g., isoproterenol or alprenolol for β$_2$AR and carbachol or atropine for m2AChR, respectively).

3. Cool the samples on ice, and then separate bound and free arrestins by gel filtration at 4°C as described in Section 3.A.a, step 4. Such studies re-

veal that nonvisual arrestins bind to the β_2AR and m2AChR with subnanomolar affinity (Gurevich and Benovic, 1993; Gurevich et al., 1995). However, it is noteworthy that the total amount of arrestin bound is often substantially lower than the total number of receptors present in the assay (even with excess arrestin present in the assay).

3.B. Effect of Arrestins on Receptor Internalization in Mammalian Cells

Depending on the cell type and receptor, nonvisual arrestins can promote agonist-induced internalization of various GPCRs. This has been best characterized for the β_2AR where nonvisual arrestin binding to the phosphorylated β_2AR mediates receptor internalization via their ability to also bind to clathrin, the major protein component of clathrin-coated pits (Krupnick et al., 1997a; Goodman et al., 1996, 1997). The receptor ultimately traffics to endosomes where it is dephosphorylated and then either recycled back to the cell surface or trafficked to lysosomes (Kallal et al., 1998; Gagnon et al., 1998). In general, COS cells, which contain relatively low endogenous arrestin levels, are the best cells in which to assay the *promotion* of internalization by arrestins (Krupnick et al., 1997a). Conversely, HEK 293 cells, which have higher endogenous arrestin levels, are the best cells in which to observe *inhibition* of internalization by dominant-negative mutants (Krupnick et al., 1997a; Ferguson et al., 1996; Gagnon et al., 1998; Orsini and Benovic, 1998). In the following sections, we describe methods to assay receptor internalization with both ligand-binding and enzyme-linked immunosorbent assay (ELISA) assays, using the β_2AR as an example. The ligand-binding assay is based on the binding of a hydrophilic radioligand, such as the β-antagonist ^{3}H-CGP-12177, that will bind to cell surface but not intracellular receptors. Because internalized receptors will not bind this hydrophilic ligand, the difference in binding before and after agonist treatment reflects the degree of receptor internalization. The ELISA is based on the presence of an N-terminal epitope tag on the receptor, such as the hemagglutinin (HA) or Flag epitope (see Chapter 1), that is no longer recognized by the cognate antibody once the receptor is internalized (Daunt et al., 1997).

3.B.a. Materials Required

Transfected cells expressing the receptor and arrestin of interest (see Section 2.B.)

14°C shaking water bath

Cell harvester (Brandel)

Whatman GF/C glass fiber filters

Disposable 15-ml and 50-ml conical tissue culture tubes (Falcon)

12 × 75 mm disposable polystyrene tubes (Fisher)

12 × 75 mm glass borosilicate tubes

Vortex mixer

Bradford reagent for protein assays (Bio-Rad)

Spectrophotometer

Disposable polystyrene cuvettes

Pipettes and disposable tips

Disposable gloves

Trypsin/EDTA (Gibco/BRL)

Dulbecco's modified Eagle's medium (DMEM), containing 10% fetal calf serum and 100 µg/ml penicillin and streptomycin

Dulbecco's PBS

Bench top centrifuge with swinging bucket rotor

Scintillation counter and scintillation vials

24-well tissue culture dishes (ELISA)

Aspirator (ELISA)

ELISA reader (ELISA)

Eppendorf repeat pipettor and combitips (ELISA)

3.B.b. *Internalization Assayed by Ligand Binding*

1. Treat transfected cells (that have been split 1:2 shortly after transfection as described in Section 2.C.c) either with or without 10 µM (–) isoproterenol (Sigma). We generally co-transfect 3 µg of receptor and 1 µg of arrestin per 60-mm dish to observe promotion of internalization in COS and 3 µg of receptor and 1–3 µg of a dominant-negative arrestin or dynamin to observe inhibition of internalization in HEK 293 cells. These amounts can be scaled up or down accordingly depending on the number of cells to be transfected. We have found that internalization assays performed in COS cells are less sensitive to the level of receptor expression because there is little internalization of the β_2AR in the absence of co-transfected arrestins. Conversely, optimal levels of internalization and effective inhibition of internalization by dominant-negative arrestins is best observed in HEK 293 cells that express 0.5–2 pmol/mg of receptor. Expression of greater numbers of receptors has been found to lower the percentage of receptors that are internalized and reduce the efficiency of inhibition by dominant-negative arrestins, most likely because the ability of the endogenous arrestins to promote internalization of these receptors is exceeded. Isoproterenol is freshly prepared as a 10-mM stock in water, containing 0.1 mM ascorbic acid (Sigma) to prevent oxidation. Dilute isoproterenol in warm DMEM containing 0.1 mM ascorbic acid. Aspirate complete medium from the cells, and wash once with warm DMEM. Add either warm DMEM containing 0.1 mM ascorbic acid (mock treated) or isoproterenol. Place cells in 37°C incubator for desired time.

2. For COS cells, aspirate the medium, and trypsinize cells using 0.25% trypsin, 0.5 mM EDTA either with or without 10 µM isoproterenol. After 1 minute, aspirate trypsin and tap the plate to dislodge cells before adding ice-cold PBS. (For HEK 293 cells, at the end of agonist treatment aspirate the medium, tap cells off the plate, and add ice-cold PBS.) At

this point it is necessary to keep the cells cold at all times because internalized receptors can recycle back to the cell surface. Gently tap or pipette cells off the plate and transfer to a 50-ml conical tube. Fill the tube with ice-cold PBS and place on ice. Repeat this procedure until all plates are harvested. Spin cells in bench-top centrifuge for 2 minutes at 2,200 rpm.

3. Decant the first PBS wash, and wash cells a second time with ice-cold PBS. Resuspend the final washed cell pellet in 1.25 (COS) or 2.5 (HEK 293) ml PBS per 60-mm dish. Adjust volumes accordingly for larger or smaller dishes. At this time, an aliquot can be removed to save for Western blot analysis to check expression levels.

4. Aliquot 50 μl of cell suspension in quadruplicate into 12 × 75 mm polystyrene tubes placed in an ice-water bath. In the fourth tube of each group, add 2.5 μl of 1 mM alprenolol as a control for nonspecific binding. Repeat this for each condition (i.e., ± isoproterenol) such that each transfection ultimately yields eight assay tubes. To each tube, add 200 μl PBS containing ^{3}H-CGP-12177 (Amersham) such that the final concentration of ligand is 1 nM and the final volume is 250 μl. Incubate for 3 hours with shaking in a 14°C water bath.

5. During the incubation in step 4, perform a protein assay on each sample, preferably in triplicate. To 12 × 75 mm glass borosilicate tubes, add 5–10 μl of cell suspension and 800 μl of water. Vortex vigorously for 10 seconds and leave at room temperature for 5 minutes to lyse cells. Prepare a standard curve using 0, 2, 4, 6, 8, and 10 μg bovine serum albumin (BSA) (Sigma). Add 200 μl of Bradford dye reagent, vortex, and read in polystyrene disposable cuvettes at 595 nm.

6. At the end of the 14°C incubation, rapidly filter the samples on Whatman GF/C filter paper with ice-cold 25 mM Tris, pH 7.4, 2 mM MgCl$_2$ (25/2 buffer). This is best performed with a Brandel cell harvester, but a Millipore filter unit will also work. 25/2 buffer can be prepared as a 50× stock but should be diluted and stored at 4°C ahead of time. If the wash buffer is not cold, some ligand dissociation from the receptor might occur during washing.

7. Remove filter circles with forceps and place in scintillation vials. Add 5 ml Scintiverse (Fisher), cap vials, and vortex. Wait at least 1 hour before counting. The counts will generally increase if recounted the next morning; however, the percentage of internalized receptors should not differ greatly from the initial values.

8. Calculate the percentage of receptor internalization as follows. Calculate the specific binding (in dpm) by averaging the triplicate values and subtracting the nonspecific counts from the alprenolol-containing tube. Calculate the specific activity of the ligand (in dpm/fmol), divide the dpm by the specific activity to obtain fmol of receptor, and then determine the fmol of receptor per ml of cells based on the volume of sample that was originally assayed. Divide this value by the protein concentration (mg/ml) to obtain fmol/mg. Divide the fmol/mg from mock-treated cells by the fmol/mg from those treated with isopro-

terenol. Subtract this value from 1 and multiply by 100 to get percent internalization.

3.B.c. *Internalization Assayed by ELISA*

1. Cells are transfected with an epitope-tagged receptor ± arrestins as described in Section 2.C.c. Before splitting the cells, treat wells of a 24-well plate with a solution of 0.1 mg/ml poly-L-lysine (PLL) (Sigma). Briefly, dissolve the PLL in sterile water, add ~250 µl to cover the well, wait 1 minute, remove the PLL, and then let the wells dry (~10 minutes). PLL can be reused up to three times. Split the cells into 24-well dishes using six wells per transfection (60-mm dish). The cell density should be ~1–3 × 10^5 cells/well. It is convenient to divide the 24-well plate in half vertically, with one side for untreated and the other side for agonist-treated conditions. Each 24-well dish can then accommodate four separate transfections horizontally.

2. Aspirate medium from the wells, wash once with DMEM, and treat triplicate wells with DMEM with or without isoproterenol at 37°C as described above in Section 3.B.b (step 1), for the desired time.

3. Aspirate medium and fix the cells with 250 µl/well of 3.7% formaldehyde (Sigma) in TBS (20 mM Tris, pH 7.5, 150 mM NaCl) for 5 minutes. The remainder of the assay can be performed at room temperature.

4. Wash the plate three times with TBS (500 µl/well) and then block each well with 250 µl of 1% BSA in TBS for 45 minutes.

5. Aspirate the blocking solution, add the primary antibody diluted 1:1,000 in TBS/1% BSA (250 µl/well), and incubate for 1 hour at room temperature. We have successfully used the 101R antihemagglutinin as raw ascites (Babco) as well as the anti-Flag antibody M1 (Sigma). The binding of the M1 antibody depends on the presence of calcium; therefore, 1 mM $CaCl_2$ must be present for this step and all subsequent steps before color development.

6. Wash the plate three times with TBS/(Ca^{2+}) (500 µl/well) and reblock with 250 µl of 1% BSA in TBS/(Ca^{2+}) for 15 minutes.

7. Aspirate the blocking solution, and add 250 µl/well of secondary antibody (alkaline phosphatase–conjugated goat antimouse; Bio-Rad) diluted 1:1,000 in TBS/(Ca^{2+}). Incubate for 1 hour at room temperature.

8. Wash plate three times with 500 µl/well of TBS/(Ca^{2+}). Develop with the alkaline phosphatase substrate kit (Bio-Rad). Dilute diethanolamine solution 1:5 in water and add 1 substrate tablet per 5 ml (dissolve the tablet completely by vigorous vortexing). The color of the solution should be colorless to a slight pale yellow. Add 250 µl of developing solution per well, and develop until a bright yellow color appears, generally 15–30 minutes depending on efficiency of transfection and receptor expression. If the signal is weak, development is aided by incubation at 37°C. It is important not to allow development to proceed beyond the linear range of the ELISA reader. Ideally, OD_{405} readings should fall between 0.5 and 1.2 after stopping the reaction with NaOH (see step 9).

9. Aliquot 100 µl/well of 0.4 M NaOH in a 96-well plate. Remove 100 µl from the developed wells and add to the NaOH to stop the development reaction. The color is now stable for several days if the plate is wrapped in Parafilm and stored at 4°C.

10. Read plate in ELISA reader at 405 nm. Subtract background from mock-transfected or untransfected control cells, and calculate percent internalization as described in Section 3.B.b. (step 8).

REFERENCES

Attramadal H, Arriza JL, Dawson TM, Codina J, Kwatra MM, Snyder SH, Caron MG, Lefkowitz RJ (1992): β-Arrestin2, a novel member of the arrestin/β-arrestin gene family. J Biol Chem 267:17882–17890.

Benovic JL, Kuhn H, Weyand I, Codina J, Caron MG, Lefkowitz RJ (1987): Functional desensitization of the isolated β-adrenergic receptor by the β-adrenergic receptor kinase: Potential role of an analog of the retinal protein arrestin (48 kDa protein). Proc Natl Acad Sci USA 84:8879–8882.

Craft CM, Whitmore DH, Wiechmann AF (1994): Cone arrestin identified by targeting expression of a functional family. J Biol Chem 269:4613–4619.

Daunt DA, Hurt C, Hein L, Kallio J, Feng F, Kobilka BK (1997): Subtype-specific intracellular trafficking of alpha2-adrenergic receptors. Mol Pharmacol 51:711–720.

Dawson TM, Arriza JL, Jaworsky DE, Borisy FF, Attramadal H, Lefkowitz RJ, Ronnett GV (1993): β-Adrenergic receptor kinase-2 and β-arrestin2 as mediators of odorant-induced desensitization. Science 259:825–829.

Donoso LA, Gregerson DS, Smith L, Robertson S, Knospe V, Vrabec T, Kalsow CM (1990): S-antigen: Preparation and characterization of site-specific monoclonal antibodies. Curr Eye Res 9:343–355.

Faure JP, Mirshahi M, Dorey C, Thillaye B, de Kozak Y, Boucheix C (1984): Production and specificity of monoclonal antibodies to retinal S antigen. Curr Eye Res 3:867–872.

Ferguson SSG, Downey WE III, Colapietro A-M, Barak LS, Menard L, Caron MG (1996): Role of β-arrestin in mediating agonist-promoted G protein–coupled receptor internalization. Science 271:363–365.

Gagnon AW, Kallal L, Benovic JL (1998): Role of clathrin-mediated endocytosis in agonist-induced downregulation of the β_2-adrenergic receptor. J Biol Chem 273:6976–6981.

Goodman OB Jr, Krupnick JG, Santini F, Gurevich VV, Penn RB, Gagnon AW, Keen JH, Benovic JL (1996): β-Arrestin functions as a clathrin adaptor to promote β_2-adrenergic receptor endocytosis. Nature 383:447–450.

Goodman OB Jr, Krupnick JG, Gurevich VV, Benovic JL, Keen JH (1997): Arrestin/clathrin interaction: Localization of the arrestin binding locus to the clathrin terminal domain. J Biol Chem 272:15017–15022.

Gurevich VV (1996): Use of bacteriophage RNA polymerase in RNA synthesis. Methods Enzymol 275:382–397.

Gurevich VV, Benovic JL (1992): Cell-free expression of visual arrestin: Truncation mutagenesis identifies multiple domains involved in rhodopsin interaction. J Biol Chem 267:21919–21923.

Gurevich VV, Benovic JL (1993): Visual arrestin interaction with rhodopsin: Sequential multisite binding ensures strict selectivity toward light-activated phosphorylated rhodopsin. J Biol Chem 268:11628–11638.

Gurevich VV, Dion SB, Onorato JJ, Ptasienski J, Kim CM, Sterne-Marr R, Hosey MM, Benovic JL (1995): Arrestin interactions with G protein–coupled receptors: Direct binding studies of wild type and mutant arrestins with rhodopsin, β_2-adrenergic, and m2 muscarinic cholinergic receptors. J Biol Chem 270:720–731.

Gurevich VV, Richardson RM, Kim CM, Hosey MM, Benovic JL (1993): Binding of wild type and chimeric arrestins to the m2 muscarinic cholinergic receptor. J Biol Chem 268:16879–16882.

Jackson RJ, Hunt T (1983): Preparation and use of nuclease-treated rabbit reticulocyte lysates for the translation of eukaryotic messenger RNA. Methods Enzymol 96:50–75.

Kallal L, Gagnon AW, Penn RB, Benovic JL (1998): Visualization of agonist-induced sequestration and down-regulation of a green fluorescent protein–tagged β_2-adrenergic receptor. J Biol Chem 273:322–328.

Kieselbach T, Irrgang K-D, Ruppel H (1994): A segment corresponding to amino acids Val170-Arg182 of bovine arrestin is capable of binding to phosphorylated rhodopsin. Eur J Biochem 226:87–97.

Knospe V, Donoso LA, Banga JP, Yue S, Kasp E, Gregerson DS (1988): Epitope mapping of bovine retinal S-antigen with monoclonal antibodies. Curr Eye Res 7:1137-1147.

Krupnick JG, Goodman OB Jr, Keen JH, Benovic JL (1997a): Arrestin/clathrin interaction: Localization of the clathrin binding domain of nonvisual arrestins to the C-terminus. J Biol Chem 272:15011–15016.

Krupnick JG, Gurevich VV, Benovic JL (1997b): Mechanism of quenching of phototransduction: Binding competition between arrestin and transducin for phosphorhodopsin. J Biol Chem 272:18125–18131.

Kuhn H (1978): Light-dependent binding of rhodopsin kinase and other proteins to cattle photoreceptor membranes. Biochemistry 17:4389–4395.

Kuhn H, Hall SW, Wilden U (1984): Light-induced binding of 48-kDa protein to photoreceptor membranes is highly enhanced by phosphorylation of rhodopsin. FEBS Lett 176:473–478.

Laemmli UK (1970): Cleavage of structural proteins during the assembly of the head of bacteriophage T4. Nature 227:680–685.

Lohse MJ, Andexinger S, Pitcher J, Trukawinski S, Codina J, Faure J-P, Caron MG, Lefkowitz RJ (1992): Receptor-specific desensitization with purified proteins: Kinase dependence and receptor specificity of β-arrestin and arrestin in the β_2-adrenergic receptor and rhodopsin systems. J Biol Chem 267:8558–8564.

Lohse MJ, Benovic JL, Caron MG, Lefkowitz RJ (1990): β-Arrestin: A protein that regulates β-adrenergic receptor function. Science 248:1547–1550.

Murakami A, Yajima T, Sakuma H, McLaren M, Inana G (1993): X-Arrestin: A new retinal arrestin mapping to the X chromosome. FEBS Lett 334:203–209.

Orsini MJ, Benovic JL (1998): Characterization of dominant negative arrestins that inhibit β_2-adrenergic receptor internalization by distinct mechanisms. J Biol Chem 273:34616–34622.

Parruti G, Peracchia F, Sallese M, Ambrosini G, Masini M, Rotilio D, DeBlasi A (1993): Molecular analysis of human β-arrestin-1: Cloning, tissue distribution, and regulation of expression: Identification of two isoforms generated by alternative splicing. J Biol Chem 268:9753–9761.

Rapaport B, Kaufman KD, Chalzenbalk GD (1992): Cloning of a member of the arrestin family from a human thyroid cDNA library. Mol Cell Endocrinol 84:R39–R43.

Razaghi A, Bonaly J, Chacun H, Faure JP, Mirshahi M, Barque JP (1997): Immunodetection of a protein related to mammalian arrestin in *Euglena gracilis*. Biochem Biophys Res Commun 233:601–605.

Shinohara T, Dietzschold B, Craft CM, Wistow G, Early JJ, Donoso LA, Horwitz J, Tao R (1987): Primary and secondary structure of bovine retinal S antigen (48-kDa protein). Proc Natl Acad Sci USA 84:6975–6979.

Smith WC, Milam AH, Dugger D, Arendt A, Hargrave PA, Palczewski K (1994): A splice variant of arrestin: Molecular cloning and localization in bovine retina. J Biol Chem 269:15407–15410.

Sterne-Marr R, Gurevich VV, Goldsmith P, Bodine RC, Sanders C, Donoso LA, Benovic JL (1993): Polypeptide variants of β-arrestin and arrestin3. J Biol Chem 268:15640–15648.

Wu G, Krupnick JG, Benovic JL, Lanier SM (1997): Interaction of arrestins with intracellular domains of muscarinic and alpha$_2$-adrenergic receptors. J Biol Chem 272:17836–17842.

Yamaki K, Takahashi Y, Sakuragi S, Matsubara K (1987): Molecular cloning of the S-antigen cDNA from bovine retina. Biochem Biophys Res Commun 142:904–910.

Yamaki K, Tsuda M, Kikuchi T, Chen K-H, Huang K-P, Shinohara T (1990): Structural organization of the human S-antigen gene. J Biol Chem 265:20757–20762.

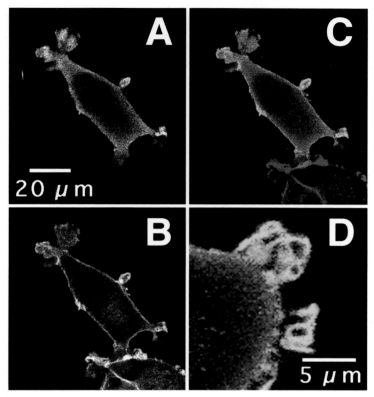

Figure 1.5. Immunocytochemical co-localization of eIF-2Bα and β$_2$-adrenergic receptor in the plasma membrane (Klein et al., 1997). HA epitope–tagged eIF-2Bα was transiently transfected into an HEK 293 cell line stably expressing the N-terminally Flag epitope–tagged β$_2$-adrenergic receptor. Cells were grown on glass coverslips, fixed, and permeabilized as described in the protocol for a staining experiment. HA epitope–tagged eIF-2Bα was detected with the monoclonal mouse antibody HA.11 (BAbCO, Richmond, CA), and β$_2$-adrenergic receptor was detected with receptor-specific rabbit antiserum (von Zastrow and Kobilka, 1992). After incubation with the secondary antibodies (FITC-conjugated donkey antimouse and Texas Red–conjugated donkey antirabbit, Jackson ImmunoResearch Laboratories, West Grove, PA), eIF-2Bα and β$_2$-adrenergic receptor were localized by dual-color confocal fluorescence microscopy with a BioRad MRC1000 and a Zeiss 100 NA1.3 objective. **A:** EIF-2Bα (green channel) localizes both in the cytoplasm and is observed in association with the plasma membrane, where it is concentrated in specialized regions. **B:** β$_2$-Adrenergic receptor (red channel) co-localizes with eIF-2Bα in these regions as shown in yellow in the two-color merged image (**C**). **D:** Another example of this co-localization is shown at higher magnification.

Figure 12.5. CLSM images of the cells expressing GFP-tagged GPCRs. **A:** Hela cells expressing CXC chemokine receptor 4 (in green) after labeling the cell surface with tetramethyl rhodamine ConA (red). **B:** Hela cells expressing CC-chemokine receptor 5 (green) with endosomes labeled with tetramethyl rhodamine transferrin (red). **C,D:** Co-localization of CCKAR-GFP with the ligand, Cy3.29–CCK-8 (red) in CHO cells 15 minutes after addition of the peptide (C) or 60 minutes after removal of the ligand (D).

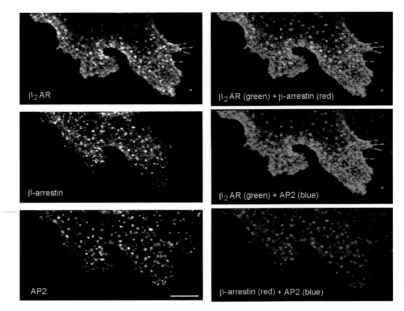

Figure 11.4. Flag-β₂AR and β-arrestin are present in the same clathrin-coated pits on agonist treatment. The punctate pattern of β₂AR and β-arrestin induced by isoproterenol coincides with coated pits, as indicated by the co-localization of AP.2, a marker for plasma membrane clathrin-coated pits. Flag-β₂AR was detected by mouse M2 monoclonal antibody followed by FL-tagged secondary antibody; β-arrestin was detected with rabbit KEE primary antibody followed by LR-tagged secondary antibody; coated pit AP.2 were revealed by mouse biotinylated-AP.6 followed by Cy5-tagged streptavidin. For details, see section 4.C. Bar = 10 μm.

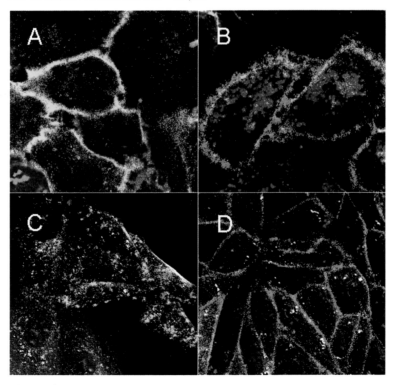

Figure 12.5.

CHAPTER 9

THE G PROTEIN–COUPLED RECEPTOR PHOSPHATASE

JULIE A. PITCHER and ROBERT J. LEFKOWITZ

1. INTRODUCTION

This chapter details the methodology used to characterize the G protein–coupled receptor phosphatase (GRP) and to investigate the regulatory mechanisms controlling the activity of this enzyme. The term *GRP* refers to the enzyme that dephosphorylates G protein–coupled receptor kinase (GRK) phosphorylated G protein–coupled receptors (GPCRs). The GRKs are a family of serine/threonine kinases (reviewed by Pitcher et al., 1998), and the GRP is thus a member of the protein serine/threonine phosphatase family (described in more detail in Section 2.C. and reviewed by Cohen [1989]). GRK-mediated phosphorylation of agonist-occupied GPCRs and the subsequent binding of an arrestin family member serves to uncouple the receptor from its cognate G protein and targets the receptor for internalization via clathrin-coated pits (reviewed by Zhang et al., 1997). These processes lead to attenuation of GPCR-mediated signaling, a

Regulation of G Protein–Coupled Receptor Function and Expression,
Edited by Jeffrey L. Benovic.
ISBN 0-471-25277-8   Copyright © 2000 Wiley-Liss, Inc.

phenomenon termed *desensitization.* Upon removal of agonist, the attenuated responsiveness is reversed in a process termed *resensitization,* which involves receptor dephosphorylation (Pippig et al., 1995).

Considerable progress has been made in identifying and characterizing the mechanisms of action of the GRKs and arrestins and in illustrating the roles of these proteins in the agonist-dependent desensitization of a vast array of GPCRs (reviewed by Pitcher et al., 1998). In contrast, the GRP(s), responsible for receptor resensitization, are less well characterized. The majority of information concerning the identity of this enzyme has come from studies using GRK phosphorylated β_2-adrenergic receptor (βAR) as a substrate. The GRP responsible for βAR dephosphorylation is a membrane-associated form of PP-2A (Pitcher et al., 1995).

The activity of this enzyme is tightly regulated in a manner that is dependent on receptor conformation. GRK-phosphorylated βAR localized in endosomes is the only substrate for this enzyme (Krueger et al., 1997). Exposure of the extracellular domains of the phosphorylated βAR to the low pH characteristic of internalized vesicles induces a change of receptor conformation, converting it to a form that can be recognized by the phosphatase (Krueger et al., 1997). In this respect, the GRP is similar to the GRKs, the kinase whose action it reverses. The activity of both the GRK and the GRP is dependent on receptor conformation. GRK-mediated βAR phosphorylation occurs only when the receptor is in an agonist-occupied conformation, and GRP-mediated βAR dephosphorylation occurs only when the receptor undergoes a pH-induced conformational change.

In this chapter, GRK phosphorylated βAR is used to illustrate the methods employed to identify and characterize the GRP. These experimental approaches are applicable, however, to the study of any phosphorylated GPCR. Methods employed to characterize receptor phosphatase activity in crude extracts and in cells as well as techniques employed for the partial purification of the GRP are described. In each case, a brief description of the rationale underlying the experimental design is followed by detailed descriptions of the experimental procedures. When appropriate, the description has been abbreviated by omitting protocols readily available from other sources or incorporated into other chapters of this volume.

2. CHARACTERIZATION OF RECEPTOR PHOSPHATASE ACTIVITY IN TISSUE EXTRACTS

The biochemical properties of protein serine/threonine phosphatases have been extensively characterized (reviewed by Cohen, 1989). This, combined with the availability of specific and potent inhibitors of these enzymes, permits their identification and quantitation in tissue extracts (Cohen, 1991). To identify GRP activity, a ^{32}P-labeled, GRK-phosphorylated receptor substrate is incubated with a source of phosphatase in the presence of phosphatase activators and inhibitors, and the loss of receptor associated ^{32}P over time is monitored. Although simple in concept, the limited number of purified reconstituted GPCRs available has limited this approach. Indeed, to date, the GRK-phosphorylated βAR is the only receptor whose dephosphorylation has been assessed in this manner (Pitcher et al., 1995).

An alternative methodology has been to use as a phosphatase substrate highly purified membranes containing GRK-phosphorylated ^{32}P-labeled GPCRs at high specific activity. In this manner, the nature of the phosphatase responsible for dephosphorylating GRK2–phosphorylated βAR and α_2C_2-adrenergic receptor has been examined with purified membranes derived from baculovirus-infected *Spodoptera frugiperda* (Sf9) cells (Pitcher et al., 1995). Similarly, the dephosphorylation of rhodopsin kinase (GRK1)–phosphorylated rhodopsin has been studied with rod outer segment membranes (King et al., 1994).

The assay of GRP activity necessitates the preparation of a suitable substrate and source of phosphatase and the use of appropriate assay conditions such that protein phosphatase activities can be distinguished. Each of these requirements is considered separately below.

2.A. Substrate Preparation

2.A.a. GRK-Phosphorylated βAR. This protocol describes the method used to produce GRK2–phosphorylated ^{32}P-labeled purified reconsituted βAR but can be adapted for use with any purified receptor or GRK family member. The purification of human βAR from Sf9 cells (Chapter 1), the reconstitution of purified receptors into phospholipid bilayers (Chapter 3), and details of the purification procedures for GRKs (Chapter 7) are described in other chapters of this volume and thus are omitted here.

Protocol: ^{32}P-Labeled Purified Reconstituted βAR Preparation

1. In a 2 ml capped Eppendorf tube place

    Purified reconstituted βAR, 100 pmol

    Purified GRK2, 62.5 pmol

    5× Reaction buffer, 100 μl

    (Reaction buffer [final concentration]: 20 mM Tris-HCl, pH 8.0; 2 mM EDTA; 10 mM MgCl$_2$; 1 mM dithiothreitol, 50 μM [−]isoproterenol)

    Make up to a final volume of 400 μl with H$_2$O

    Note: The presence of the GPCR agonist ([−]isoproterenol for the βAR) is essential for GRK-mediated receptor phosphorylation

2. Start phosphorylation reaction by addition of 100 μl of 300 μM ATP containing ~3,000 cpm/pmol γ^{32}P-ATP. Vortex briefly, and incubate at 30°C for 20 minutes.

3. Stop the phosphorylation reaction by addition of 1 ml of ice-cold 50 mM Tris-HCl (pH 7.0), 0.1 mM EDTA, 50 mM 2-βmercaptoethanol (buffer A); and transfer tubes to ice.

4. Divide reaction equally between two pre-cooled Beckman TLA 100.2 tubes, reserving 10 μl of the diluted reaction mix for determination of receptor phosphorylation stoichiometry (see step 9).

5. Centrifuge in a TLA 100.2 rotor at 80,000 rpm (~200,000g) for 15 minutes at 4°C. A 2–3-mm pellet will become apparent.

6. Carefully aspirate γ^{32}P-ATP–containing supernatant, and resuspend each pellet in 1 ml ice-cold buffer A.

7. Repeat steps 5 and 6 twice to remove free ATP. Finally, resuspend the pellet of phosphorylated βAR substrate in 125 μl of buffer A and pool in a single screw cap tube.

8. Determine receptor concentration of ^{32}P-labeled βAR preparation by radioligand binding using [^{125}I](–)-iodocyanopindolol (Samama et al., 1993). Because βAR is radioactively labeled, a control incubation lacking [^{125}I](–)-iodocyanopindolol should be included in the binding assay to determine counts present in the absence of added ligand. In practice, because small amounts of receptor are assayed (~15 fmol receptor/binding assay) and the receptor is labeled with ^{32}P rather than ^{125}I, negligible counts are observed under these conditions.

9. To determine the stoichiometry of GRK2–mediated βAR phosphorylation, 1 pmol of labeled receptor is electrophoresed on a 10% polyacrylamide gel. The gel is subsequently dried on paper and subjected to phosphorimager analysis. Before phosphorimager exposure with the reaction mixture reserved at step 4, spot volumes equivalent to 0.5, 1.0, 3.0, and 5.0 pmol ATP on paper adjacent to the gel to determine pixels/pmol ATP and thus quantitate the number of pmol ^{32}P incorporated/pmol βAR.

2.A.b. ^{32}**P-Labeled, Membrane–Incorporated GPCRs.** Phosphorylated membrane-incorporated receptor substrates represent an alternative, more readily accessible source of labeled substrate for *in vitro* phosphatase assays. GRK1–phosphorylated rhodopsin in purified rod outer segment membranes (King et al., 1994), together with GRK2–phosphorylated βAR and $\alpha_2 C_2$ receptor in purified membranes derived from baculovirus-infected Sf9 cells (Pitcher et al., 1995), have been used for this purpose. Because recombinant baculovirus infection of Sf9 cells represents an approach applicable to the study of any GPCR, this methodology, illustrated using the βAR, is described below.

A detailed description of the methods used to produce recombinant baculovirus and the expression of recombinant proteins in Sf9 cells is beyond the scope of this chapter, and readers are referred to additional resources (Gruenwald and Heitz, 1993). In brief, for the βAR, expression in Sf9 cells is induced by infection of 1.5×10^6 cells/ml with a recombinant baculovirus at a multiplicity of infection of 5. Cells are grown for 48 hours after infection and harvested by centrifugation. The cell pellets are subsequently resuspended in 20 mM Hepes, pH 7.2, 5 mM EDTA supplemented with a mixture of protease inhibitors (PIs) (10 μg/ml benzamidine, 100 mM phenylmethylsulfonyl fluoride, 4 μg/ml leupeptin, and 1 μg/ml pepstatin) (buffer B) and stored frozen (–70°C) until required.

Protocol: Preparation of βAR Containing Membranes of High Specific Activity
All procedures are performed at 4°C.

1. Fifty milliliters of frozen Sf9 cell lysate in buffer B (approximately 10 liters of Sf9 cell culture) expressing recombinant βAR (~10 pmol βAR/mg total cell protein) is thawed on ice and supplemented with additional PIs, and cells are lysed by dounce homogenization (approximately 12 strokes with a tight-fitting pestle).

2. Lysate is centrifuged at 300g for 5 minutes to remove unbroken cells and nuclei, and a crude membrane preparation is obtained by centrifugation of the resulting supernatant at 300,000g for 30 minutes.

3. Membrane pellets are resuspended in 5 ml buffer B and layered carefully on top of two 35 ml 5%–50% continuous, nonlinear, sucrose gradients made in the same buffer. Gradient composition: 10 ml 5%–25% sucrose, 15 ml 25% sucrose, and 10 ml 25%–55% sucrose (Kassis and Sullivan, 1986); 40 ml polycarbonate tubes are convenient for this purpose.

4. Samples are centrifuged in a swinging bucket rotor at 100,000g for 100 minutes.

5. Following centrifugation gradient fractions are collected and assayed for [^{125}I](–)-cyanopindolol binding (Samama et al., 1993) to determine the location of βAR. Two separated peaks of βAR-binding activity are detected corresponding to vesicle and plasma membrane associated receptor. Fractionation of the 37.5 ml gradient into 2 ml aliquots is sufficient to adequately separate these two membrane fractions. Numbering fractions from the top of the gradient (5% sucrose), vesicle-associated βAR is eluted at approximately 18% sucrose, fraction numbers 4–7, while plasma membrane associated βAR is eluted at approximately 40% sucrose, fraction numbers 14–18.

6. Fractions containing vesicle-associated βARs (fractions 4–7) are pooled, diluted approximately fivefold with buffer B, and subjected to centrifugation at 300,000g for 30 minutes.

7. Membrane pellets are resuspended using gentle pipetting in approximately 1 ml of buffer A and protein concentration, and βAR content is determined to obtain the specific activity of βAR present. This technique provides an approximately 20–40-fold purification of the βAR such that final specific activities are in the range of ~300 pmol βAR/mg protein. A preparation of this size generally yields around 2–3 mg of total protein. Once prepared, the βAR-containing membranes can be stored at –70°C before phosphorylation with GRKs.

Protocol: Phosphorylation of Membrane-Incorporated βAR

The phosphorylation protocol utilized is identical to that described in Section 2.A.a. with the exception that purified reconstituted βAR is replaced with βAR containing purified Sf9 cell membranes.

2.B. Source of Protein Phosphatases

Tissue extracts prepared in 50 mM Tris (pH 7.0), 5 mM EDTA, 50 mM 2-βmercaptoethanol + PIs (buffer C) are used as the source of phosphatase. The pattern of expression of the GPCR to be used as substrate determines the choice of tissue. In the case of phosphorylated βAR, the GRP present in bovine brain extracts was characterized (Pitcher et al., 1995). Bovine cerebral cortex (100 g) was homogenized with a polytron tissue disruptor in 500 ml buffer C and subjected to low-speed centrifugation (300g, 15 minutes) to obtain a crude extract (the supernatant). Fractionation of this extract by centrifugation at 300,000g for 45 minutes yielded a supernatant containing soluble phosphatases and a pellet of membrane-enriched particulate material. Characterization of GRP activity in these fractions revealed this enzyme to be a membrane-associated phosphatase (Pitcher et al., 1995).

2.C. Phosphatase Assays

Four types of protein serine/threonine phosphatase catalytic subunits account for most of the activity of these enzymes detectable in crude extracts. These enzymes, PP1, PP2A, PP2B, and PP2C, can be readily distinguished biochemically based on substrate specificity and on their sensitivity to a variety of inhibitors or activators (reviewed by Cohen, 1989, 1991). Type 1 protein phosphatases specifically dephosphorylate the β subunit of phosphorylase kinase and are inhibited by nanomolar concentrations of two small heat- and acid-stable proteins, termed *inhibitor 1* (I-1) and *inhibitor 2* (I-2). In contrast, the type 2 protein phosphatases (PP2A, PP2B, and PP2C) dephosphorylate preferentially the α subunit of phosphorylase kinase and are insensitive to inhibition by I-1 and I-2.

PP2A, PP2B, and PP2C activities are most easily distinguished by their metal ion sensitivity. PP2B and PP2C exhibit absolute requirements for Ca^{2+} and Mg^{2+}, respectively, while PP2A, like PP1, is at least partially active toward most substrates in the absence of divalent cations. In addition to their cation dependence and I-1 sensitivity, a useful criterion for distinguishing phosphatase activities is their sensitivity to okadaic acid, a complex fatty acid polyketal first extracted from the marine sponge *Halichondria okadaii*. When crude extracts are diluted so as to obtain a phosphorylase-a phosphatase activity of <0.1 mU/ml, 1 nM okadaic acid inhibits PP2A activity completely (IC_{50} ~ 0.1 nM); in contrast, PP1 activity is unaffected. PP1 activity is, however, completely inhibited by 5 μM okadaic acid (IC_{50} ~ 10–15 nM). PP2C and PP2B are insensitive to okadaic acid inhibition. Thus in a crude extract assayed at the appropriate phosphorylase-a phosphatase activity, an unknown phosphatase activity can be identified as

PP1, if dephosphorylation of the receptor substrate is observed in the absence of divalent metal ions in the presence of 1 nM okadaic acid (to block PP2A) and is inhibited by addition of 0.2 μM I-1 or I-2

PP2A, if dephosphorylation is observed in the absence of divalent cations in the presence of 0.2 μM I-1 or I-2, but is inhibited by addition of 1 nM okadaic acid

PP2B, if in the presence of 5 μM okadaic acid (to inhibit PP1 and PP2A) Ca^{2+}-dependent dephosphorylation is observed

PP2C, if Mg^{2+}-dependent dephosphorylation is observed in the presence of 5 µM okadaic acid (to inhibit PP-1 and PP-2A) and 0.1 mM EGTA (to inhibit PP2B).

If multiple protein serine/threonine phosphatases dephosphorylate a phosphorylated substrate, the use of the appropriate inhibitors can be used to distinguish the relative contributions of each.

The procedure outlined above measures active forms of protein phosphatases; inactive forms of PP1 and PP2 are, however, also present in mammalian tissue extracts. Indeed, the GRP responsible for dephosphorylating ^{32}P-labeled GRK-phosphorylated βAR is latent when assayed at neutral pH (i.e., exhibits no spontaneous activity; Pitcher et al., 1995). In these situations, artificial activators of PP2A in combination with the inhibitors and activators outlined above are used to identify phosphatase activity. PP2A is activated at low (submicromolar) concentrations of basic proteins such as polylysine and protamine. Freeze–thaw treatment of tissue extracts under reducing conditions has additionally been shown to potently actvate PP2A (reviewed by Cohen, 1989). Basic protein addition or freeze–thaw treatment combined with the use of okadaic acid, I-1 and I-2, and metal ions was used to identify the GRP as a latent form of PP2A. It is important to note that activation by basic proteins is by itself insufficient evidence for the identification of a protein phosphatase activity as PP2A because PP1-mediated dephosphorylation of certain phosphorylated proteins, including glycogen synthase, pyruvate kinase, and hormone sensitive lipase, is also stimulated under these conditions (reviewed by Cohen, 1991).

Agents that block βAR internalization block receptor resensitization (Pippig et al., 1995; Yu et al., 1993; Sibley et al., 1986). Similarly, βAR mutants that are impaired in their ability to internalize fail to resensitize (Barak et al., 1994). These observations, together with earlier findings that indicated that receptors in a vesicular fraction are in a less phosphorylated state than receptors in the plasma membrane (Sibley et al., 1986), led to a model proposing that dephosphorylation of the receptor occurs upon its sequestration into a vesicle population (Yu et al., 1993). Consistent with this hypothesis, treatment of agonist-stimulated cells with concanavalin A, which blocks GPCR internalization, prevents receptor resensitization and dephosphorylation (Pippig et al., 1995). Because internalization of the βAR into vesicles appears to be required for their dephosphorylation and because these vesicles have been identified as endosomes (von Zastrow et al., 1992, 1993), an acidified vesicle population (Gruenberg and Howell, 1989), a hypothesis was proposed suggesting that the uniquely low pH of the vesicles is the key regulator of GRP-mediated receptor dephosphorylation. To test this hypothesis, GRP assays were performed under conditions of varying pH (Krueger et al., 1997). Although, as described above, GRP activity is latent at pH 7.0, significant spontaneous GRP activity is observed at lower pHs. The pH optimum for GRP-mediated dephosphorylation of GRK2–phosphorylated βAR is 4.85. The effect of low pH is specific for the βAR substrate and is dramatically attenuated if GRP activity is assayed with heat-denatured βAR (Krueger et al., 1997). These results suggest that the effect of low pH is to modulate βAR conformation, converting it to a form that can be dephosphorylated by GRP rather than to directly activate the GRP itself. Protocols for characterizing GRP activity at neutral pH and for examining the effect of pH-induced changes of receptor conformation on GRP activity are described below.

Protocol: Identification of GRP in Tissue Extracts at Neutral pH

1. In an Eppendorf tube, place

    10 µl 50 mM Tris-HCl (pH 7.0), 0.1 mM EDTA, 50 mM 2-βmercaptoethanol + PIs (± 3 × phosphatase activators/inhibitors; see assay conditions)

    10 µl bovine brain extract containing <0.1 mU/ml phosphorylase-a phosphatase activity (see phosphorylase-a phosphatase assay, below)

2. Start reaction by addition of 10 µl of ^{32}P-labeled βAR substrate (1.0 pmol of βAR-phosphorylated to a stoichiometry of between 2 and 4 mol P_i/mol βAR).

    Assay Conditions

    a. Buffer alone

    b. Buffer + 3 nM okadaic acid (PP1 activity)

    c. Buffer + 3 nM okadaic acid + 0.6 µM I-1 (decreased activity compared with b = PP1 activity)

    d. Buffer + 0.6 µM I-1 (PP2A activity)

    e. Buffer + 0.6 µM I-1 + 1 nM okadaic acid (decrease in activity compared with d = PP2A activity)

    f. Buffer + 5 µM okadaic acid + 0.1 mM Ca^{2+} or 0.1 mM Ca^{2+}/0.2 µM calmodulin (PP2B activity)

    g. Buffer + 5 µM okadaic acid + 0.1 mM EGTA + 5 mM Mg^{2+} (PP2B activity)

    h–n. As for a–g using extract subjected to freeze–thaw treatment in the presence of 0.2 M 2-βmercaptoethanol or as for a–g + protamine (3 mg/ml)

3. Incubate reaction at 30°C for the appropriate time; for βAR this ranged between 20 and 60 minutes.

4. Stop reaction by addition of SDS-PAGE sample loading buffer and subject samples to electrophoresis on 10% polyacrylamide gels.

5. Dry gels and quantitate phosphatase-mediated loss of ^{32}P label from the βAR substrate by phosphorimager analysis.

Negative control: ^{32}P-labeled βAR incubated in the absence of divalent cations + 5 µM okadaic acid or in the absence of divalent cations + 1 nM okadaic acid and 0.2 µM I-1 (final concentrations in the assay) is used to eliminate the possibility that proteolysis contributes to the loss of ^{32}P from the receptor. The amount of βAR-incorporated ^{32}P under these conditions should equal that observed when receptor is incubated in the absence of extract and is equivalent to 0% dephosphorylation.

Positive control: Phosphorylase-a phosphatase and/or casein phosphatase activities are determined under the same conditions as for the βAR. This is of particular importance in the case of a latent phosphatase because it establishes and quantifies the amount of spontaneous phosphatase activity present in the extract.

Protocol: Assessing the pH Dependence of GRP Activity

Assays are performed as described above (see Identification of GRP in Tissue Extracts at Neutral pH) with the exception that 50 mM acetic acid–acetate buffer is used to replace 50 mM Tris-HCl (pH 7.0). Varying the proportion of acetic acid relative to acetate in the 50 mM buffer is used to adjust the pH of the assay between 4.0 and 6.5.

Protocol: Phosphorylase-a Phosphatase Assays

The production of ^{32}P-labeled phosphorylase-a (Shenolikar and Ingebritsen, 1984) and phosphorylase-a phosphatase assays (Cohen et al., 1988) have been described in detail elsewhere. One unit of phosphorylase-a phosphatase activity is that amount that catalyzes the dephosphorylation of 1 μmol of substrate in 1 minute.

3. MONITORING RECEPTOR DEPHOSPHORYLATION IN CELLS

Transient transfection of HEK 293 cells with tagged GPCRs and the subsequent immunoprecipitation of receptor from ^{32}P-labeled cells following agonist exposure represents a powerful technique for studying GRK-mediated phosphorylation of a wide array of GPCRs. This methodology, which is described in detail in Chapter 2, is readily modified to study receptor dephosphorylation. Cells are treated with agonist to promote GRK-mediated receptor phosphorylation, the agonist is subsequently removed, and the cells are incubated in the absence of agonist to allow receptor resensitization and dephosphorylation. The ^{32}P content of receptor immunoprecipitated from labeled cells at various times post-agonist treatment is compared with the ^{32}P content of receptor immunoprecipitated immediately following agonist treatment, thus allowing a time course of receptor dephosphorylation to be determined.

In this fashion, the effect of vesicle acidification on the rate of receptor dephosphorylation has been investigated for the βAR (Krueger et al., 1997). Incubation of cells with NH_4Cl, a weak base used to raise the pH of acidic cellular compartments such as endosomes and lysosomes (Maxfield, 1982; Beaumelle et al., 1992), significantly impairs βAR dephosphorylation (Fig. 9.1). These experiments demonstrate in a cellular system the requirement for vesicular acidification for GRP activity and provide a molecular explanation for the observation that inhibition of βAR internalization prevents βAR resensitization and dephosphorylation (Pippig et al., 1995). As the methodology for immunoprecipitation of GPCRs from ^{32}P-labeled cells has been described previously (Chapter 2), an abbreviated protocol, highlighting modifications re-

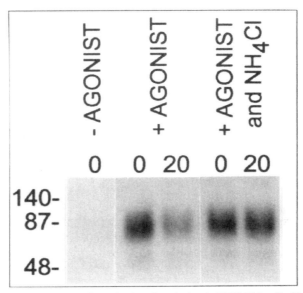

Figure 9.1. NH$_4$Cl inhibits βAR dephosphorylation in HEK 293 cells. HA-tagged βAR was phosphorylated by exposure of transfected HEK 293 cells to 10 μM (–)-isoproterenol either alone or with 20 mM NH$_4$Cl. Cells were thoroughly washed to remove agonist and were either harvested at this time (time 0) or incubated an additional 20 minutes with medium alone or medium containing 20 mM NH$_4$Cl before harvesting (time 20). The HA-tagged βAR was immunoprecipitated and subjected to SDS-PAGE. The autoradiogram of a representative of four experiments is shown. (From Krueger et al. [1997], with permission of the publisher.)

quired for monitoring receptor dephosphorylation and for modulating vesicular pH using NH$_4$Cl is described here.

Protocol

HEK 293 cells are transiently transfected with cDNA encoding tagged βAR (for the experiments described here, HA-tagged βAR was used). The intracellular ATP pool is labeled using ^{32}Pi, and the cells are incubated with 10 μM isoproterenol either alone or in the presence of 20 mM NH$_4$Cl for 10 minutes at 37°C. Cells are washed with phosphate-buffered saline and either harvested immediately (0% dephosphorylation) or incubated an additional 20 minutes with medium alone or medium containing 20 mM NH$_4$Cl. ^{32}P-labeled βAR is immunoprecipitated using the monoclonal 12CA5 antibody and protein A Sepharose, and equivalent amounts of protein are subjected to SDS-PAGE on 10% polyacrylamide gels. Dried gels are exposed to film and subsequently quantified by phosphorimager analysis. The amount of receptor dephosphorylation is indicated by the difference in ^{32}P content of receptor immunoprecipitated immediately following agonist treatment (100% phosphorylated) and that immunoprecipitated after the additional 20 minutes incubation (dephosphorylated sample). In the absence of NH$_4$Cl treatment, approximately 80% of the receptor is dephosphorylated over this

20-minute period. A representative example of results obtained using this protocol is shown in Figure 9.1.

4. PURIFICATION OF GRP

4.A. Partial Purification of GRP From Bovine Brain Extract

The GRP is a latent membrane-associated enzyme that biochemically displays characteristics consistent with its identification as a form of PP2A (see Section 2.C.). PP2A holoenzymes exist as heterotrimeric complexes composed of a 36-kD catalytic subunit (C), a 65-kD structural subunit (A), and a third subunit (B) ranging in molecular weight between 52 and 74 kD (reviewed by Wera and Hemmings, 1995). Two isoforms of C and A and at least six different gene products encoding B subunits have been identified.

In an attempt to gain information concerning the structure of the GRP and how this compares with the structure of previously characterized forms of PP2A, the GRP was partially purified by gel filtration. Detergent-solubilized preparations derived from the particulate fraction of bovine brain were fractionated on Superose-12 in either the presence or absence of detergent (Fig. 9.2). Gel filtration of solubilized GRP in the presence of detergent revealed a major peak of phosphorylase-a phosphatase and GRP activity, that gel filtered with a molecular weight of 35 kD (Fig. 9.2), was inhibited by low concentrations of okadaic acid, and was stimulated by protamine. These observations are consistent with the identification of the catalytic subunit of GRP as PP2A.

In contrast, when solubilized extracts are gel filtered in the absence of detergent, two peaks of GRP activity are observed with apparent molecular masses of 150 and 35 kD. Notably, although the spontaneous phosphorylase-a phosphatase activity of the 35 kD peak is similar when gel filtered in either the presence or absence of detergent, the GRP exhibits very different characteristics when gel filtered under these conditions. In the absence of detergent, GRP activity is latent, that is, it is only observed in the presence of protamine, while in the presence of detergent spontaneous GRP activity is observed. These results suggest that detergent, much like a change in pH, induces a conformational change in the phosphorylated βAR substrate such that it becomes a substrate for the GRP.

The high molecular weight (150 kD) latent form of GRP observed in the absence of detergent had an approximately fourfold higher activity per unit of phosphorylase-a activity than did the isolated catalytic subunit. Immunoblot analysis of this form of the enzyme using antibodies directed against the C, A, and three distinct B subunits (B_α [55 kD], B_β [55 kD], and B′[53 kD]) of PP2A suggested a potential GRP subunit structure of $AB_\alpha C$ (Pitcher et al., 1995). Notably, PP2A of this subunit composition is the most prevalent form of soluble PP2A; however, negligible soluble GRP activity is detected.

Several potential explanations exist for this observation. The B_α subunit antibodies are directed against a short peptide sequence from this protein. The GRP may thus contain a regulatory B subunit highly related to but distinct from B_α. Alternatively, because two isoforms (α and β) of the C and A subunits of

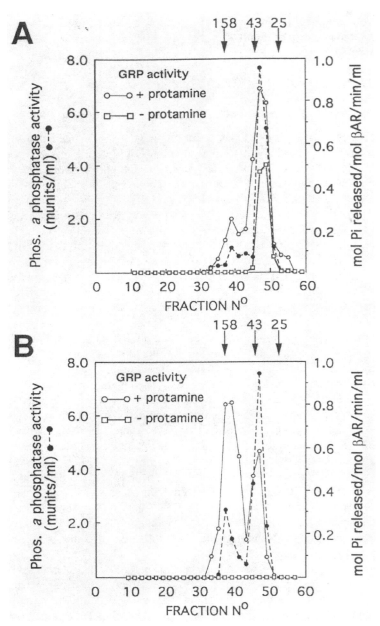

Figure 9.2. Gel filtration of GRP. A detergent solubilized particulate fraction from bovine brain was subjected to gel filtration in the presence of detergent (0.05% dodecyl β-maltoside) **(A)**. Alternatively, before gel filtration, detergent was removed from the solubilized fraction, and gel filtration was performed in the absence of detergent **(B)**. Fractions were assayed for phosphorylase-a phosphatase (●) and GRP activity. GRP activity was measured in the presence (○) or absence (□) of protamine sulfate (1 mg/ml). The elution positions of aldolase (158 kD), ovalbumin (43 kD), and chymotrypsinogen A (25 kD) are indicated. (From Pitcher et al. [1995], with permission of the publisher.)

PP2A have been identified, the GRP and soluble PP2A may have different iso-form compositions. Additionally, covalent modifications of the catalytic sub-unit of PP2A have been reported, including phosphorylation (Chen et al., 1992; Guo and Damuni, 1993) and carboxyl methylation (Xie and Clarke, 1994). Dif-ferential modification of one or more of the constituent subunits of the GRP may account for the altered properties of this enzyme. The exact subunit com-position of the GRP remains somewhat obscure and awaits the purification of this enzyme. The partial purification of GRP by gel filtration is likely to prove a useful step in this purification procedure, and a protocol for this chromato-graphic step is included here.

Protocol: Partial Purification of GRP From Bovine Brain

All procedures are performed at 4°C.

1. Homogenize 100 g of bovine cerebral cortex in 500 ml of 50 mM Tris-HCl, pH 7.0, 5 mM EDTA, 50 mM 2-βmercaptoethanol + PIs (buffer C) using a polytron homogenizer (approximately 5 × 30 second bursts) and pass extract through cheesecloth to remove unhomogenized tissue.

2. Centrifuge extract at 300g for 15 minutes.

3. Carefully remove supernatant, and centrifuge at 300,000g for 45 minutes to obtain a soluble and particulate fraction.

4. Resuspend pellet in 2 volumes buffer C containing 0.25% dodecyl-β-maltoside, and incubate with constant stirring for 45 minutes.

5. Recover soluble proteins by centrifugation at 350,000g for 10 minutes.

6. Remove detergent from the supernatant by passing this fraction over an extractigel column (25 ml volume).

7. Load small aliquots of sample (250 μl) on a Superose-12 column (10 × 300 mm) equilibrated in 20 mM Tris-HCl, pH 7.0, 2 mM EDTA, 5% (vol/vol) glycerol, 100 mM NaCl. Develop column at a flow rate of 0.3 ml/min and collect 0.3 ml fractions. Assay fractions for the presence of GRP and phosphorylase-a phosphatase activities as described in Section 2,C.

4.B. Co-Immunoprecipitation of GRP With Phosphorylated GPCRs

The immunoprecipitation of overexpressed tagged receptors and the subse-quent identification of associated phosphatase catalytic subunits by Western blot analysis represents an alternative approach for purifying the GRP. For the βAR, this approach has been used to both confirm the identity of the catalytic subunit of the GRP and to investigate the effect of vesicular acidification on βAR/GRP association (Krueger et al., 1997). HEK 293 cells overexpressing Flag-tagged βAR are treated either with or without isoproterenol, harvested, and subjected to fractionation on nonlinear sucrose gradients. As described in Section 2.A. this fractionation procedure separates vesicular and plasma mem-brane localized βAR.

Immunoprecipitation of βAR from both membrane fractions and the subse-quent blotting of immunoprecipitates for the catalytic subunit of PP2A reveal that, although the catalytic subunit of PP2A is present in all fractions, its association with βAR is agonist dependent and occurs exclusively in vesicles (Fig. 9.3). Agonist treatment promotes GRK-mediated receptor phosphorylation. The agonist-dependent association of GRP and βAR suggests that the phosphatase binds exclusively to its phosphorylated βAR substrate. Furthermore, the vesicular location of this interaction supports the hypothesis (outlined in Sections 2 and 3) that vesicular acidification is a necessary prerequisite for GRP/βAR binding. This model is further supported by the observation that GRP fails to associate with phosphorylated βAR immunoprecipitated from the vesicular fraction of cells treated with NH_4Cl, an agent that inhibits vesicular acidification (Maxfield, 1982) (Fig. 9.3).

The co-immunoprecipitation of receptor and GRP is a useful technique both for the identification of GRP and for studying the regulatory mechanisms controling the activity of this enzyme. Because epitope tagging of GPCRs is a widely applicable approach (see Chapter 1), this methodology can be used to study the dephosphorylation of potentially any cloned GPCR.

Protocol: Co-Immunoprecipitation of GRP and βAR From HEK 293 Cells

Fractionation of HEK 293 cell lysates by sucrose density gradient fractionation. This procedure is essentially identical to that used to fractionate Sf9 cell lysates described in Section 2.A. The description of this method included here is thus limited to a description of those changes incorporated when using HEK cell lysates.

HEK 293 cells (1×150 mm plate) transiently transfected with Flag-tagged βAR are incubated for 15 minutes in medium alone, medium containing 10 μM isoproterenol, or medium containing 20 mM NH_4Cl and isoproterenol. Cells are placed on ice, washed, and incubated with phosphate-buffered saline containing 250 μg of concanavalin A/ml to minimize vesicularization of the plasma membrane. Cells are subsequently harvested and lysed in 50 mM Tris-HCl (pH 8.0), 5 mM EDTA containing 5% sucrose + PIs using a dounce homogenizer. The preparation of crude membranes, fractionation on sucrose gradients, and detection of βAR are performed as described previously (Section 2.A.b.).

Immunoprecipitation of Flag-tagged βAR. Fractions from the sucrose gradients containing either vesicle or plasma membrane associated βAR are pooled, diluted with 50 mM Tris-HCl (pH 8.0), 5 mM EDTA containing 5% sucrose + PIs, and centrifuged at 100,000g for 30 minutes. Pellets are resuspended in 10 mM Hepes (pH 7.4), 1 mM EDTA, 1 mM dithiotreitol, 100 mM NaCl, and 1% CHAPS, and receptor complexes are immunoprecipitated with 15 μg of monoclonal antibody M2 in the presence of protein G–coupled agarose beads. Immunoprecipitates are washed in the same buffer, resuspended in SDS sample loading buffer, and subjected to polyacrylamide gel electrophoresis on a 10% polyacrylamide gel. Gels are transferred to nitrocellulose and probed using an antibody directed against the catalytic subunit of PP2A. Blots are developed by incubation with a horseradish peroxidase–conjugated secondary antibody followed by processing of the blot as recommended by the manufacturer (ECL by Amersham Corp.).

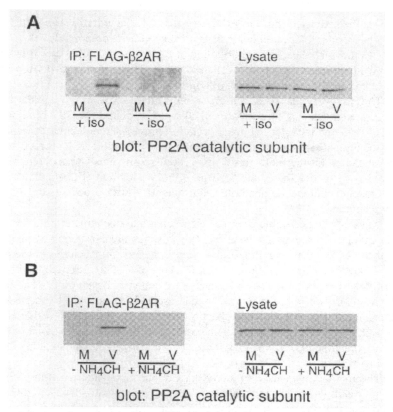

Figure 9.3. Agonist and vesicular acidification-dependent association of the GRP subunit with internalized βAR. **A:** Agonist-dependent association of GRP with internalized βAR. **B:** Acidification-dependent association of GRP with βAR. HEK 293 cells transiently expressing Flag-tagged βAR were incubated with isoproterenol either alone (A and B) or in the presence of 20 mM NH$_4$Cl (B). The expressed βAR present in either the vesicles (V) or plasma membrane (M) was immunoprecipitated. Immunoprecipitated protein complexes (left panels), as well as samples of the fractions taken before immunoprecipitation (right panels), were subjected to SDS-PAGE followed by immunoblotting with antibody against the PP2A catalytic subunit. Immunoblots representative of three separate experiments are shown. (From Krueger et al. [1997], with permission of the publisher).

The results of a representative experiment with this protocol are shown in Figure 9.3.

5. SUMMARY

The identification of the GRP activity responsible for dephosphorylating GRK-phosphorylated βAR as a latent, membrane-associated form of PP2A initiated a series of experiments designed to elucidate the regulatory mechanisms controling the activity of this enzyme. Assays of GRP activity *in vitro* using purified βAR substrates and bovine brain tissue extracts, combined with cellular studies in which the dephosphorylation of βAR was monitored by ^{32}P-labeling or by directly assessing βAR/GRP association, have revealed that GRP activity is intimately dependent on βAR conformation.

The discovery that βAR conformation is altered following vesicular acidification and that this conformational change is required for GRP binding and GRP-mediated receptor dephosphorylation provides a molecular explanation for the observation that inhibiting receptor internalization prevents receptor resensitization. Furthermore, this finding reveals a previously unappreciated parallel between GRK-mediated βAR phosphorylation and GRP-mediated βAR dephosphorylation, the absolute requirement of both processes on receptor conformation.

Although considerable progress has been made in characterizing the GRP, several questions remain to be addressed, most notably the subunit composition of the GRP and the generality of this dephosphorylation mechanism. Members of the GRK family phosphorylate a vast array of GPCRs. It remains to be determined if the GRP activity responsible for dephosphorylating βAR similarly dephosphorylates other receptors. In this respect it is important to note that dephosphorylation of the GRK2–phosphorylated $\alpha_2 C_2$ receptor is mediated by a phosphatase activity with biochemical properties identical to those of GRP, suggesting that this enzyme may indeed play a general role in regulating GPCR resensitization.

The methodology used to study βAR dephosphorylation outlined in this chapter is in large part applicable to the study of any phosphorylated GPCR. Studying the phosphatases responsible for dephosphorylating a variety of GPCRs will prove useful in elucidating the substrate specificity of the GRP.

REFERENCES

Barak LS, Tiberi M, Freedman NJ, Kwatra MM, Lefkowitz RJ, Caron MG (1994): A highly conserved tyrosine residue in G protein–coupled receptors is required for agonist-mediated β_2-adrenergic receptor sequestration. J Biol Chem 269:2790–2793.

Beaumelle B, Bensammar L, Bienvenue A (1992): Selective translocation of the A chain of diphtheria toxin across the membrane of purified endosomes. J Biol Chem 267:11525–11531.

Chen J, Martin BL, Brautigan DL (1992): Regulation of protein serine-threonine phosphatase type-2A by tyrosine phosphorylation. Science 257:1261–1264.

Cohen P (1989): The structure and regulation of protein phosphatases. Annu Rev Biochem 58:458–508.

Cohen P (1991): Classification of protein–serine/threonine phosphatases: Identification and quantitation in tissue extracts. Methods Enzymol 201:389–399.

Cohen P, Alemany S, Hemmings BA, Resnick TJ, Stralfors P, Tung HYL(1988): Protein phosphatase 1 and protein phosphatase-2A from rabbit skeletal muscle. Methods Enzymol 107:102–129.

Gruenberg J, Howell KE (1989): Membrane traffic in endocytosis: Insights from cell-free assays. Annu Rev Cell Biol 5:453–481.

Gruenwald S, Heitz JD (1993): Baculovirus Expression Vector System: Procedures and Methods Manual, 1st ed. San Diego: Pharmingen.

Guo H, Damuni Z (1993): Autophosphorylation-activated protein kinase phosphorylates and inactivates protein phosphatase 2A. Proc Natl Acad Sci USA 90:2500–2504.

Kassis S, Sullivan M (1986): Desensitization of the mammalian beta-adrenergic receptor: Analysis of receptor redistribution on non-linear sucrose gradients. J Cyclic Nucleotide Protein Phosphor Res 11:35–46.

King AJ, Andjelkovic N, Hemmings BA, Akhtar M (1994): The phospho-opsin phosphatase from bovine rod outer segments. An insight into the mechanism of stimulation of type-2A protein phosphatase activity by protamine. Eur J Biochem 225:383–394.

Krueger KM, Daaka Y, Pitcher JA, Lefkowitz RJ (1997): The role of sequestration in G protein–coupled receptor resensitization. J Biol Chem 272:5–8.

Maxfield FR (1982): Weak bases and ionophores rapidly and reversibly raise the pH of endocytic vesicles in cultured mouse fibroblasts. J Cell Biol 95:676–681.

Pippig S, Andexinger S, Lohse MJ (1995): Sequestration and recycling of beta-2-adrenergic receptors permit receptor resensitization. Mol Pharmacol 47:666–676.

Pitcher JA, Freedman NJ, Lefkowitz RJ (1998): G protein–coupled receptor kinases. Annu Rev Biochem 67:653–692.

Pitcher JA, Payne ES, Csortos C, Depaoli-Roach AA, Lefkowitz RJ (1995): The G protein–coupled receptor phosphatase: A protein phosphatase type 2A with a distinct subcellular distribution and substrate specificity. Proc Natl Acad Sci USA 92:8343–8347.

Samama P, Cotecchia S, Costa T, Lefkowitz RJ (1993): A mutation-induced activated state of the β_2-adrenergic receptor. J Biol Chem 268:4625–4636.

Shenolikar S, Ingebritsen TS (1984): Protein (serine and threonine) phosphate phosphatases. Methods Enzymol 159:390–408.

Sibley DR, Strasser RH, Benovic JL, Daniel K, Lefkowitz RJ (1986): Phosphorylation/dephosphorylation of the β-adrenergic receptor regulates its functional coupling to adenylate cyclase and subcellular distribution. Proc Natl Acad Sci USA 83:9408–9412.

von Zastrow M, Kobilka BK (1992): Ligand-regulated internalization and recycling of human β_2-adrenergic receptors between the plasma membrane and endosomes containing transferrin receptors. J Biol Chem 267:3530–3538.

von Zastrow M, Link M, Daunt D, Barsh G, Kobilka BK (1993): Subtype specific differences in the intracellular sorting of G protein–coupled receptors. J Biol Chem 268:763–766.

Wera S, Hemmings BA (1995): Serine/threonine protein phosphatases. Biochem J 311:17–29.

Xie H, Clarke S (1994): Protein phosphatase 2A is reversibly modified by methyl esterification at its C-terminal leucine residue in bovine brain. J Biol Chem 269:1981–1984.

Yu SS, Lefkowitz RJ, Hausdorf WP (1993): β-Adrenergic receptor sequestration. A potential mechanism of receptor resensitization. J Biol Chem 268:337–341.

Zhang J, Ferguson SS, Barak LS, Aber MJ, Giros B, Lefkowitz RJ, Caron MG (1997): Molecular mechanisms of G protein–coupled receptor signaling: Role of G protein–coupled receptor kinases and arrestins in receptor desensitization and resensitization. Receptors Channels 5:193–199.

PART 3

REGULATION OF RECEPTOR TRAFFICKING AND EXPRESSION

CHAPTER 10

RADIOLIGAND-BINDING ASSAYS FOR G PROTEIN–COUPLED RECEPTOR INTERNALIZATION

MYRON L. TOEWS

Regulation of G Protein–Coupled Receptor Function and Expression,
Edited by Jeffrey L. Benovic.
ISBN 0-471-25277-8   Copyright © 2000 Wiley-Liss, Inc.

I. INTRODUCTION

I.A. Desensitization, Internalization, and Down-Regulation

G protein–coupled receptors (GPCRs) are integral membrane proteins that normally reside in the plasma membrane. Considerable evidence indicates that these receptors span the plasma membrane seven times, with their N terminus and their ligand-binding domain exposed on the extracellular face of the membrane and their C terminus and their G protein interaction domains exposed on the intracellular face of the membrane. This orientation allows them to detect the presence of their physiological ligands in the extracellular milieu and to respond to those ligands by activating G protein–mediated signal transduction pathways within the cell.

The localization of GPCRs in the plasma membrane is dynamic rather than static, with several highly regulated processes determining the overall number and the distribution of these receptors among various plasma membrane and intracellular membrane domains. GPCRs are synthesized in the endoplasrnic reticulum and then further processed in the Golgi before their initial transport to the plasma membrane. Agonist binding to the receptor in the plasma membrane leads not only to generation of second messengers and intracellular responses but also to a series of adaptive changes that decrease the subsequent responsiveness of the receptor, a process referred to as *desensitization*. Desensitization is often accompanied by *internalization* of the desensitized receptors, and the mechanisms and significance of this agonist-induced receptor internalization have been the subject of extensive investigation. Following their internalization, the desensitized receptors can be *resensitized* and *recycled* back to the plasma membrane to once again function in ligand recognition and signal transduction. Alternatively, some of the internalized receptors may be transferred to other intracellular compartments and further processed by poorly defined mechanisms leading to receptor "down-regulation," defined as a decrease in the total number of radioligand-binding sites. These regulatory processes are the topic of numerous other chapters in this volume, and their mechanisms are discussed only briefly here and in terms of their relationship to receptor internalization.

Desensitization and the accompanying internalization of GPCRs have been most extensively characterized for the prototypical β_2-adrenergic receptor (β_2AR). Progress in understanding the cellular pathways and molecular mechanisms involved in internalization for β_2ARs and other GPCRs has been covered in numerous review articles over the years (Perkins et al., 1991; Hein and Kobilka, 1995; Freedman and Lefkowitz, 1996; Bohm et al., 1997; Koenig and Edwardson, 1997; Krupnick and Benovic, 1998), and the reader is referred to these reviews for citations of the individual studies leading to the current picture of internalization summarized here.

Internalization of β_2ARs was originally viewed as part of the desensitization pathway for these receptors. However, current evidence indicates that functional desensitization of β_2ARs is primarily due to a rapid "uncoupling" of the receptors from G protein activation, a process mediated by receptor phosphorylation and binding of β-arrestin to the phosphorylated receptors in the plasma membrane. This functional desensitization can occur without internalization. β-Arrestin also serves as an adaptor protein to target the phosphorylated and desensitized β_2ARs to clathrin-coated pits for subsequent internalization into intracellular vesicles. Recent studies suggest that internalization is critical for *resensitization* of β_2ARs, with intracellular vesicle acidification promoting receptor dephosphorylation and preparing the receptors for recycling back to the plasma membrane. Receptor internalization is widely thought to be an intermediate step on the pathway to receptor down-regulation, but this remains to be definitively demonstrated. Similar pathways and mechanisms for internalization have been documented for other GPCRs, although there is some evidence that both the cellular pathways and the functional significance of internalization for some GPCRs may be different from those that have been established for β_2ARs.

I.B. History of GPCR Internalization Assays

Most of our current understanding of the cellular pathways and molecular mechanisms of receptor internalization have come from studies of "nutrient receptors" that are really transport proteins, primarily the low density lipoprotein (LDL) receptor for uptake of cholesterol into the cell and the transferrin receptor for uptake of transferrin and its associated iron, and from studies with receptors for peptide growth factors, primarily the epidermal growth factor (EGF) receptor and the insulin receptor (Trowbridge et al., 1993; Carpentier, 1994). These systems were amenable to study because the ligand was tightly bound and internalized along with the receptor and because useful antibodies were available for immunohistochemical localization of both the receptors and their ligands. The well-characterized pathway for internalization of these receptors involves their interaction with adaptor proteins that lead to their accumulation in clathrin-coated pits in the plasma membrane. These coated pits invaginate and pinch off from the plasma membrane to form coated vesicles, which then lose their coats to become structures called *endosomes*. Acidification of the endosomes leads to dissociation of the ligand from the receptor, with the ligand delivered to intracellular domains and the receptor generally recycled to the plasma membrane for another round of uptake.

Considerable evidence now indicates that β_2ARs and some other GPCRs, but perhaps not all GPCRs, are internalized by the same clathrin-mediated pathway (Freedman and Lefkowitz, 1996; Bohm et al., 1997; Koenig and Edwardson, 1997; Krupnick and Benovic, 1998). However, direct morphological studies of the cellular pathways and molecular mechanisms for GPCR internalization have been more difficult than for the receptors discussed above for several reasons. The agonist ligands for GPCRs, particularly for the prototypical β_2ARs, typically dissociate from the receptor before its internalization, thus requiring tools for *receptor* localization rather than relying on studies of *ligand* uptake. Furthermore, immunohistochemical studies of GPCR localization were not feasible for many years because of the difficulty of generating useful antibodies to GPCRs.

Thus early work on GPCR internalization was limited to more indirect studies, such as radioligand-binding assays following subcellular fractionation to identify the various compartments containing the receptors and assays of cell surface accessibility of the receptors for radioligand binding under various assay conditions. Most of these radioligand-binding assays for GPCR internalization were adapted from similar assays used to study internalization of LDL, transferrin, EGF, and insulin receptors, although some were developed specifically for GPCRs.

I.C. The Radioligand-Binding Assays for GPCR Internalization

This chapter covers six types of assays that rely on radioligand binding to monitor GPCR internalization. They can be divided into two distinct groups that provide quite different types of information regarding receptor location and accessibility. The first group utilize *subcellular fractionation* to identify the specific cellular membrane compartments in which the receptors reside. The second group take advantage of the physical properties of cellular membranes and/or the ligands used in the assay to determine the *cell surface accessibility* of the receptors. These internalization assays based on radioligand binding have provided the bulk of our current knowledge regarding GPCR internalization. Furthermore, they continue to be important tools and to provide valuable new information, even though more elegant "morphological" assays of internalization are now becoming widely available (see Chapters 1, 12, and 13).

The rationales and some of the advantages and disadvantages of each of these assays are discussed first, in Section 2 for the subcellular fractionation assays and in Section 3 for the cell surface accessibility assays. References to historically important and more recent typical studies using each of these assays are also presented in these sections, with an emphasis on the adrenergic and muscarinic receptors, which have generally been the prototype systems for the development and application of these assays. Section 4 describes how results from combinations of these assays have led to a two-step model and a proposed nomenclature for the multiple steps of internalization. The remaining sections then present the general experimental considerations and specific experimental protocols for each of the six types of assays in more detail.

2. OVERVIEW OF SUBCELLULAR FRACTIONATION ASSAYS

The subcellular fractionation assays rely on centrifugation of cell lysates to separate and partially purify the various subcellular fractions, with the plasma membrane, the various intracellular membranous compartments, and the cytosol being of greatest interest. These compartments can be separated by multiple-step differential centrifugation protocols or by density gradient centrifugation using any of several centrifugation media, with sucrose being the most widely used.

Assays based on subcellular fractionation following a detergent extraction step are also used to isolate plasma membrane compartments called *caveolae*. Although there is recent evidence suggesting the possible involvement of caveolae in the internalization of some GPCRs (Roettger et al., 1995; de Weerd and Leeb-Lundberg, 1997; Feron et al., 1997), procedures for isolation of caveolae are not included here. The reader is referred to recent reviews for information on methods for the isolation and study of caveolae (Anderson, 1993; Smart et al., 1995; Lisanti et al., 1995).

2.A. Sucrose Density Gradient Centrifugation

Sucrose density gradient centrifugation relies on the different buoyant densities of the various subcellular membrane fractions. Cells are lysed, and the lysates are centrifuged at high speed through either continuous or step gradients of sucrose concentration, generally ranging from 15% to 30% (weight/volume) sucrose at the top of the gradient to 40%–60% at the bottom. Unbroken cells and nuclei are the most dense and pass through even the 60% sucrose and pellet at the bottom of the centrifuge tube. Plasma membrane fragments pass through the lighter sucrose fractions and are retained in the 40%–50% sucrose range. Light intracellular membrane vesicles containing internalized receptors are retained in the 25%–35% sucrose range. Soluble cytosolic cellular constituents do not enter the sucrose and remain with the applied sample at the top of the gradient.

An advantage of continuous sucrose gradient separations vs. step gradients and differential centrifugation is that they allow the assessment of the entire subcellular distribution of receptors, not just of a few specific fractions. By the same token, however, these assays also generate a large number of fractions, each of which must be assayed for the receptor of interest, whereas step gradients and differential centrifugation generally involve assaying only two fractions, the plasma membrane and light vesicle fractions. The preparation of the gradients is somewhat time consuming, with continuous gradients being more difficult to prepare than step gradients.

Sucrose density gradient centrifugation assays with β_2-adrenergic receptors (β_2ARs) provided the first clear evidence for the occurrence of GPCR internalization (Harden et al., 1980), and they continue to be widely used in ongoing mechanistic studies. As examples, we have used these assays extensively in our studies of internalization for β_2ARs (Toews et al., 1984, 1986; Toews, 1987), α_{1B}ARs (Toews, 1987; Cowlen and Toews, 1988; Zhu et al., 1996; Wang et al., 1997), muscarinic receptors (Hoover and Toews, 1990), and vasoactive intestinal peptide receptors (Turner et al., 1988), including both continuous and step gradients.

2.B. Differential Centrifugation

Differential centrifugation assays are widely used for subcellular fractionation in general, and they are sometimes used to monitor receptor internalization as well. Cells are lysed, and the lysates are then centrifuged sequentially at increasing speeds, with various subcellular constituents "pelleting" to the bottom of the centrifuge tube at different centrifugal forces. Larger and/or more dense particles pellet at lower centrifugal forces, and higher forces are required to pellet smaller or less dense particles. A low-speed spin, typically 1,000g, is used to pellet unbroken cells and nuclei. A higher speed spin, typically 30,000g to 40,000g, is used to pellet plasma membranes. Finally, a high-speed spin in an ultracentrifuge, typically 150,000g to 200,000g, is then used to pellet the "light" membrane fractions containing internalized receptors. Cellular constituents remaining in the supernatant from this final high-speed spin are considered "soluble."

Differential centrifugation assays are somewhat simpler than gradient assays because they do not require the preparation of gradients; however, multiple centrifugation steps are involved rather than the single spin for gradient assays. Generally only two fractions need to be assayed, the medium-speed pellet containing the plasma membrane receptors and the high-speed pellet containing the internalized vesicles. These assays also have the advantage that the receptors are concentrated in the process of their isolation because the pellets can be dissolved in small volumes to give high receptor concentrations. Many of these advantages are shared with step gradient assays.

Differential centrifugation assays were used in several early studies of β_2AR internalization (Stadel et al., 1983; Strader et al., 1984; Strasser et al., 1984). Some laboratories continue to use these assays routinely, such as in recent mechanistic studies of muscarinic receptor internalization (Slowiejko et al., 1996; Sorensen et al., 1997) and of β_2AR internalization (Lin et al., 1997).

3. OVERVIEW OF CELL SURFACE ACCESSIBILITY ASSAYS

This group of assays rely on the ability of ligands with different physical solubility properties to bind to receptors under various assay conditions, and the results are interpreted in terms of the accessibility of the receptors to the ligand. The variables upon which these assays rely are the lipophilicity of the ligands and the fluidity and permeability of cellular membranes. Lipophilic ("lipid-loving") ligands, also referred to as hydrophobic ("water-fearing") ligands, are relatively more soluble in cellular membrane lipid bilayers, and thus are more membrane permeable, than hydrophilic ("water-loving"), also referred to as lipophobic ("lipid-fearing"), ligands.

At warmer temperatures, such as the physiological 37°C, cellular membranes are relatively fluid and relatively permeable to lipophilic ligands but are relatively impermeable to hydrophilic ligands. At lower temperatures, such as with cells on ice (4°C), cellular membranes are markedly less fluid and relatively impermeable even to lipophilic ligands. Both the lipophilicity/hydrophilicity of the ligands and the permeability of cellular membranes can vary over a broad range and thus provide useful tools for assessing receptor distribution within the cell. In practice, the most hydrophilic and the most lipophilic ligands available tend to be the most useful; the temperature extremes of 37° and 0°C are commonly used, although assays at intermediate temperatures that retain selectivity of binding are also used. Based on different combinations of these variables, four types of assays are the most widely used, each of which is summarized below.

3.A. Hydrophilic Radioligand Binding

For those receptors for which a very hydrophilic radioligand is available, the number of receptors labeled by the hydrophilic radioligand can be taken as a direct indicator of the number of receptors accessible on the surface of the cell. Thus the decrease in binding of a hydrophilic radioligand to intact cells following agonist pretreatment provides an indication of the number of receptors that have redistributed to domains that are not accessible at the surface of the cell. These assays are generally done with cells on ice to prevent reversal of internalization during the time required for the binding assay. Binding of a hydrophilic radioligand alone has the disadvantage that it cannot differentiate receptor internalization from receptor down-regulation; both processes lead to a decrease in the number of cell surface receptors. Thus separate assays to determine the possible contribution of down-regulation to the loss of binding may be required. In most cases, however, internalization is measured on a time frame of less than 60 minutes, during which there is usually minimal down-regulation, and thus hydrophilic radioligand binding alone provides a convenient tool for estimating receptor internalization.

A limitation on the use of these assays is that a suitably hydrophilic radioligand may not be available for all receptors. The classic examples for GPCRs whose ligands are small molecules are the use of [^{3}H]CGP-12177 for β_2AR internalization (Hertel et al., 1983; Toews et al., 1984; Gagnon et al., 1998) and of N-methyl-[^{3}H]scopolamine for muscarinic receptor internalization (Feigenbaum and El-Fakahany, 1985; Hoover and Toews, 1990; Slowiejko and Fisher, 1997; Pals-Rylaarsdam et al., 1997; Shockley et al., 1997). Use of these two hydrophilic radioligands provided critical support to the early evidence from subcellular fractionation assays for the occurrence of GPCR internalization, and they continue to be used in ongoing mechanistic studies. [^{125}I]- or [^{3}H]-labeled sulpiride have been suggested for studies of dopamine receptor internalization but have been used less extensively (Barton et al., 1991; Itokawa et al., 1996). Most peptides are sufficiently membrane impermeable, and these assays can be used to study internalization of GPCRs whose ligands are peptides, as in our study with receptors for vasoactive intestinal peptide (Turner et al., 1988). However, the acid/salt stripping assays, which often include an initial low temperature binding step to label cell surface receptors, are more commonly used with receptors for peptide ligands.

3.B. Low Temperature Assays of Lipophilic Radioligand Binding

Low temperature assays of lipophilic radioligand binding take advantage of the relative impermeability of cellular membranes at low temperature to even relatively lipophilic ligands. Thus either hydrophilic or lipophilic radioligands can be used in these assays. In the case of hydrophilic radioligands, these assays are essentially identical to those described in the section above because those assays are routinely carried out at low temperature to prevent reversal of internalization during the assay. Cells are treated in the absence or presence of agonist to induce internalization, the agonist is washed away, and radioligand binding to control and agonist-treated intact cells is measured at a low temperature, generally on ice. The decrease in radioligand binding to agonist-treated cells is taken as a reflection of receptor internalization.

A major advantage of the low temperature assays is that they can be used to monitor internalization even for those receptors for which no suitable hydrophilic radioligand is available. These assays share with the hydrophilic radioligand-binding assays described in the previous section the limitation that the decrease in radioligand binding could arise from either internalization or from down-regulation, although distinguishing between these two processes is generally straightforward.

These assays have been used for a variety of GPCRs. Studies of $[^{125}I]$iodopindolol binding to β_2ARs of intact cells on ice was important in early studies characterizing the nature of β_2AR internalization (Toews et al., 1986). We and others have used low temperature binding assays with $[^3H]$prazosin to study $\alpha_{1B}AR$ internalization (Toews, 1987; Leeb-Lundberg et al., 1987; Cowlen and Toews, 1988; Wang et al., 1997).

3.C. Hydrophilic Ligand Competition for Lipophilic Radioligand Binding

Hydrophilic ligand competition for lipophilic radioligand binding, like the hydrophilic radioligand-binding assay, takes advantage of the relative impermeability of cellular membranes to hydrophilic ligands. In this case, the assays are conducted at 37°C and utilize a lipophilic radioligand that is assumed to have full access to the entire population of receptors; a hydrophilic nonradioactive competing ligand is included in the assay to selectively inhibit binding of the lipophilic radioligand to those receptors still exposed on the cell surface. The hydrophilic competing ligand can be used at multiple concentrations to generate competition curves or at a single concentration previously determined to optimally differentiate between cell surface and intracellular receptors. In these assays, changes in the amount of lipophilic radioligand bound in the absence of competing ligand are assumed to reflect changes in the total number of receptor-binding sites (down-regulation), whereas changes in the number or percentage of receptors labeled in the presence of the hydrophilic competing ligand are assumed to reflect receptor internalization into domains that are inaccessible to the hydrophilic ligand.

A potential advantage of these assays is that they can provide information about down-regulation as well as internalization in a single type of assay, all with the same preparation. A disadvantage of these assays is that they are conducted at 37°C (to maximize accessibility of the entire population of receptors to the lipophilic radioligand) and with intact cells (to ensure that membrane permeability barriers are not altered). Because internalization can reverse rapidly at 37°C, these assays are limited to relatively short assay times, ideally much shorter than the half-time for reversal of internalization. Alternatively, if the hydrophilic competing ligand that is used is an agonist, it can induce internalization of receptors during the binding assay, and again short assay times are required to reduce the extent to which this may contribute to the measured internalization. The requirement for short assay times makes it difficult to assay a large number of samples at the same time because the reactions must be stopped shortly after they are started. The short assay time, significantly shorter than the half-time for radioligand binding, also means that only a small percentage of the total receptor population will be labeled unless very high concentrations of radioligand are used. Thus a high specific activity radioligand with very low nonspecific binding is generally required to generate reliable data at these short assay times.

The requirements for a sufficiently lipophilic radioligand and a sufficiently hydrophilic competing ligand together with the difficulties of obtaining reliable data in short-time assays have limited the extent to which these assays are routinely employed. However, our short-time assays of competition by the agonists isoproterenol and epinephrine and the antagonist sotalol for binding of [125I]iodopindolol to β_2ARs provided important early evidence in support of β_2AR internalization (Toews and Perkins, 1984). This assay was also used in studies comparing internalization for β_2ARs and α_{1B}ARs (Toews, 1987). We also used short-time assays of epinephrine competition for [3H]prazosin binding to assess α_{1B}AR internalization in these studies (Toews, 1987; Cowlen and Toews, 1988). Variations of these assays in which hydrophilic ligand competition for lipophilic radioligand binding is measured with longer assay times at lower temperatures (4° or 16°C, to prevent induction or reversal of internalization during the assay) have been used for β_2ARs (Gabilondo et al., 1996) and for opioid receptors (Trapaidze et al., 1996).

3.D. Acid/Salt Stripping

Acid/salt stripping assays are somewhat different from the three described above in that they utilize a treatment *after* the radioligand-binding step to differentiate cell surface from internalized receptors. Following pretreatment with agonist to induce receptor internalization, receptors are incubated with the radioligand under conditions in which both cell surface and internalized receptors will be labeled.

Following the labeling step, free (unbound) radioligand is washed away. Radioligand binding associated with cell surface receptors is then "stripped" from the cells by incubation at low pH and/or high salt concentration to cause radioligand dissociation. Radioligand that is bound to internalized receptors is presumed to be protected from the acid/salt environment outside the cell and thus remains bound to the receptor. This stripping step is conducted at low temperature to further ensure that internalized receptors are protected from the acid/salt medium. The low temperature also prevents any radioligand that may dissociate from internalized receptors during the stripping step from being washed away during the subsequent wash. Radioactivity that dissociates from the cells into the acid/salt medium provides a measure of cell surface receptors, whereas radioligand that remains cell associated following the acid/salt incubation provides a measure of internalized receptors.

A variation of the procedure described above is actually the most commonly used and is particularly useful for peptide receptors where the peptide agonist is also a useful high-affinity radioligand. In this case, cells are incubated with a radiolabeled agonist peptide, which both binds to the receptors and induces internalization of the receptors along with the bound radioligand. Free radioligand is washed away, and the acid/salt stripping step then follows the combined labeling/internalization step. Because of the difficulty of removing high-affinity peptide agonists from their receptors after a pretreatment step to induce internalization, acid/salt stripping assays following a combined labeling/internalization step with an agonist radioligand are often the internalization assays of choice for GPCRs whose ligands are peptides.

A few examples of the use of these assays with peptide ligands include studies with receptors for substance P (Garland et al., 1996), angiotensin (Chaki et al., 1994; Thomas et al., 1995), gonadotropin-releasing hormone (Arora et al., 1995), and gastrin-releasing peptide (Slice et al., 1994). One early study used acid/salt stripping to further characterize the nature of the apparent internalization of β_2ARs (Mahan et al., 1985).

4. INTEGRATION OF RESULTS FROM MULTIPLE ASSAYS

Although each of the assays described above can be used alone as a useful tool for monitoring GPCR internalization, various combinations of these assays can provide a much more comprehensive assessment of both the extent and the nature of GPCR internalization. Several points related to the value of using multiple assays for receptor internalization, including evidence for multiple steps in the internalization pathway and a proposed new nomenclature, are discussed here.

4.A. Selective Labeling of Cell Surface Versus Internalized Receptors

The hydrophilic radioligand-binding assays and the low temperature binding assays with lipophilic radioligands both selectively label the cell surface–accessible population of receptors. In contrast, the acid/salt stripping assays label both internalized and cell surface receptors before stripping but leave radioligand selectively bound to the internalized receptors after stripping. Similarly, the hydrophilic ligand competition binding assays also label both populations of receptors in the absence of hydrophilic competitor but selectively label the internalized receptors in the presence of hydrophilic competitor. Thus either the cell surface–accessible or the cell surface–inaccessible receptors can be selectively labeled, or both populations of receptors can be labeled depending on which combination of assays is chosen.

4.B. Combining and Comparing Subcellular Fractionation Assays With Cell Surface Accessibility Assays

Additional critical information can be obtained from combining subcellular fractionation assays with cell surface–accessibility assays. These two types of assays are conceptually different, one indicating whether receptors are still physically associated with the plasma membrane after cell lysis and the other indicating whether receptors are accessible for ligand binding at the surface of intact cells. It is possible to perform subcellular fractionation assays following radioligand binding under the various conditions described above to characterize the cellular membrane compartments containing the surface-accessible and surface-inaccessible receptors (Toews et al., 1984, 1986). Alternatively, the surface accessibility assays can be performed following cell lysis and subcellular fractionation to determine the accessibility of the receptors in the plasma membrane and vesicular compartments thus isolated (Hertel et al., 1983; Sorensen, et al, 1997)

Although the subcellular fractionation and cell surface accessibility assays generally agree at least qualitatively, in some cases both qualitative and quantitative differences between the assays may be observed. In particular, several studies have now identified a population of receptors that appears inaccessible for ligand binding at the cell surface but that is still physically associated with the plasma membrane (Toews et al., 1984, 1986; Strader et al., 1984; Cowlen and Toews, 1988; Wang et al., 1997; Sorensen et al., 1997). These receptors are most likely present in an intermediate compartment on the receptor endocytosis pathway, perhaps a deeply invaginated clathrin-coated pit in the process of being pinched off to form a clathrin-coated vesicle. The neck of these vesicles may not allow extracellular ligands access to the receptors, even though they are still attached to the plasma membrane.

Alternatively, a covalent modification or unique conformation of the receptor-binding site could occur that would make the receptors appear inaccessible in intact cell binding assays even though they are still attached to the plasma membrane; this modification or conformation might then be lost upon cell lysis, accounting for the detectability of these receptors in the plasma membrane fraction in subcellular fractionation assays. Studies using other types of internalization assays have also provided evidence for multiple steps and compartments in the GPCR internalization pathway (von Zastrow and Kobilka, 1994; Mostafapour et al., 1996). The precise identities and relationships among the multiple compartments and steps identified by these diverse assays remain to be established.

4.C. A Two-Step Model and Proposed Nomenclature for GPCR Internalization

The studies described above that document a population of cell surface–inaccessible receptors still associated with the plasma membrane bring up the issue of nomenclature for the apparently multiple steps and compartments involved in GPCR internalization. The terms *sequestration* and *internalization* have often been used synonymously. Similarly, most investigators have assumed that GPCR "internalization" assessed by the various types of assays described above would be essentially the same process previously defined as "endocytosis" for nutrient and growth factor receptors, and these two terms have also been used synonymously, particularly with the more recent evidence for involvement of the clathrin-mediated endocytosis pathway in GPCR internalization. What terms do we then use to distinguish the receptors that are surface inaccessible but plasma membrane associated from those that are physically separated from the plasma membrane in intracellular vesicles, and what terms do we use to describe the processes involved? We have proposed the use of the term *sequestration* to refer to the conversion of receptors to the plasma membrane–associated but surface-inaccessible form and the term *endocytosis* to refer to the physical redistribution of these receptors from the plasma membrane to intracellular vesicles (Wang et al., 1997). We utilize the term *internalization* to refer to the sum of both processes. We hypothesize that sequestration and endocytosis thus defined constitute two sequential steps in the overall internalization pathway; however, it remains possible that they represent distinct pathways rather than separate steps of a single pathway. A model presenting the two-step pathway and the proposed nomenclature is presented in Figure 10.1.

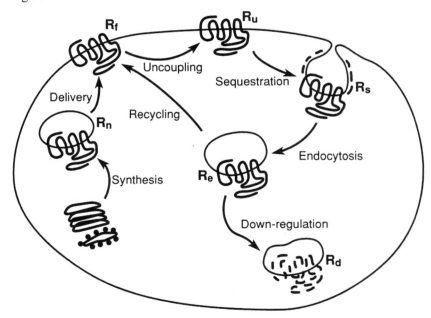

Figure 10.1. Two-step model and proposed nomenclature for GPCR internalization. R_n, *n*ewly synthesized receptors not yet delivered to the plasma membrane; R_f, *f*unctional receptors on the cell surface; R_u, receptors that are *u*ncoupled from G proteins, functionally desensitized but still exposed on the cell surface; R_s, *s*equestered receptors that are inaccessible to ligands at the cell surface but still physically associated with the plasma membrane; R_e, receptors that have been *e*ndocytosed into intracellular vesicles; and R_d, receptors that have been *d*own-regulated, presumably degraded by proteolysis.

In this proposed nomenclature, both sequestered and endocytosed receptors appear as inaccessible in the cell surface accessibility assays, but only the endocytosed receptors appear in the light vesicle fraction in subcellular fractionation assays. Subcellular fractionation assays thus provide direct information about the extent of endocytosis; in contrast, the extent of sequestration is determined as the difference between the fraction of receptors that are cell surface inaccessible and the fraction that migrate in the light vesicle fraction. Alternatively, sequestered receptors can be identified as those receptors that are labeled on intact cells by lipophilic radioligands in the presence of hydrophilic competing ligands or that remain labeled following acid/salt stripping, but that nonetheless migrate in the plasma membrane fraction in subcellular fractionation assays.

5. GENERAL CONSIDERATIONS FOR INTERNALIZATION ASSAYS

5.A. Growth of Cells

The first step in all of the assays is to grow a sufficient quantity of cells in vessels appropriate for the type of assay to be performed. All of the assays can be used with cells grown in monolayer culture or with cells grown in suspension culture. Thus these assays are applicable to virtually all cell types. An advantage of monolayer culture is that the various treatment medium changes and washes required for the assays can be quickly and easily accomplished by simply aspirating with vacuum and replacing with the next solution. In contrast, cells in suspension culture require centrifugation for medium changes and washing. This procedure takes longer and may require that cells be maintained at low temperature to prevent changes in the extent of internalization from occurring during the time required for the centrifugation steps.

A disadvantage of monolayer culture is that cells must be grown in as many separate vessels as the number of assays that are to be performed. This is not a significant factor for the subcellular fractionation assays, but it necessitates a large number of separate dishes or wells for the cell surface accessibility assays with intact cells. For example, an experiment with 16 different conditions, each to be assayed in triplicate, requires 48 individual dishes or a 48-well plate. Details of the specific experiment to be conducted need to be known at the time of plating cells so that the proper number of dishes or wells are available.

Each of these dishes or wells must then be treated separately for all steps in the assay. Our standard protocols for intact cell-binding assays treat the dishes in sets of six, because six dishes is a convenient number to manipulate within a 1-minute time frame (1 every 10 seconds). Because multiple washes plus treatment steps are involved at various points in the protocol, the sets of six dishes are treated at 5- to 6-minute intervals depending on the specific experiment. In contrast, suspension culture cells can be grown in a single vessel and can often be pretreated in a relatively small number of vessels; only for the final assay steps are separate tubes or wells required for each individual replicate. For suspended cells, the assay tubes can generally be prepared on ice, all started simultaneously by placing the rack of tubes in a water bath, and all terminated simultaneously by filtration with a cell harvester.

The methods used for cell lysis are also different depending on how cells are grown. Suspension cultures are generally lysed by use of a homogenizer or sonicator, whereas monolayer cultures can often be lysed by scraping with a rubber policeman after incubation in a hypotonic medium to cause cell swelling; however, subsequent further homogenization may be required with monolayer cells also. If monolayer culture is most appropriate for cell growth but suspended cells are found to be more convenient for assays, it is also possible to detach monolayer cells from the culture dish and then to perform the internalization assays with the cells in suspension. For cells assayed in suspension, the final collection of samples for counting is by filtration. For cells assayed as monolayers, the final collection of samples for counting is by dissolving the cell sheet with bound radioligand in NaOH or detergent.

5.B. Agonist Pretreatment and Removal

All of the assays involve a pretreatment step in which cells are exposed to agonist or appropriate vehicle followed by an assay step to assess changes in the state of the receptors. Variables during the agonist pretreatment step include the agonist or other activators used, for example to assess specificity; the concentration of agonist used, either a single concentration or for concentration-effect studies; the time of pretreatment, either a single time point or for time course studies; and the absence or presence of various additional agents, such as antagonists, enzyme inhibitors, or other modulators, whose effects on internalization are being assessed.

The agonist pretreatment is generally conducted at 37°C, although incubation at other temperatures may be used for specific experiments; for example, incubation at 16°C has been used to demonstrate the occurrence of multiple steps in internalization based on their differential temperature dependence (von Zastrow and Kobilka, 1994). Internalization assays typically utilize a 20- to 30-minute exposure to agonist; this time is generally long enough for internalization to reach a steady-state value but short enough to eliminate any significant contribution from receptor down-regulation.

All of the assays also involve a wash step to remove the agonist following the pretreatment step. If these washes can be performed quickly, it may be best to conduct the washes with 37°C medium because this is more likely to allow any intracellular agonist to pass through cell membranes and be washed away before the assay step. If more than 1–2 minutes are required to complete the washes, low temperatures should be used to prevent reversal of internalization during the wash step.

5.C. Use of Concanavalin A To Improve Plasma Membrane Isolation

The initial studies of β_2AR internalization using sucrose density gradient centrifugation included a treatment with the plant lectin concanavalin A before cell lysis (Harden et al., 1980). This treatment was found to shift the plasma membrane fraction to higher sucrose densities and to markedly improve separation of the plasma membrane fraction from the light vesicle fraction. Concanavalin A apparently maintains the plasma membrane fragments as larger sheets that have improved separation properties (Lutton et al., 1979). Accordingly, many subsequent gradient studies with these and other receptors have also utilized concanavalin A. However, many studies also appear to obtain good separation of plasma membrane from light vesicle fractions without the use of concanavalin A, particularly with appropriate step gradients. Concanavalin A should only be used if it provides a significant improvement in gradient profiles because it introduces an additional step and additional time between the agonist pretreatment and the final assay.

Concanavalin A has also been used to improve plasma membrane separation in differential centrifugation assays (Slowiejko et al., 1996; Lin et al., 1997). This use of concanavalin A *after* the agonist pretreatment to improve membrane separations in subcellular fractionation assays should not be confused with the use of concanavalin A *during* the agonist pretreatment step as an inhibitor of receptor internalization (Wakshull et al., 1985). Thus, concanavalin A has proved to be useful in two completely different ways for studies of GPCR internalization.

6. SUCROSE DENSITY GRADIENT CENTRIFUGATION

6.A. Choice of Continuous Versus Step Gradients

The advantages and disadvantages of step versus continuous gradients were discussed above. Ideally, continuous gradients should be utilized in initial experiments to quantitate the sucrose densities at which the plasma membrane and light vesicle fractions migrate for the system under study. Appropriate sucrose concentrations for step gradients can then be chosen for subsequent studies. These studies should include verification that the receptor distribution obtained from the simpler step gradients agrees with that obtained from assaying the entire continuous gradient.

A typical continuous gradient would be from 30% to 60% sucrose, with the light vesicle receptors banding near the top of the gradient and the plasma membrane receptors banding nearer the bottom of the gradient. A typical step gradient would include steps of 15%, 30%, and 60%, with plasma membrane receptors banding at the 30%:60% interface and light vesicle receptors banding at the 15%:30% interface. For both step and continuous gradients, unbroken cells and nuclei pass through the 60% sucrose and pellet to the bottom of the tube. A small cushion of 5% sucrose is often included at the top of continuous gradients to prevent mixing of cytosolic components with the upper fractions of the gradient.

6.B. Gradient Preparation, Centrifugation, and Fractionation

Continuous gradients can be prepared with any standard gradient-forming apparatus, ranging from a simple two column mixing chamber to more elaborate programmable mechanical models. Step gradients are usually formed manually by careful layering of each step into the centrifuge tube. Mixing of the interfaces is best prevented by layering the lightest sucrose fraction first and then adding succeeding heavier layers under the previous layers using a disposable syringe with a long needle.

The amount of cell lysate required to give an adequate number of receptors for accurate quantitation will determine the size of the centrifuge tube and rotor to be used. Swinging bucket rotors are used in all cases. For cells expressing moderate levels of receptors, and particularly for transfected cells expressing high levels of receptors, a typical choice would be to layer 1–2 ml of cell lysate on a 10-ml gradient in a 12-ml centrifuge tube for use in a Beckman SW41 or comparable ultracentrifuge rotor. For cells expressing low levels of receptors, 10 ml of cell lysate could be layered onto a 30-ml gradient in a 40-ml centrifuge tube for use in a Beckman SW28 or comparable rotor.

The cell lysate is carefully added to the top of the gradient. The samples are then centrifuged, typically for 60 minutes at 100,000g. The gradient fractions are then collected; again, this can be done manually or with mechanical gradient fractionators and fraction collectors. The number of fractions collected from continuous gradients will determine the resolution of the receptor distribution. For 12-ml gradients, 1-ml fractions are usually adequate; for 40-ml gradients, 2–4-ml fractions are collected. For step gradients, the cellular material from the interfaces is collected manually either with a syringe and needle through the side of the tube or by carefully aspirating the sucrose above the membrane band and then collecting the membranes with a pipettor. These samples are then used in the standard membrane radioligand-binding assay for the receptor of interest, generally assaying both total and nonspecific binding to each fraction in triplicate. A portion of the original cell lysate may also be retained and assayed to ensure that the receptors in the light vesicle and plasma membrane fractions account for all of the receptors initially applied to the gradient.

6.C. Protocol for Step Gradients

1. *Cell growth.* Grow cells to near confluence in 10 ml medium on the appropriate number of 100-mm culture dishes, typically two dishes per treatment condition.

2. *Gradient preparation.* Using a disposable 2-ml plastic syringe, add 1.5 ml of 15% (w/v) sucrose to the bottom of each 14 × 89 mm ultracentrifuge tube. Gently inject 5.0 ml of 30% sucrose to the bottom of each tube under the 15% sucrose and then 2.5 ml of 60% sucrose under the 30% sucrose. Be sure that the gradients are identical in height and weight. The stock sucrose solutions should be prepared in a buffered physiological salt solution and stored refrigerated, and gradients should be kept cold after preparation.

3. *Pretreatment to induce internalization.* To control dishes, add 100 μl of the vehicle in which the agonist is dissolved, and to treated dishes add 100 μl of agonist dissolved at 100 times the desired final concentration. Return the dishes to the incubator at 37°C for 30 minutes. For some systems, it may be beneficial to feed the cells with fresh medium the day before the experiment. If the presence of serum might interfere with the

action of the agonist, the pretreatment may be done in serum-free medium, with the medium changed at least 1 hour before adding agonist.

4. *Cell washing and lysis.* Place the dishes on ice, and aspirate the pretreatment medium. Add 10 ml of ice-cold lysis buffer (20 mM Tris, pH 7.4, 2 mM EDTA); aspirate this first wash and add a second 10 ml of lysis buffer; and, finally, aspirate most of this second wash, leaving about 0.5 ml lysis buffer on each dish. Incubate for 10–20 minutes on ice to allow cell swelling. Then scrape the cells from the dishes with a rubber policeman or cell scraper. Pool the lysate from the two control dishes and from each pair of treated dishes and adjust all lysates to equal volume (about 1.2 ml) with lysis buffer.

5. *Centrifugation.* Gently layer 1.0 ml of lysates onto the top of the gradients. Carefully place the tubes in ultracentrifuge buckets, fasten the bucket caps, carefully attach the buckets to a Beckman SW41 swinging bucket rotor, and properly seat the rotor in a Beckman L8-70 refrigerated ultracentrifuge. Centrifuge at 28,000 rpm (100,000g) for 60 minutes at 4°C.

6. *Fraction collection.* Collect the material at the 15%:30% interface as the light vesicle fraction and at the 30%:60% interface as the plasma membrane fraction. The membranes banded at each interface are usually visible. Carefully remove the 15% sucrose with a Pasteur pipette or digital pipettor down to a few millimeters above the 15%:30% interface and discard. Then carefully collect the membranes from the interface, including the bottom few millimeters of the 15% sucrose and the top few millimeters of the 30% sucrose. Repeat these steps for the 30%:60% interface, discarding the bulk of the 30% sucrose and collecting the 30%:60% interface and a few millimeters above and below. Alternatively, the two interfaces can be collected with a syringe by inserting a needle through the side of the tube at the level of the interface and carefully removing the banded membranes together with a minimal volume of the sucrose above and below the band.

7. *Binding assays.* Dilute each fraction 1:1 with the appropriate binding buffer. Distribute equal aliquots of each fraction to tubes for radioligand-binding assays according to standard procedures for the receptor of interest.

6.D. Protocol for Continuous Gradients

1. *Steps common to step and continuous gradients.* Steps 1, 3–5, and 7 are identical to those above for step gradients and are not repeated here.

2. *Gradient preparation.* Place 0.5 ml of 70% sucrose in the bottom of each centrifuge tube. Using a commercial or homemade two-column gradient former, prepare an 8-ml linear gradient from 30% to 60% sucrose. Begin with 4 ml of 60% sucrose in the mixing chamber and 4 ml of 30% sucrose in the reservoir chamber. Begin stirring the mixing chamber, open the stopcock separating the chambers, and use a peristaltic pump to slowly pump the sucrose solution from the mixing chamber into the centrifuge tube. Have the tube tilted at a 45° angle, and let the sucrose solution gen-

tly flow down the bottom side of the tube to prevent mixing. Then carefully add a 0.5-ml pad of 5% sucrose to the top of each gradient. Again, be sure the gradients are equal in height and weight.

6. *Fraction collection.* Manually or with a commercial gradient fractionator, carefully collect 12 fractions of 0.8 ml each.

6.E. Results and Interpretation

Typical results from a continuous gradient centrifugation experiment demonstrating endocytosis of α_{1B}ARs is shown in Figure 10.2. The majority of the receptors (80%–90%) are generally localized in the plasma membrane fraction in

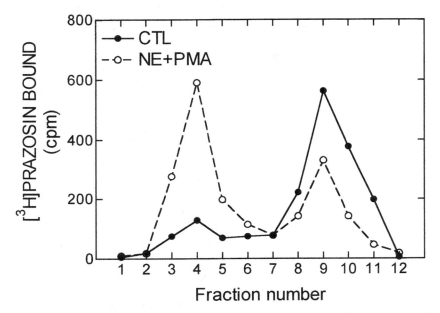

Figure 10.2. Sucrose density gradient centrifugation. DDT$_1$ MF-2 cells were incubated for 30 minutes in the absence (●, CTL) or presence (○) of the α_{1B}AR agonist norepinephrine (NE, 10 μM) plus the protein kinase C activator phorbol 12-myristate, 13-acetate (PMA, 1 μM), washed and incubated on ice with 0.5 mg/ml concanavalin A, and lysed. Cell lysates were centrifuged for 60 minutes at 100,000g on 30% to 50% continuous sucrose density gradients. Internalization (endocytosis) of α_{1B}ARs was then assessed by measuring [^{3}H]prazosin binding to the gradient fractions. The data show that pretreatment with agonist plus PMA induces a shift of about half of the plasma membrane receptors (fractions 8–11) to the light vesicle fraction (fractions 3–5), indicating α_{1B}AR endocytosis. This study also includes data comparing results from step gradient assays, assays of lipophilic radioligand binding to intact cells at low temperature, and short-time assays of hydrophilic ligand competition for lipophilic radioligand binding. Together, the data in this study suggested that α_{1B}ARs in these cells undergo sequestration but not endocytosis when treated with agonist alone, whereas treatment with agonist plus PMA leads to both sequestration and endocytosis of these receptors. In contrast, β_2ARs in the same cells undergo both sequestration and endocytosis when treated with agonist alone (Toews, 1987), and adding PMA has no further effect (Cowlen and Toews, 1987). (From Cowlen and Toews [1988], with permission of the publisher.)

control cells not exposed to agonist, although some receptors (10%–20%) are often found in the light vesicle fraction even in control cells. These receptors are presumably newly synthesized and en route to the plasma membrane or may represent receptors undergoing a low level of constitutive internalization and recycling even in the absence of agonist. Agonist exposure induces a decrease in the number of receptors migrating in the heavier sucrose fractions containing the plasma membrane and a corresponding increase in the number of receptors migrating in the light vesicle fraction containing internalized receptors. The decrease in binding to the plasma membrane fraction between control and agonist-treated cells and the increase in binding to the light vesicle fraction between control and agonist-treated cells are expressed as a percentage of the binding to the plasma membrane fraction from control cells; this value represents the percentage of the originally plasma membrane-localized receptors that became internalized to the light vesicle fraction during the agonist treatment. If no down-regulation has occurred, the amount of binding lost from the plasma membrane fraction should be essentially identical to that gained in the light vesicle fraction.

When establishing sucrose gradient assays for a new cell system, the distribution of subcellular marker enzymes on the gradient should ideally be established to ensure that the "plasma membrane" fraction of receptors co-migrates with other plasma membrane markers and that the "light vesicle" receptors co-migrate with other markers for intracellular vesicles. Ideally, initial experiments should also include saturation assays of radioligand binding to receptors in the plasma membrane and light vesicle fractions to ensure that both populations of receptors have similar affinities for the radioligand and that the changes in binding to these fractions reflect changes in receptor number rather than alterations in binding affinity.

7. DIFFERENTIAL CENTRIFUGATION

7.A. Experimental Considerations

The primary variables to consider with differential centrifugation assays are the number of subcellular fractions to be isolated and the speed of centrifugation (centrifugal force) to be used for each fraction. With appropriate conditions, it is possible to generate separate fractions reasonably highly enriched in a wide variety of subcellular organelles, including but not limited to nuclei, mitochondria, plasma membranes, lysosomes, endoplasmic reticulum and Golgi apparatus, endosomes, and caveolae. Receptor internalization assays typically focus on separating only two fractions, one containing plasma membrane receptors and the other containing intracellular vesicles with internalized receptors. An initial low-speed spin to remove intact cells may precede the isolation of the two membrane fractions. The final spin is typically relatively long and at very high centrifugal force to ensure that all of the light vesicles containing the internalized receptors are pelleted.

Some of the standard differential centrifugation protocols lyse cells in 0.25 M sucrose and use this as the medium from which the various membrane fractions are pelleted, but most receptor internalization protocols utilize hypotonic medium without sucrose for cell lysis. As with gradient protocols, it is also important to conduct marker enzyme assays on the fractions collected with differential centrifugation to ensure that the desired subcellular fractions are in fact being separated and isolated in reasonably pure form and to compare binding with the two isolated fractions to that with the original lysate to ensure that all receptors are being measured.

7.B. Protocol

1. *Cell growth, treatment, and lysis.* Grow cells, pretreat to induce internalization, and wash and lyse cells as described above for gradient assays. Again, it is useful to save an aliquot of the lysate to assay for total receptors along with the assays of the plasma membrane and vesicle fractions to ensure that these two fractions account for all of the receptor binding activity of the original lysate.

2. *Centrifugation steps.* Centrifuge the lysate at 16,000 rpm (30,000g) for 10 minutes at 4°C in a Sorvall SS34 rotor in a Sorvall RC-5B centrifuge to pellet the plasma membrane fraction. Adaptors and small tubes can be used for small volumes of lysate. An initial low-speed spin (1,000g) can be added before the 30,000g spin to pellet intact cells if cell lysis is incomplete and to pellet nuclei if this is desired. Then centrifuge the supernatant from the 30,000g spin at 54,000 rpm (200,000g) for 90 minutes at 4°C in a Beckman Ti80 fixed-angle rotor in a Beckman L8-70 ultracentrifuge to pellet the plasma membrane fraction.

3. *Binding assays.* Resuspend the 30,000g pellet and the 200,000g pellet in binding buffer to achieve the desired membrane and receptor concentration (typically 1–2 mg protein/ml) by homogenization with a Teflon-glass homogenizer. The 30,000g pellets can be resuspended during the 200,000g spin and then kept on ice. Assay radioligand binding to each fraction according to standard procedures for the receptor of interest.

7.C. Results and Interpretation

Interpretation of results from these assays is essentially identical to that for the step gradient assays described above, with agonist treatment inducing a decrease in the amount of binding to the plasma membrane fraction and a corresponding increase in binding to the light vesicle fraction.

8. HYDROPHILIC RADIOLIGAND BINDING

8.A. Availability of Hydrophilic Radioligands

A limiting factor in the use of hydrophilic radioligand-binding assays is the availability of a good radioligand that is sufficiently hydrophilic to differentiate cell surface receptors from intracellular receptors. Peptide ligands are generally large enough and hydrophilic enough to not penetrate cell membranes, so theoretically these assays could be particularly useful for GPCRs whose ligands are peptides (e.g., endothelin, angiotensin). However, a complication arises because of the fact that agonist peptides generally bind to their receptors with high affinity. This makes it difficult to ensure removal of all of the pretreatment agonist from the receptors before initiating the binding assay with radiolabeled peptide ligand. Thus, if a decrease in radiolabeled peptide binding is observed following agonist pretreatment, additional precautions must be taken to ensure that the apparent decrease in cell surface–accessible receptors is not an artifact due to retained agonist from the pretreatment step. For this reason, inducing internalization with a radiolabeled agonist and using the acid/salt stripping assay to remove cell surface–bound radioligand is more often the assay of choice for high-affinity peptide ligands.

Most of the endogenous amine neurotransmitters (e.g., epinephrine, dopamine, serotonin) that act on GPCRs are sufficiently hydrophilic to monitor cell surface accessibility of their receptors, but they do not bind to their receptors with sufficiently high affinity to be useful as radioligands. However, these ligands can be useful as competing ligands for assessing receptor internalization in hydrophilic competition binding assays. Most of the useful nonpeptide radioligands for GPCRs are antagonists, and thus an appropriate hydrophilic radiolabeled antagonist must be identified. As mentioned earlier, [³H]CGP-12177 for β_2ARs (Hertel et al., 1983; Toews et al., 1984) and N-methyl-[³H]scopolamine for muscarinic receptors (Feigenbaum and El-Fakahany, 1985) are the two classic examples. Similarly useful hydrophilic radioligands may be available for a few other receptors, but for many receptors the lack of a suitable ligand may preclude use of this type of assay. However, as pointed out above, these assays are conducted at low temperatures to prevent reversal of internalization, and at low temperatures even lipophilic radioligands can selectively label cell surface receptors. In many ways, then, these assays with a hydrophilic radioligand are identical to the low temperature binding assays with lipophilic radioligands, but with the added advantage that the hydrophilicity of the radioligand provides an additional factor to enhance the selectivity of the labeling.

8.B. Binding Assay Parameters

The assays are conducted at a low temperature, generally with the dishes or tubes on ice, to prevent reversal of internalization (or further internalization if the radioligand is an agonist) during the time required for the radioligand-binding assay. This low temperature presumably also helps to ensure the selectivity of labeling of cell surface–accessible receptors. The length of the binding assay is another variable factor. Ideally, the binding reaction would be allowed to proceed to equilibrium, which will generally require many hours of binding at the low temperatures used for these assays. Initial experiments should document the time courses for binding of the hydrophilic radioligand to intact control and agonist-treated cells at low temperature, and the incubation time for subsequent experiments should be determined based on these results. If a decrease in hydrophilic radioligand binding is observed following agonist treatment, limited saturation binding experiments should be conducted to ensure that the decrease is due to a change in the number of accessible receptors and not due to a change in the binding affinity for the radioligand.

The final step in the binding assay is the wash to remove unbound radioligand before counting. For adrenergic receptors, we have found that rapid washing at 37°C is more effective at reducing nonspecific binding than washing with ice-cold wash buffer. A receptor antagonist is included in the wash buffer to prevent any further binding to receptors during the 37°C washes. Having the βAR antagonist propranolol present at 100 μM during the wash steps also reduces nonspecific binding of various ligands by an unknown mechanism, so we routinely include 100 μM propranolol in our final wash buffers for adrenergic receptor assays as well as for muscarinic receptor assays with intact cells (Toews, 1987; Hoover and Toews, 1990). We routinely use Hepes-buffered serum-free growth medium as the physiological wash buffer for intact cell-binding assays, but any other isotonic buffered physiological salt solution should be satisfactory.

8.C. Protocol

Because the manipulations required for assays with monolayer cells are somewhat more involved than with cells in suspension, a detailed protocol for monolayer cells is presented here. The assay presented is the simplest possible, with only two treatment conditions, control and agonist. The protocol is easily expanded to include multiple agonists, agonist concentration curves, effects of antagonist blockade, effects of inhibitors, or other experimental variables. Very similar protocols but with appropriate modifications are also used for the assays of lipophilic radioligand binding to intact cells at low temperature described in Section 9 and for the assays of hydrophilic ligand competition for lipophilic radioligand binding described in Section 10.

1. *Cell growth.* Grow cells to near confluence on 12 35-mm culture dishes (three for total and three for nonspecific binding for both control and agonist-pretreated cells) in 2 ml growth medium per dish.

2. *Preparation of solutions and assay setup.* Prepare 15 ml of Hepes-buffered serum-free growth medium containing the appropriate concentration of hydrophilic radioligand. Transfer 3.5 ml of this solution to each of four 12-ml polypropylene tubes. These tubes need to have a large enough diameter to allow easy use of a 1-ml pipettor tip. To two of these tubes (for total binding), add 35 μl of the vehicle for the agent used to define nonspecific binding and label as "T*." To the other two tubes (for nonspecific binding), add 35 μl of 100× concentrated solution of the agent used to define nonspecific binding and label as "N*." Place these tubes in a 37°C water bath to equilibrate at the same time that cells are pretreated with agonist. (Note that slightly more solution is prepared at each step than is needed, to ensure that the appropriate number of identical aliquots can be recovered from each tube). Place one beaker of wash medium (Hepes-buffered serum-free growth medium) in a 37°C water bath and another on ice. Place an appropriate-sized steel or aluminum

cake pan upside down in an appropriate-sized container of ice (we use plastic dishpans) to serve as an ice-cold platform for manipulating dishes. Prepare a Pasteur pipette connected by Tygon tubing to a vacuum flask connected to a vacuum pump to use for aspirating medium from dishes.

3. *Time course for treatment and binding assay.* A detailed time course of a simple experiment that allows a single investigator to conduct all steps is presented below. Times are expressed as t = hr:min. A 30-minute pretreatment and a 4-hour binding assay are used. All solution additions are done with a Pipetman or similar hand-held adjustable pipettor. Solutions should be gently added against the inside wall of the dish, not directly to the cell monolayers, to avoid detaching cells from the dish during the multiple medium changes.

Agonist Pretreatment

t = 0:00: At 10-second intervals, add 20 µl of vehicle for the pretreatment agonist to each of six dishes (three labeled "CT" for control cells, total binding; and three labeled "CN" for control cells, nonspecific binding). Return dishes to CO_2 incubator.

t = 0:05: At 10-second intervals, add 20 µl of pretreatment agonist at 100× final concentration to each of six dishes (three labeled "AT" for agonist-treated cells, total binding; and three labeled "AN" for agonist-treated cells, nonspecific binding). Return dishes to CO_2 incubator.

t = 0:10, 15, 20, 25: For experiments with additional pretreatment or assay conditions, additional sets of six dishes would be started as above.

Initiation of Binding

t = 0:30: At 10-second intervals, aspirate pretreatment medium and add 2 ml 37°C wash buffer to the three "CT" dishes and the three "CN" dishes. This step is to wash away the pretreatment agonist.

t = 0:31: At 10-second intervals, aspirate wash buffer and add 2 ml ice-cold wash buffer to the three "CT" dishes and the three "CN" dishes. Place on ice tray. This step is to further wash the cells and to cool them before adding the radioligand.

t = 0:32: At 10-second intervals, aspirate wash buffer and add 1 ml "T*" solution to the three "CT" dishes and 1 ml "N*" solution to the three "CN" dishes. If the samples are treated in this order, it is not necessary to change pipette tips between the total and nonspecific binding solutions. However, the tip must be changed before the total binding solution is added to the next set of dishes (unless separate pipettors are available for use with the total and nonspecific binding solutions).

t = 0.33: Move dishes to a second ice tray in a refrigerated chamber to keep cells at low temperature for the duration of the binding assay.

t = 0:35, 36, 37, 38: Repeat the above steps for the dishes labeled "AT" and "AN."

t = 0:40, 45, 50, 55: For experiments with additional pretreatment or assay conditions, additional sets of six dishes would be treated as above.

Termination of binding

t = 4:30: At 10-second intervals, aspirate the binding medium and add 2 ml 37°C wash buffer containing 100 μM propranolol to the three "CT" dishes and the three "CN" dishes. This step is to stop the binding reaction and wash away unbound radioligand.

t = 4:31: Repeat the step above for a second wash.

t = 4:32: At 10-second intervals, aspirate the wash buffer and add 1 ml 0.2 M NaOH to the three "CT" dishes and the three "CN" dishes. Stack dishes and set aside.

t = 4:35, 36, 37, 38: Repeat the above steps for the dishes labeled "AT" and "AN".

t = 4:40, 45, 50, 55: For experiments with additional pretreatment or assay conditions, additional sets of six dishes would be treated as above.

Transfer and Quantitation of Bound Radioactivity

After the binding and wash steps are completed for all dishes, transfer the dissolved cells and the associated radioactivity to vials for scintillation counting or to tubes for gamma counting, depending on the radioligand used. The NaOH may cause problems with chemiluminescence in some scintillation cocktails; this can be avoided by neutralizing the NaOH after transfer or by choosing a different scintillation cocktail.

8.D. Results and Interpretation

The decrease in binding to agonist-treated cells compared with control cells is expressed as a percentage of the binding to control cells to obtain the percentage of initially cell surface–accessible receptors that have become inaccessible. Again, these assays do not distinguish whether these inaccessible receptors have been endocytosed into intracellular vesicles or are only sequestered within the plasma membrane.

9. LOW TEMPERATURE BINDING WITH A LIPOPHILIC RADIOLIGAND

The protocols involved in these assays are essentially identical to those used in the hydrophilic radioligand-binding assays above. Thus the details of the assay protocols are not repeated here. Because these assays rely solely on the low temperature of the assay and the consequent impermeability of cellular membranes to prevent radioligand binding to internalized receptors, it is even more critical in these assays that the cells have been cooled before addition of the radioligand and that antagonist is present in the final wash buffer after the binding step to prevent the lipophilic radioligand from crossing cell membranes and labeling internalized receptors.

10. HYDROPHILIC LIGAND COMPETITION FOR LIPOPHILIC RADIOLIGAND BINDING

10.A. Experimental Considerations

The combination of ligand properties that is required for these assays and the relative difficulty in performing short-time assays has limited their usefulness. These assays require a suitably hydrophilic ligand to use as competing ligand. However, this ligand does not need to be available in radiolabeled form nor does it need to be of particularly high affinity. Thus the constraints on the competing ligand are often not a significant problem. As mentioned above, many of the endogenous hormones and neurotransmitters that act on GPCRs satisfy these conditions.

Constraints on the properties of the radioligand are more limiting. These assays require a lipophilic radioligand that will very rapidly equilibrate across cell membranes and among all cellular compartments containing receptors. The radioligand should also have high specific radioactivity and very low nonspecific binding because experiments are generally limited to short times of binding to avoid changes in the extent of internalization during the assay, and thus only a small fraction of the receptors are labeled. The concentration of radioligand can be increased to increase the amount of binding that is obtained; however, this increases cost and generally increases nonspecific binding as well.

The difficulty of conducting the short-time assays and obtaining reliable data have also limited the use of this type of assay. The time allowed for binding can be increased if lower temperatures are used; however, this increases the likelihood that internalized receptors may not be labeled as effectively as cell surface receptors. Despite these constraints, these assays have proved useful for studying internalization of several receptors. We have used assay times as short as 15 seconds (Toews and Perkins, 1984) but more commonly use 1-minute assays (Toews, 1987; Cowlen and Toews, 1988) so that a larger number of dishes can be assayed simultaneously. Others have used longer assay times at lower temperatures (Gabilondo et al., 1996; Trapaidze et al., 1996).

Full competition curves should be performed initially to identify the optimal concentration for distinguishing between the cell surface receptors, which show a high apparent affinity for the hydrophilic competing ligand, and the internalized receptors, which show a very low apparent affinity for the hydrophilic competing ligand. This single concentration can then be used in subsequent experiments to differentiate cell surface from internalized receptors. It should be noted that equilibrium binding assumptions do not apply in short-time assays, and thus the half-maximal inhibitory concentrations for the competing ligands in these nonequilibrium assays generally do not reflect the true equilibrium binding affinity of these ligands. Rather, the ratio of the apparent affinities of the competing ligand for the high- and low-affinity binding sites is a reflection of its selectivity for cell surface versus internalized receptors, and the proportions of receptors exhibiting high and low affinity reflect the proportions of cell surface versus internalized receptors, respectively. Results from a full competition curve with the hydrophilic agonist epinephrine competing for binding of the lipophilic radioligand [^{125}I]iodopindolol to β_2ARs are shown in Figure 10.3.

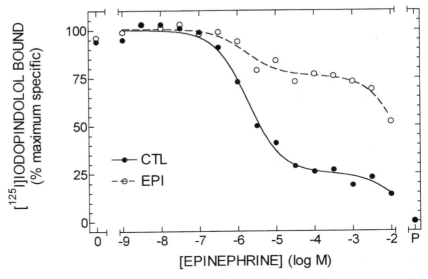

Figure 10.3. Hydrophilic ligand competition for lipophilic radioligand binding. DDT$_1$ MF-2 cells were incubated in the absence (●, CTL) or presence (○, EPI) of the β$_2$AR agonist epinephrine (10 μM) for 30 minutes. Cells were rapidly washed, and then binding of the lipophilic radioligand ^{125}I-iodopindolol to β$_2$AR on intact cells was measured in 1-minute assays in the absence (0) or presence of the indicated concentrations of the hydrophilic competing ligand epinephrine or 1 μM propranolol (P) to define nonspecific binding. Computerized curve-fitting indicated two-site fits in both cases, with 74% high affinity and 26% low affinity in control cells but 24% high affinity and 76% low affinity in the agonist-pretreated cells. The high-affinity component of binding is presumed to be to cell surface receptors readily accessible to epinephrine, and the low-affinity component is presumed to be to internalized receptors that are relatively inaccessible to epinephrine. The data are interpreted to indicate that 50% of the total cellular receptors (74%–24%) or about two thirds of the original cell surface receptors ((74%–24%)/74%) became sequestered and/or endocytosed during the agonist pretreatment. These data also indicate that competition by 10^{-4} M epinephrine would be a good choice for routine experiments to assess cell surface accessibility using only a single concentration of hydrophilic competing ligand. (From Toews [1987], with permission of the publisher.)

10.B. Protocol

The general procedures for these assays are very similar to those presented in detail above for hydrophilic radioligand binding and are therefore presented only briefly. One major difference is that in these assays binding is measured under a minimum of three conditions—total binding, binding in the presence of the hydrophilic competing ligand, and nonspecific binding—as opposed to only total and nonspecific binding for the assays in the previous two sections. If full comptetion curves with the hydrophilic ligand are performed, then binding is measured under a larger number of conditions. The second major difference is that the binding step is only 1 minute long in this protocol, and thus the "termination of binding" washes immediately follow the "initiation of binding" step. For a typical protocol with 1-minute binding assays preceded by a single wash to remove pretreatment agents and followed by two washes to remove nonspecifically bound radioligand, this allows sets of six dishes to be assayed at 6-minute intervals. If additional washes are needed, the interval between sets of assays must be increased accordingly.

Solutions containing radioligand and the appropriate concentration(s) of hydrophilic competing ligand, the agent to define nonspecific binding or vehicle are prepared. Sets of six dishes are pretreated with vehicle or agonist at 10-second intervals every six minutes. After 30 minutes of incubation with agonist, the binding assays are performed on one set of six dishes, again at 10-second intervals every 6 minutes. For these assays, it is most convenient to assay binding under one set of conditions to three control and three agonist-pretreated dishes within each set of six dishes so that the pipettor tip does not have to be changed during the series of 10-second intervals.

During the first minute, the dishes are washed to remove pretreatment agents. During the second minute, solution containing radioligand and appropriate competing agent is added and binding occurs. During the third and fourth minutes, the dishes are washed to remove unbound radioligand. Finally, during the fifth minute, 0.2 M NaOH is added to dissolve cells and associated radioactivity and the dishes are set aside. The next set of dishes is then assayed beginning with the sixth minute and additional sets at successive 6-minute intervals. When all of the assays are completed, the samples are transferred to vials or tubes for counting. It is important to note that, because of the short time of the binding step and the fact that binding is far from steady state and is changing rapidly with time, it is much more critical in these assays than in the other types of intact cell binding assays that the initiation and termination of binding be done at exactly 10-second intervals so that the binding time is exactly identical for each dish or well.

10.C. Results and Interpretation

The difference between total and nonspecific binding is the specific binding in the absence of hydrophilic competing ligand. The difference between binding in the presence of the hydrophilic competing ligand and nonspecific binding is the specific binding in the presence of hydrophilic competing ligand. The specific binding in the presence of hydrophilic competing ligand divided by the specific binding in the absence of competing ligand is the fraction of receptors that are internalized (inaccessible to hydrophilic ligand). The increase in this fraction with agonist pretreatment indicates the extent of agonist-induced internalization. These calculations are illustrated in Figure 10.3 for the data presented there.

Ideally, agonist pretreatment should not decrease the specific binding in the absence of hydrophilic competing ligand. If a decrease in specific binding in the absence of hydrophilic competing ligand does occur, this could indicate (1) that pretreatment agonist was not adequately washed away and is inhibiting radioligand binding, (2) that some agonist-induced down-regulation of binding sites has occurred, or (3) that the internalized receptors in the agonist-pretreated cells are not as rapidly accessible to the radioligand as are the cell surface receptors. The first possibility can be tested and avoided by additional rapid washes. The second possibility can be tested in binding assays with isolated membrane preparations and can perhaps be avoided by using a shorter pretreatment time. The last possibility is the most difficult to deal with; in general, it means that a more lipophilic radioligand is needed or that only qualitative and not quantitative assessment of internalization will be possible with this type of assay. It should also be pointed out that these assays do not distinguish whether the receptors that are inaccessible to the hydrophilic competing ligand have been endocytosed into intracellular vesicles or are only sequestered within the plasma membrane.

II. ACID/SALT STRIPPING

II.A. Experimental Considerations

As mentioned above, these assays are often the most convenient approach for monitoring internalization of those GPCRs whose ligands are peptides that bind to the receptor with high affinity. For these receptors, the binding of the agonist, which induces receptor internalization, is of sufficiently high affinity (pM range) that the agonist will stay bound to the receptor during its internalization, and the inducing ligand will thus be internalized along with the receptor. In this situation, *ligand* internalization can be taken as an indicator of *receptor* internalization, with the assumption generally being made that one molecule of agonist internalizes with each molecule of receptor. The most common version of these assays uses a radioactive agonist peptide as the ligand to both induce receptor internalization and to label the receptors. In some cases the entire incubation is conducted at 37°C, allowing internalization to take place along with equilibration of agonist radioligand with the receptor.

Another common protocol employs an initial binding step conducted at 4°C, a condition that generally allows steady-state labeling of the cell surface population of receptors without internalization, which is effectively blocked at this low temperature. Free radioligand is then removed, and the cells are incubated for an additional period of time at 37°C, during which the pre-bound agonist drives internalization of the receptors along with the pre-bound agonist radioligand. In either case, after the internalization step, the cells are placed on ice and washed to remove any unbound ligand. Cells are then incubated in a low pH/high salt solution, generally 0.2 M acetic acid/0.5 M NaCl, which promotes dissociation of the bound agonist radioligand from the receptors that remain on the cell surface. This medium is then washed from the cells and counted as an indication of the number of cell surface (acid-labile) receptors. The radioactivity remaining associated with the cells is also counted and taken as an indication of the number of internalized (acid-resistant) receptors. Two sets of dishes must be taken through the entire protocol, one labeled under total binding conditions and one under nonspecific binding conditions; the difference between the two then provides the specific binding to the cell surface and internalized receptors. Variables to consider in developing these assays are the radiolabeled agonist ligand to be used and its concentration, the time to be used for labeling of the receptors, the time to be used for the internalization step, the precise composition of the acid/salt solution used in the stripping step, and the time to be used for the stripping step.

II.B. Protocol

1. *Cell growth.* Grow cells to near confluence on 35-mm dishes or multi-well plates.

2. *Receptor labeling.* Wash away growth medium and incubate cells for 2 hours on ice with 1 ml serum-free Hepes-buffered medium containing ^{125}I-labeled agonist peptide at two to five times its K_d. For the total binding dishes, use only the radiolabeled agonist ligand; for the nonspecific binding dishes, include a 100-fold excess of competing nonradioactive ligand. Protease inhibitors are often included to prevent degradation of the agonist peptide.

3. *Receptor internalization.* Leaving the dishes on ice, wash twice with 2 ml ice-cold culture medium or phosphate-buffered saline to remove unbound radioligand. Replace the ice-cold medium with 1 ml 37°C medium and incubate at 37°C for 30 minutes to allow internalization of receptor and ligand. Note that some radiolabeled agonist may dissociate from the prelabeled receptors during this step, either on the cell surface or in the internalized vesicles.

4. *Acid/salt stripping.* Aspirate the medium and wash to remove any dissociated radioligand. Then add 0.5 ml of ice-cold 0.2 M acetic acid/0.5 M NaCl. Incubate on ice for 5 minutes to dissociate cell surface radioligand.

5. *Radioligand quantitation.* Transfer the acetic acid/salt solution to tubes for counting. Rinse dishes with an additional 0.5 ml of ice-cold acid/salt solution to quantitatively remove all of the dissociated radioligand from the dishes. Then add 1 ml of 0.2 M NaOH to each dish to dissolve the cells and the internalized radioligand. Transfer to tubes for counting.

II.C. Results and Interpretation

Total minus nonspecific binding in the acid/salt solution is taken as the amount of cell surface–associated receptors. Total minus nonspecific binding in the dissolved cell fraction is taken as the amount of internalized receptors. The fraction of receptors internalized is calculated as the internalized receptors divided by the sum of the cell surface plus internalized receptors, which represents the amount of receptors that were initially present at the cell surface during the labeling step.

REFERENCES

Anderson RG (1993): Plasmalemmal caveolae and GPI-anchored membrane proteins. Curr Opin Cell Biol 5:647–652.

Arora KK, Sakai A, Catt KJ (1995): Effects of second intracellular loop mutations on signal transduction and internalization of the gonadotropin-releasing hormone receptor. J Biol Chem 270:22820–22826.

Barton AC, Black LE, Sibley DR (1991): Agonist-induced desensitization of D_2 dopamine receptors in human Y-79 retinoblastoma cells. Mol Pharmacol 39:650–658.

Bohm SK, Grady EF, Burnett NW (1997): Regulatory mechanisms that modulate signaling by G-protein–coupled receptors. Biochem J 322:1–18.

Carpentier JL (1994): Insulin receptor internalization: Molecular mechanisms and physiopathological implications. Diabetologia 37 (suppl 2):S117-S124.

Chaki S, Guo DF, Yamano Y, Ohyama K, Tani M, Mizukoshi M, Shirai H, Inagami T (1994): Role of carboxyl tail of the rat angiotensin II type 1A receptor in agonist-induced internalization of the receptor. Kidney Int 46:1492–1495.

Cowlen MS, Toews ML (1987): Effects of agonist and phorbol ester on adrenergic receptors of DDT$_1$ MF-2 cells. J Pharmacol Exp Ther 243:527–533.

Cowlen MS, Toews ML (1988): Evidence for α_1-adrenergic receptor internalization in DDT$_1$ MF-2 cells following exposure to agonists plus protein kinase C activators. Mol Pharmacol 34:340–346.

de Weerd WF, Leeb-Lundberg LM (1997): Bradykinin sequesters B2 bradykinin receptors and the receptor-coupled Gα subunits Gα_q and Gα_i in caveolae in DDT$_1$ MF-2 smooth muscle cells. J Biol Chem 272:17858–17866.

Feigenbaum P, El-Fakahany EE (1985): Regulation of muscarinic cholinergic receptor density in neuroblastoma cells by brief exposure to agonist: Possible involvement in desensitization of receptor function. J Pharmacol Exp Ther 233:134–140.

Feron O, Smith TW, Michel T, Kelly RA (1997): Dynamic targeting of the agonist-stimulated m2 muscarinic acetylcholine receptor to caveolae in cardiac myocytes. J Biol Chem 272:17744–17748.

Freedman NJ, Lefkowitz RI (1996): Desensitization of G protein–coupled receptors. Recent Prog Horm Res 51:319–351; discussion 352.

Gabilondo AM, Krasel C, Lohse MJ (1996): Mutations of Tyr^{326} in the β_2-adrenoceptor disrupt multiple receptor functions. Eur J Pharmacol 307:243–250.

Gagnon AW, Kallal L, Benovic JL (1998): Role of clathrin-mediated endocytosis in agonist-induced down-regulation of the β_2-adrenergic receptor. J Biol Chem 273:6976–6981.

Garland AM, Grady EF, Lovett M, Vigna SR, Frucht MM, Krause JE, Bunnett NW (1996): Mechanisms of desensitization and resensitization of G protein–coupled neurokinin$_1$ and neurokinin$_2$ receptors. Mol Pharmacol 49:438–446.

Harden TK, Cotton CU, Waldo GL, Lutton JK, Perkins JP (1980): Catecholamine-induced alteration in sedimentation behavior of membrane bound β-adrenergic receptors. Science 210:441–443.

Hein L, Kobilka BK (1995): Adrenergic receptor signal transduction and regulation. Neuropharmacology 34:357–366.

Hertel C, Staehelin M, Perkins JP (1983): Evidence for intravesicular β-adrenergic receptors in membrane fractions from desensitized cells: Binding of the hydrophilic ligand CGP-12177 only in the presence of alamethicin. J Cyclic Nucleotide Protein Phosphor Res 9:119–128.

Hoover RK, Toews ML (1990): Activation of protein kinase C inhibits internalization and downregulation of muscarinic receptors in 1321N1 human astrocytoma cells. J Pharmacol Exp Ther 253:185–191.

Itokawa M, Toru M, Ito K, Tsuga H, Kameyama K, Haga T, Arinami T, Hamaguchi H (1996): Sequestration of the short and long isoforms of dopamine D$_2$ receptors expressed in Chinese hamster ovary cells. Mol Pharmacol 49:560–566.

Koenig JA, Edwardson JM (1997): Endocytosis and recycling of G protein–coupled receptors. Trends Pharmacol Sci 18:276–287.

Krupnick JG, Benovic JL (1998): The role of receptor kinases and arrestins in G protein–coupled receptor regulation. Annu Rev Pharmacol Toxicol 38:289–319.

Leeb-Lundberg LM, Cotecchia S, DeBlasi A, Caron MG, Lefkowitz RJ (1987): Regulation of adrenergic receptor function by phosphorylation. I. Agonist-promoted desensitization and phosphorylation of α_1-adrenergic receptors coupled to inositol phospholipid metabolism in DDT$_1$ MF-2 smooth muscle cells. J Biol Chem 262:3098–3105.

Lin FT, Krueger KM, Kendall HE, Daaka Y, Fredericks ZL, Pitcher JA, Lefkowitz RJ (1997): Clathrin-mediated endocytosis of the β-adrenergic receptor is regulated by phosphorylation/dephosphorylation of β-arrestin1. J Biol Chem 272:31051–31057.

Lisanti MP, Tang Z, Scherer PE, Sargiacomo M (1995): Caveolae purification and glycosylphosphatidylinositol-linked protein sorting in polarized epithelia. Methods Enzymol 250:655–668.

Lutton JK, Frederich RCJ, Perkins JP (1979): Isolation of adenylate cyclase–enriched membranes from mammalian cells using concanavalin A. J Biol Chem 254:11181–11184.

Mahan LC, Motulsky HJ, Insel PA (1985): Do agonists promote rapid internalization of β-adrenergic receptors? Proc Natl Acad Sci USA 82:6566–6570.

Mostafapour S, Kobilka BK, von Zastrow M (1996): Pharmacological sequestration of a chimeric beta 3/beta 2 adrenergic receptor occurs without a corresponding amount of receptor internalization. Recept Signal Transduct 6:151–163.

Pals-Rylaarsdam R, Gurevich VV, Lee KB, Ptasienski JA, Benovic JL, Hosey MM (1997): Internalization of the m2 muscarinic acetylcholine receptor. Arrestin-independent and -dependent pathways. J Biol Chem 272:23682–23689.

Perkins JP, Hausdorff WP, Lefkowitz RJ (1991): Mechanisms of ligand-induced desensitization of beta-adrenergic receptors. In Perkins JP (ed): The Beta-Adrenergic Receptors. Clifton, NJ: Humana Press, pp 73–124.

Roettger BF, Rentsch RU, Pinon D, Holicky E, Hadac E, Larkin JM, Miller LJ (1995): Dual pathways of internalization of the cholecystokinin receptor. J Cell Biol 128:1029–1041.

Shockley MS, Burford NT, Sadee W, Lameh J (1997): Residues specifically involved in down-regulation but not internalization of the m1 muscarinic acetylcholine receptor. J Neurochem 68:601–609.

Slice LW, Wong HC, Sternini C, Grady EF, Bunnett NW, Walsh JH (1994): The conserved NPX$_n$Y motif present in the gastrin-releasing peptide receptor is not a general sequestration sequence. J Biol Chem 269:21755–21761.

Slowiejko DM, Fisher SK (1997): Radioligand binding measurement of receptor sequestration in intact and permeabilized cells. Methods Mol Biol 83:243–250.

Slowiejko DM, McEwen EL, Ernst SA, Fisher SK (1996): Muscarinic receptor sequestration in SH-SY5Y neuroblastoma cells is inhibited when clathrin distribution is perturbed. J Neurochem 66:186–196.

Smart EJ, Ying YS, Mineo C, Anderson RG (1995): A detergent-free method for purifying caveolae membrane from tissue culture cells. Proc Natl Acad Sci USA 92:10104–10108.

Sorensen SD, McEwen EL, Linseman DA, Fisher SK (1997): Agonist-induced endocytosis of muscarinic cholinergic receptors: Relationship to stimulated phosphoinositide turnover. J Neurochem 68:1473–1483.

Stadel JM, Strulovici B, Nambi P, Lavin TN, Briggs MM, Caron MG, Lefkowitz RJ (1983): Desensitization of the β-adrenergic receptor of frog erythrocytes. Recovery and characterization of the down-regulated receptors in sequestered vesicles. J Biol Chem 258:3032–3038.

Strader CD, Sibley DR, Lefkowitz RJ (1984): Association of sequestered beta-adrenergic receptors with the plasma membrane: A novel mechanism for receptor down regulation. Life Sci 35:1601–1610.

Strasser RH, Stiles GL, Lefkowitz RJ (1984): Translocation and uncoupling of the β-adrenergic receptor in rat lung after catecholamine promoted desensitization in vivo. Endocrinology 115:1392–1400.

Thomas WG, Thekkumkara TJ, Motel TJ, Baker KM (1995): Stable expression of a truncated AT_{1A} receptor in CHO-K1 cells. The carboxyl-terminal region directs agonist-induced internalization but not receptor signaling or desensitization. J Biol Chem 270:207–213.

Toews ML (1987): Comparison of agonist-induced changes in β- and α_1-adrenergic receptors of DDT_1 MF-2 cells. Mol Pharmacol 31:58–68.

Toews ML, Perkins JP (1984): Agonist-induced changes in β-adrenergic receptors on intact cells. J Biol Chem 259:2227–2235.

Toews ML, Waldo GL, Harden TK, Perkins JP (1984): Relationship between an altered membrane form and a low affinity form of the β-adrenergic receptor occurring during catecholamine-induced desensitization. Evidence for receptor internalization. J Biol Chem 259:11844–11850.

Toews ML, Waldo GL, Harden TK, Perkins JP (1986): Comparison of binding of ^{125}I-iodopindolol to control and desensitized cells at 37° and on ice. J Cyclic Nucleotide Protein Phosphor Res 11:47–62.

Trapaidze N, Keith DE, Cvejic S, Evans CJ, Devi LA (1996): Sequestration of the δ opioid receptor. Role of the C terminus in agonist-mediated internalization. J Biol Chem 271:29279–29285.

Trowbridge IS, Collawn JF, Hopkins CR (1993): Signal-dependent membrane protein trafficking in the endocytic pathway. Annu Rev Cell Biol 9:129–161.

Turner JT, Bollinger DW, Toews ML (1988): Vasoactive intestinal peptide receptor/adenylate cyclase system: Differences between agonist– and protein kinase C–mediated desensitization and further evidence for receptor internalization. J Pharmacol Exp Ther 247:417–423.

von Zastrow M, Kobilka BK (1994): Antagonist-dependent and -independent steps in the mechanism of adrenergic receptor internalization. J Biol Chem 269:18448–18452.

Wakshull E, Hertel C, O'Keefe EJ, Perkins JP (1985): Cellular redistribution of β-adrenergic receptors in a human astrocytoma cell line: A comparison with the epidermal growth factor receptor in murine fibroblasts. J Cell Biochem 29:127–141.

Wang J, Zheng J, Anderson JL, Toews ML (1997): A mutation in the hamster α_{1B}-adrenergic receptor that differentiates two steps in the pathway of receptor internalization. Mol Pharmacol 52:306–313.

Zhu SJ, Cerutis DR, Anderson JL, Toews ML (1996): Regulation of hamster α_{1B}-adrenoceptors expressed in Chinese hamster ovary cells. Eur J Pharmacol 299:205–212.

CHAPTER II

CHARACTERIZATION OF RECEPTOR SEQUESTRATION BY IMMUNOFLUORESCENCE MICROSCOPY

FRANCESCA SANTINI and JAMES H. KEEN

Regulation of G Protein–Coupled Receptor Function and Expression,
Edited by Jeffrey L. Benovic.
ISBN 0-471-25277-8   Copyright © 2000 Wiley-Liss, Inc.

1. INTRODUCTION

Immunofluorescence techniques have been instrumental in studying the cell and tissue distribution of many G protein–coupled receptors (GPCRs) as well as their fate following agonist stimulation. There is no substitute for being able to visualize the focus of ones studies when possible, and immunofluorescence provides a morphological counterpart to other, more functional, approaches. Whereas radioligand-binding assays can reveal the disappearance of agonist-binding sites from the plasma membrane of agonist-treated cells, they do not reveal if this is due to loss of receptor affinity for the ligand (receptor desensitization) or to the physical removal of the receptor from the plasma membrane (internalization).

With immunofluorescence it is possible to more directly monitor GPCR trafficking and distribution before, during, and after stimulation. This can provide important information with regard to the intracellular compartments involved in receptor endocytosis, down-regulation, and resensitization, yielding a spatiotemporal correlation between events. Also, co-localization experiments can potentially reveal the locations of several species of interest at once, particularly of value in the GPCR signaling system in which multiple players contribute to activation, signaling, and attenuation phases. Like other techniques, the power of immunofluorescence microscopy is accompanied by real and potential limitations. Careful experimental design and internal controls can make these limitations recognizable, if not always surmountable.

2. IMMUNOSTAINING

The steps involved in performing an immunostaining protocol are highlighted in Table 11.1. These include a fixation step to maintain the integrity of cell structure, permeabilization to allow antibody penetration, and blocking to minimize nonspecific interactions. These steps are followed by incubation with primary antibody directed against the protein of interest and, generally, addition of a second, fluorescently tagged antibody. After fixation, it is this second antibody that is actually visualized in the microscope. In the following sections, we discuss each step in more detail. At the end of this chapter (Section 4), we provide the reader with several examples, as well as modifications of this protocol to achieve specific goals.

TABLE 11.1. Immunofluorescence Staining Flow Chart

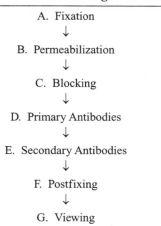

A. Fixation
↓
B. Permeabilization
↓
C. Blocking
↓
D. Primary Antibodies
↓
E. Secondary Antibodies
↓
F. Postfixing
↓
G. Viewing

2.A. Preparing Cells for Labeling

2.A.a. Coverslips. Cells to be processed for immunofluorescence are generally grown on glass coverslips. For use with high magnification, high numerical aperture objectives, No. 1.5 coverslips should be used. Unfortunately, these are not readily available in 12-mm diameter circles suitable for use in 24-well plates described here. Fisher No. 1 coverslips can generally be substituted without adverse results. Glass coverslips can be cleaned and sterilized by dipping them in ethanol (95%) and setting them to dry under a sterile hood. They can also be briefly flamed, but should not be allowed to become misshapen.

2.A.b. Plates. Cells grown on coverslips can be processed for staining in a 24-well plate to employ minimal reagent volumes (0.3–0.5 ml/step). Changing solution in the well is also easily accomplished by gentle aspiration of the solution through a pipette tip connected to a house vacuum. Unless otherwise noted (see next section), it is important to never let the cell layer on the coverslip dry out.

2.A.c. Adherent Cells. Many adherent cells grow fairly well when plated directly on glass coverslips. Cells should be plated so that they are approximately 40%–60% confluent at the time of staining, unless a different cell density is required for the aim of the study. In general, $4–8 \times 10^5$ cells per coverslip is a good working density when dealing with cells of 10–25 μm diameter.

2.A.d. Cells Growing in Suspension and Cells That Require Special Substrata for Growth. Sometimes microscopy has to be performed on cells that grow in suspension or adhere poorly to the glass coverslip. In other cases, cells may require special substrata in order to grow or differentiate. For these instances, specially prepared coverslips or physical sedimentation of the cells onto the substrate needs to be used. In addition to purchasing precoated coverslips directly from commercial sources (e.g., BioCoat Cell Environments, Bedford, MA), we suggest one of the following protocols.

2.A.d.1. Polylysine-Coated Coverslips. In many cases, when the goal is to induce increased cell adherence to the glass, coverslips can be coated with polylysine. Cells plated on polylysine-treated coverslips are generally adherent in about 30 minutes, but they can be grown on treated coverslips overnight.

1. Place coverslips in a 24-well plate and rinse twice with distilled water.
2. Incubate coverslips with 1 mg/ml of poly-L-lysine (55 kD, Sigma) in distilled water for 30 minutes at room temperature.
3. Rinse coverslips twice with distilled water and either use immediately or let dry and store at room temperature.

2.A.d.2. Matrix Protein–Coated Coverslips. A simple method that works well with several proteins (e.g., fibronectin, collagen, laminin) is to incubate a 1 mg/ml solution of the coating protein in phosphate-buffered saline (PBS), pH 7.2, or appropriate buffer with the coverslips overnight at 37°C.

2.A.d.3. Special Coatings. When a more specific coating is required (e.g., with immunoglobulin, hormones), glutaraldehyde cross-linking of the protein can be performed (Michl et al., 1979; Santini and Keen, 1996). The following steps are carried out at room temperature:

1. Coat coverslips with poly-L-lysine as specified above.
2. Incubate coverslips with freshly prepared 2.5% glutaraldehyde (1 ml of 50% glutaraldehyde, EM grade; Polyscience Cat. No. 18428) in 19 ml PBS for 15 minutes. Rinse three times with PBS.
3. Incubate coverslips for 30 minutes in the dark with a 0.5–2 mg/ml solution of the coating protein in PBS.
4. Incubate coverslips for 5 minutes with 10 mg/ml sodium-borohydride solution made in PBS, prepared immediately before use. This step reduces autofluorescence by helping to quench unreacted glutaraldehyde (see Section 2.B.a.3).
5. Rinse coverslips with PBS and incubate with 0.2 M glycine in 10 mM sodium phosphate, pH 7.2, in the dark overnight.
6. Store coverslips at 4°C in the dark in 10 mM sodium phosphate, pH 7.2, with 0.02% sodium azide.

2.A.d.4. Cell Sedimentation Onto Coverslips. Suspension cells can be attached to the substrate by using a cytocentrifuge (Cytospin 3, Shandon).

1. Wash cells in PBS and resuspend at $1-2 \times 10^6$/ml.
2. Immobilize a coverslip on the slide under the opening in the paper mask using double-stick tape.

3. Insert microscope glass slide in paper mask and place in cytocentrifuge holder.

4. Add 0.1–0.5 ml of cell suspension to the funnel.

5. Spin at 1,200*g* for 8 minutes.

6. Remove coverslip and proceed as in section 2.H., Step 2.

2.B. Fixing Cells

The fixation step is a critical one. Ideally, while immobilizing the cellular components in a state as close as possible to their native state, the fixative used in preparing the specimen should fulfill three criteria: (1) it should not alter the distribution of the antigen; (2) it should not alter the morphology of the specimen; and (3) it should preserve the antigenicity of the protein of interest. While no fixative entirely fulfills all of these criteria, the fixatives successfully used in immunofluorescence light microscopy fall into two categories: cross-linking agents and organic solvents.

2.B.a. Cross-Linking Fixatives. The most commonly used cross-linking agents are formaldehyde (2%–3.7%), paraformaldehyde (2%–3%), and glutaraldehyde (2%–4%), alone or in combination. These reagents work by cross-linking proteins to each other and to the cytoskeleton while generally leaving the cell membrane intact. The intracellular structure is well preserved and, because cell membranes are quite permeable to these agents, cell monolayers are rapidly fixed at room temperature. These fixatives are generally not strongly denaturing. As a result, they may not be useful if the antibodies available recognize only a denatured form of the antigen. Also, care should be taken not to "overfix" the sample because the cross-linking process could mask the antigen or prevent antibody from reaching it.

Ideally, cross-linking reagents should be stored under nitrogen and working solutions prepared in PBS or Tris-buffered saline solution (TBS) just before use.

2.B.a.1. Formaldehyde. Formaldehyde is conveniently prepared by dilution in PBS or TBS of the 37% solution commercially available (Sigma, St. Louis, MO).

2.B.a.2. Paraformaldehyde. Paraformaldehyde is prepared from the dry chemical (EM grade, Sigma, Cat. No. P6148). To prepare a 5% (w/v) solution, 3 g of paraformaldehyde is dissolved into 50 ml of PBS at 80°C with constant stirring. After adjusting pH to 7.3 if necessary and bringing the volume to 60 ml with PBS, the solution can be aliquoted and stored at –20°C.

2.B.a.3. Glutaraldehyde. Glutaraldehyde is prepared from the commercially available 50% solution (EM grade from Polyscience). Because glutaraldehyde produces autofluorescence, the fixed samples should be incubated in a sodium-borohydride solution (10 mg/ml $NaBH_4$ in water) to quench unreacted groups before proceeding with the immunostaining.

All of these reagents have been implicated as suspected or known carcinogens, and all have some volatility at room temperature. Accordingly, reagents should be prepared in a fume hood and other appropriate precautions employed.

2.B.b. Organic Fixatives.

Other methods of fixing cells employ organic solvents in nonaqueous media, the protein components of the cell being precipitated in place while the more hydrophobic components (e.g., most lipids) are extracted. This technique does not preserve the subcellular architecture of the cell with sufficient precision for ultrastructural studies by electron microscopy at nanometer resolution, but it is often adequate for light microscopic studies with its much lower resolution (250 nm).

On the other hand, fixation with organic fixatives has the advantage of simultaneously accomplishing both fixation and permeabilization of the cell (see next section). Finally, because macromolecules tend to be more completely unfolded by this treatment than by cross-linking reagents, organic fixatives may be more useful with antibodies directed toward epitopes that are buried in the native antigen.

Most commonly used are methanol, ethanol, and acetone either alone or sequentially or in combination. The fixatives are kept cold ($-20°C$), and fixation is carried out at $-20°C$ for 5–10 minutes. After treatment, the solution is withdrawn, and the sample is air-dried before proceeding.

2.C. Permeabilization Step

2.C.1. Permeabilization of Fixed Samples.

Detergent extraction of the sample allows partial or complete solubilization of the cell membranes, thus rendering intracellular antigens available to the antibody. This step is generally not necessary if an organic fixative has been used (see above). Often a detergent is also present in all the solutions used in the subsequent steps of the immunostaining in order to keep background staining to a minimum (see next step).

Cells are usually permeabilized by brief incubation with nonionic detergents. Among the most commonly used detergents are Triton-X 100 (0.05%–0.3% in PBS for 10–15 minutes at room temperature [RT]) or Nonidet P-40 (0.2%). Saponin (0.1–0.5 mg/ml, 30 minutes, on ice) is also used because it solubilizes cholesterol, leaving the membrane structure relatively intact (Jacob et al., 1991; Willingham and Pastan, 1985).

Permeabilization of the sample is not required if the goal is to visualize cell surface proteins (e.g., receptors) on the plasma membrane. Fixed cells, after a blocking step, can be immediately incubated with an antibody directed against an extracellular protein epitope. Omission of the permeabilization step can also be used to differentiate between plasma membrane and intracellular pools of the same antigen in the sample. Receptors that are present on both the plasma membrane and intracellular compartment(s) can be separately resolved. First the plasma membrane receptors are revealed by immunostaining the sample while avoiding the use of detergents. For example, mouse antibody "A" is followed by fluorescein (FL)–conjugated antimouse antibody (steps 1–3 and then 6–12 of the protocol in Section 2.H. performed without adding Triton X-100 to the buffers) to label surface receptors. Second, the intracellular pool of receptors is detected by performing another immunostaining after permeabilizing the cells:

for example, rabbit antibody "B" followed by lissamine-rhodamine (LR)–conjugated antirabbit antibody (steps 4–13 of the protocol in section 2.H.) is used.

Some cells (e.g., HeLa) suffer membrane breakage upon formaldehyde fixation, and internal components become accessible to exogenous antibodies. A sample stained with an antibody to an internal protein (e.g., tubulin) should be included as a control to eliminate this possibility.

2.C.b. Permeabilization of Unfixed Samples. The permeabilization step can also be performed before the sample has been fixed. Proteins that constitute or are associated with the cytoskeleton are resistant to mild detergent extraction. The permeabilization of unfixed cells, therefore, releases soluble components of the cytosol while retaining the detergent-resistant proteins. Immunostaining of a sample that has been permeabilized before fixation reveals interesting features that can be obscured by soluble antigen, including an association with the cytoskeleton. Endocytic structures like clathrin-coated pits, for example, are generally detergent resistant and are retained in permeabilized and unfixed cells.

2.D. Blocking Step

Nonspecific interaction between the antibody and the specimen can severely compromise visualization of the antigen of interest. The blocking agent occupies sites on the sample that could nonspecifically bind proteins (e.g., by charge interaction) or immunoglobulins (e.g., samples rich in Fc receptors). The most common blocking agents are serum proteins (e.g., 1%–10% nonimmune serum like normal goat serum (NGS); 1%–5% bovine serum albumin), preferably from the same animal in which the second antibody has been raised. Fat-free dry milk (5%–10%) and gelatin (0.3%–1%) can also be used.

2.E. Antibody

2.E.a. Primary Antibody. The prerequisite of an immunofluorescence staining protocol, as well as that of any immunodetection method, is the availability of suitable antibodies. Antibodies are serum immunoglobulins induced by injecting host animals (mice, rats, rabbits, goats) with an immunogen of choice (the antigen).

Polyclonal antibodies, often used as whole serum or partially purified IgG fractions, are the product of multiple B-cell populations and generally recognize multiple sites (epitopes) on a macromolecular antigen. Monoclonal antibodies, on the other hand, represent a homogeneous population of immunoglobulins obtained from the clonal amplification of a B cell by fusion with a myeloma cell line. Monoclonal antibodies specifically recognize and bind a single antigenic epitope.

Several sources are available with information on how to produce, characterize, purify and handle antibodies (e.g., (Harlow and Lane, 1988).

Primary antibodies have to be thoroughly characterized to be certain that they (1) recognize the antigen of interest after fixation and (2) do not cross-react with specimen components other than the antigen.

The first thing that needs to be established when working with antibodies is to determine the minimum dilution at which the antibodies give the best signal without excessive background. Useful dilutions vary greatly depending on the concentration, affinity, and avidity of the antibody used. As a general guideline for preliminary screening, the following dilution ranges are suggested:

For polyclonal antisera: 1/50, 1/100, 1/250, 1/500, 1/1,000, 1/2,000
For purified monoclonal preparations (μg/ml): 1, 2, 5, 10, 20, 50, 100

Antibodies are always diluted and incubated with the sample in a buffer containing the same blocking agent used in the blocking step to help reduce nonspecific interactions. The antibody solution should always be briefly (10 minutes) spun in a microcentrifuge (13,500g) at 4°C and the supernatant carefully removed for incubation with the sample. The time of incubation with primary antibodies (from 10 minutes to overnight) as well as the temperature at which the incubation is carried out (RT or 37°C for relatively short incubations; 4°C for longer intervals) will depend on the antibody characteristics: we generally start with a 30-minute incubation at 37°C.

An important aspect of working with antibody, especially with primary antisera, is the limited supply. In trying to overcome this limitation we employ a procedure that minimizes the amount of reagent used. The incubation of the sample with the antibody is performed inside a sealed chamber lined with moistened filter paper to maintain a humidified atmosphere. A piece of Parafilm (3 × 3 cm/coverslip) is then appropriately labeled and inverted on the moistened paper, ensuring that a smooth surface is obtained. The paper back of the Parafilm is then carefully peeled away, and a drop of antibody solution ($\approx$20 μl/coverslip) is applied in the center of the square. With forceps, the coverslip containing the cells is then removed from the 24-well plate (a 21-gauge needle with bent tip makes this task easier), and the excess buffer is gently removed by touching the edge of the coverslip on a clean tissue. Without letting the sample dry out, the coverslip is carefully inverted on the drop so that the cell layer is now in contact with the antibody. The coverslip is transferred back to the 24-well plate at the end of the incubation for the subsequent washing procedures. Needless to say, it is important to keep the orientation of the cell surface on the coverslip in mind during these manipulations.

For the specific task of studying receptor sequestration following agonist challenge, several "approaches" can be taken.

2.E.a.1. Antibody Directed Against Receptors. An antibody that recognizes the receptor extracellular epitope can be used in the standard immunostaining protocol both with and without cell permeabilization (see above), thus allowing the differential visualization of surface versus total cell receptors. Also, such an antibody can be used as a marker to track receptor sequestration and trafficking following stimulation. For example, living cells can be incubated with the antibody at 4°C for 1 hour to load all the receptors present on the plasma membrane. After washing off unbound antibody, the cells are incubated at 37°C with or without agonist. Internalization of the receptors is then monitored by fixing cells at different times, permeabilizing, and revealing antibody-bound receptors with fluorescently tagged second antibodies (see below). An important point to bear in mind when using antireceptor antibody in this way, though, is to make sure that they do not interfere with the receptor affinity for its ligand or with receptor signaling.

2.E.a.2. Antibody Against Tagged-Receptors. Very often the study of receptor trafficking employs transient or stable transfection of receptors in a characterized cell line. If this is the case, it is very useful to transfect an epitope-tagged form of the receptor of interest. A variety of such epitope tags are available. Their use confers several advantages: (1) the receptor is easily visualized by using commercially available tag-specific antibody; (2) species specificity of antireceptor antibodies is no longer a problem; and (3) in conjunction with the use of a receptor-directed antibody, the anti-tag antibody allows differential tracking of transfected versus total receptors.

2.E.a.3. Antibody Against Receptor Ligand. In a few cases, it is possible to study receptor trafficking using antibodies specific for their ligands. To use this approach, one must ascertain that the ligand–receptor complex is stable and is resistant to the method of fixation used. Commercially available are antibodies against neuropeptides for several cholecystokinin receptors like substance P, substance K, somatostatin, gastrin, and bombesin (Accurate, Westbury, NY; Cappel Research Reagents, ICN, Costa Mesa, CA).

2.E.a.4. Conjugated Primary Antibody. It is sometimes necessary to directly label primary antibodies. This is often done when

1. Two or more antigens need to be detected but all the available primary antibodies are from the same species and thus cannot be differentially detected by means of species-specific fluorescently tagged secondary antibodies (see Section 2.E.b.).
2. The antigen concentration in the sample is low and the signal needs to be amplified.

Several companies provide the necessary reagents for direct labeling (e.g., Molecular Probes, Eugene, OR; Sigma; Pierce, Rockford, IL). A widely used protein-conjugation system involves biotin. Biotinylated primary antibody can then be revealed by fluorescent-streptavidin or streptavidin-conjugated IgG. This system can be useful in amplifying a weak antigen signal. A second alternative is to conjugate the antibody directly to a fluorophore (e.g., fluorescein-5-isothiocyanate from Molecular Probes as described by Menon et al. [1984]) in which case there is no requirement for a further detection system.

2.E.a.5. Controls. When working with antibodies, an extremely important aspect is to ascertain the specificity of the staining pattern observed. Several controls are always suggested.

1. An important control to detect artifacts is to stain a sample using preimmune serum in place of the primary antibodies in Step 7 of the protocol

in Section 2.H. It is preferable that the preimmune serum comes from the same animal in which the primary antibody was generated, obtained before or at the time of initial immunization. For monoclonal antibodies, an unrelated antibody, preferably of the same isotype, should be used.

Ideally, no significant signal should be obtained with the preimmune serum. In practice, a detectable background is frequently observed. Often, this can be lowered by inclusion of detergents in all solutions, by increasing washing times or temperatures (Steps 8 and 10, Section 2.H.), or by lowering the amount of antibody used. The last option may also proportionally decrease the positive signal obtained with the immune serum. This often necessitates affinity purification of the antibody of interest, for which several protocols are available (Harlow and Lane, 1988). Affinity purification is also the remedy of choice to remove unrelated, contaminating antibodies that cross-react with the specimen.

2. An accurate method for establishing the specificity of the staining observed with a given primary antibody is to process a parallel sample using neutralized antibody. This is done by incubating the primary antibody with a 5–10-fold molar excess of the antigen (1 hour to overnight). A sample is then stained using the neutralized primary antibody and compared to one stained with regular primary antibody: lack of signal in the former will be a good indication of specificity.

3. While the above controls are all considered "negative" in that they should produce little or no staining, a positive control can sometimes be obtained by staining a sample with a different primary antibody directed against the same protein (polyclonal) or against a different epitope of the protein (monoclonal). A similar pattern of antigen distribution should be obtained.

4. When preparing multiple labeled samples, it is important to ascertain that the two primary antibodies do not cross-react with each other due to immunoreactivity between the two species in which the antibodies were raised. This can lead to two kinds of artifacts: (a) the two antibodies form an immunocomplex that precipitates on the surface of the sample, producing fluorescent speckles upon examination; and (b) one primary antibody binds to the antigen against which it was raised *and* to the other primary antibody, resulting in false co-localization of the two antigens. To identify these problems, samples should always be labeled with each of the primary antibodies alone and the staining pattern compared with that of samples stained with both primary antibodies. Sometimes it is possible to overcome primary antibody cross-reactivity by using the primary antibodies sequentially rather than together (Steps 7 and 8 of protocol 2.H. are repeated twice).

2.E.c. Secondary Antibody. Generally, immunodetection of a protein is accomplished by indirect immunofluorescence. In this technique the antigen–primary antibody complex is detected by using fluorescently tagged secondary antibody (secondary antibody) raised against the immunoglobulins of the animal in which the primary antibodies were produced. For example, antigen X is recognized by a mouse anti-X primary antibody, which in turn is recognized by a goat antimouse conjugated secondary antibody. The fluorescently tagged secondary antibody is the reagent actually visualized. Many of the points discussed for primary antibodies apply also to secondary antibodies. As for the primary antibodies, secondary antibodies need to be checked for potential artifacts too.

Good sources of fluorescently tagged secondary antibodies are Jackson ImmunoResearch Laboratories and Amersham Corporation (Arlington Heights, IL). It is desirable to use secondary antibodies that have been affinity purified to ensure minimal species cross-reactivity for multilabeling techniques.

2.E.b.1. Controls. When using secondary antibodies, a control should be done to detect possible reactivity between the sample and the antibody (e.g., species cross-reactivity). To do so, a sample is stained using only the secondary antibodies (that is, skipping Steps 7 and 8 in Section 2.H.). If the antibody gives a reaction, switching to secondary antibodies raised in a different species might solve the problem. When planning a multilabeling experiment it is also important to determine if the labeled second antibodies cross-react with each other. This is done by incubating a sample with only one of the primary antibodies (e.g., rabbit), followed by the two fluorescently labeled second antibodies that are to be used (e.g., FL-antirabbit and LR-antimouse). By using the appropriate set of filters, immunoreactivity should be observed only from the FL channel.

2.F. Choice of Fluorophores

The choice of the fluorophore depends on the light source, the filter sets, and the detection system available. Also important is the separation between emission wavelengths in the case of multiply labeled samples. Table 11.2 lists the most widely used fluorophores and their excitation and emission wavelengths. Among those listed, fluorescein (FL), lissamine-rhodamine (LR), and Texas red (TR) provide good color separation for double labeling and can be used by conventional epifluorescence microscopes. Another relatively new fluorophore, not visible by eye, is the indodicarbocyanine dye Cy5. This is particularly valuable in multiple labeling as Cy5 emission is widely separated from that of FL.

TABLE 11.2. Fluorophores Commonly Used in Immunofluorescence Microscopy

Immunofluorescent Probes	Excitation Max (nm)	Emission Max (nm)
Indodicarbocyanine (Cy) 2	489	506
Fluorescein (FL)	492	528
Tetramethyl rhodamine (TRITC)	520–554	582
Indodicarbocyanine (Cy) 3	553–555	568–574
Lissamine-rhodamine B (LR)	539–574	602
Texas Red (TR)	558–594	623
Indodicarbocyanine (Cy) 5	649–651	666–674

2.F.a. Autofluorescence. When choosing a fluorophore, an important thing to check is the degree of intrinsic fluorescence (autofluorescence) present in the sample and its potential interference with the fluorophore. All cells show a certain degree of autofluorescence due to the presence of flavanoids and lipofuscin pigments, and for some cells (e.g., hepatocytes) this can be particularly high. The endogenous fluorophores that contribute to autofluorescence generally have an excitation maximum in the violet/blue region (400–450 nm) and therefore interfere particularly with FL detection. To assess the degree of sample autofluorescence, unstained specimens are simply fixed and mounted (steps 11–13 of Section 2.H.) and examined under the microscope using the filter sets that will be used to view stained samples. This problem can sometimes be eliminated by incubating samples with 10 mg/ml sodium-borohydride for 10 minutes at RT after fixing.

2.F.b. Bleed Through. With multiple labeled samples and depending on the characteristics of the light sources and filters, emission on one channel can "bleed through" into another. This can often be a problem when samples labeled with both FL and LR are examined, leading to false-positive signals on the LR channel. To estimate the degree of "bleed through" in an FL/LR double-labeled sample, a parallel sample should be stained with only the reagents used to detect the FL-labeled antigen; the sample is then viewed and analyzed with the LR filter set channel. Another way to avoid ambiguous results is to use FL in conjunction with Cy5 because the Cy5 excitation band lies much further from the FL emission peak.

2.G. Postfixing, Mounting, and Storing Stained Samples

After the last wash, a final short fixing step with formaldehyde in PBS can be beneficial in that it helps to prevent release of the labeled antibody. The step should be kept short to avoid quenching of the signal and should not involve glutaraldehyde due to its intrinsic autofluorescence.

Cells grown on coverslips can be mounted on slides using a drop of PBS/glycerol (1:9; v/v) as a mounting medium. All fluorophores have a tendency to fade on illumination so the use of mounting medium containing antifading agents such as *n*-propyl gallate (Giloh and Sedat, 1982) or *p*-phenylenediamine (Johnson and Nogueira Araujo, 1981; Johnson et al., 1982) is recommended. Commercial preparations such as SlowFade (Molecular Probes Inc.) or FluoroGuard (Bio Rad) are also available.

To preserve the thickness of the cells, an especially important consideration if the sample is to be analyzed by confocal sectioning, three small dots of nail polish are applied to the glass slide and allowed to dry to create a support for the round coverslip. A drop of mounting fluid is applied on the slide in the area delimited by the dots. After excess liquid is drained from it, the coverslip is inverted on the mounting fluid. The excess fluid seeping from under the coverslip is wiped, and the coverslip is sealed to the slide with nail polish.

Analysis of the samples should take place shortly after preparation, but, if necessary, slides can be kept at 4°C in the dark and are generally good for viewing for several days.

2.H. Immunostaining Protocol

1. *Wash:* 3× with PBS at room temperature (RT) or 37°C
2. *Fix:* 3.7% formaldehyde in PBS, 10 minutes at RT
3. *Wash:* 3× with PBS at RT
4. *Permeabilization:* 0.05% Triton X-100 in PBS, 10 minutes at RT
5. *Wash:* 3× with 0.05% Triton X-100 in PBS
6. *Block:* 4% normal goat serum in PBS with 0.05% Triton X-100, 10 minutes at RT
7. *Primary antibodies:* 30 minutes at 37°C
8. *Wash:* 3× with PBS with 0.05% Triton X-100 then PBS with 0.05% Triton X-100, 30 minutes at 37°C
9. *Secondary antibodies:* 30 minutes at 37°C
10. *Wash:* 3× with PBS with 0.05% Triton X-100 then PBS with 0.05% Triton X-100, 30 minutes at 37°C
11. *Postfix:* 3.7% Formaldehyde in PBS, 10 minutes at RT
12. *Wash:* 3× with PBS at RT
13. *Mount:* Use a drop of SlowFade and then seal the edges of coverslip with nail polish

3. SAMPLE ANALYSIS

3.A. Conventional Light Microscopy

Samples are generally analyzed on microscopes equipped with epifluorescence optics. There are two essential features of such a system. First, the microscope must be equipped with a light source that is capable of providing a relatively broad spectrum of intense light from about 400–700 nm (mercury or xenon lamps are commonly used). Second, the system must have filter sets capable of providing exciting light of specific wavelengths and collecting the desired emitted light (see Table 11.2), while rejecting both incident/scattered light and light of other, unwanted wavelengths. Fortunately, manufacturers now routinely provide their microscopes with the necessary equipment for such studies; for special applications, independent companies can provide more varied filter sets (e.g., Chroma, Brattleboro, VT; Molecular Probes; Omega Optical, Brattleboro, VT). Finally, images can be recorded on film (35-mm TMAX 400 ASA is suitable) or with video or other camera systems that can provide digitized output. An advantage of the latter approach is that data can more easily be analyzed, quantitated, and formatted for publication purposes using readily available software (e.g., NIH Image, Adobe Photoshop). For further details on instrumental setup, the reader is referred to Wang and Herman (1996) and Cheng (1994).

3.B. Confocal Light Microscopy

Conventional light microscopy can provide lateral resolution of about 0.25 μm (depending on wavelength), which can permit direct observation of many intracellular structures in cells (e.g., mitochondria, elements of the endoplasmic reticulum, lysosomes, endosomes, and Golgi). However, out-of-focus light from above and below the plane of focus can greatly degrade the image. Confocal microscopy systems use a laser to provide powerful, collimated illumination. This permits the use of narrow, "confocal" apertures in both the exciting and emission light paths to reject this out-of-focus light. As a result, the confocal microscope provides much sharper images than a conventional system in a single plane, as well as resolution in the vertical "z" axis of about 0.5–0.7 μm. The image signal is generally detected by a photomultiplier with an A → D readout or by direct photon counting, in either case providing a digital readout. Most systems provide two or more photomultipliers to permit simultaneous detection of more than one signal. For further general information, the reader should consult Gratton and van de Ven (1990).

Confocal microscopy systems can be obtained with a variety of laser sources: a Kr/Ar mixed gas laser is desirable for multilabel studies. A wide variety of filter sets can also generally be obtained or constructed as needed. These can permit imaging of one, two, or three labels simultaneously. Digitized images can then be processed as described above.

4. EXAMPLES

4.A. Single Labeling

We use immunofluorescence to visualize expression levels, distribution, and trafficking of β_2-adrenergic receptors (β_2AR) that had been transiently cotransfected with β-arrestin in COS-1 cells. Our strategy was to transfect an epitope-tagged receptor so that we could take advantage of a commercially available monoclonal antibody specific for the epitope. As shown in Figure 11.1, in unstimulated COS-1 cells (Fig. 11.1A) the expressed Flag-β_2AR appears diffuse and localized mainly on the plasma membrane with a fraction of receptors also found in the perinuclear region before agonist treatment. Ten minutes after stimulation with (−)-isoproterenol, the β_2AR distribution is dramatically altered, with the majority of the receptors now found in small intracellular vesicles (Fig. 11.1B).

Experimental Details

COS-1 cells were cotransfected with plasmids containing sequences coding for the β_2AR with the Flag epitope (N-Asp-Tyr-Lys-Asp-Asp-Asp-Asp-Lys-C) fused to the N terminus, and a plasmid containing the ß-arrestin coding sequence using lipofectamine (Life Technologies, Inc.) according to the manufacturer's instructions. Seven hours after transfection cells were trypsinized and plated on 12-mm round glass coverslips in a 24-well plate, 0.5 ml medium per well at 3×10^4 cells per well. Analysis was performed 24 hours after transfection. Cells grown on glass coverslips were washed twice with Hepes-buffered medium (Hepes-DMEM) and either left undisturbed or stimulated for 10 minutes with 10 μM (−)-isoproterenol.

Cells were processed for immunofluorescence according to the protocol described above (Section 2.H.). We detected Flag-β_2ARs with a mouse monoclonal

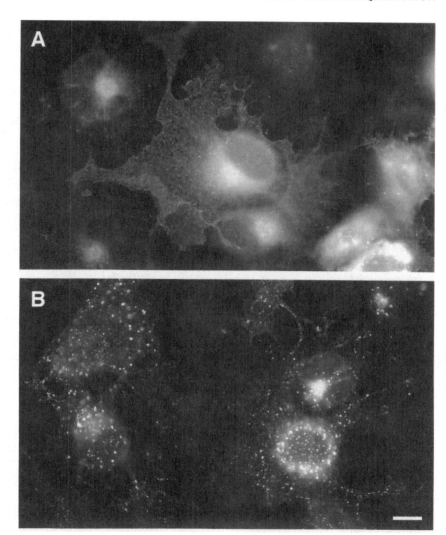

Figure 11.1. Distribution of expressed Flag–β_2-adrenergic receptors in resting **(A)** and (–)-isoproterenol–stimulated **(B)** COS-1 cells. In resting cells the transfected receptors, detected by mouse M2 primary antibody followed by FL–secondary antibody appear to be largely present on the plasma membrane and in a perinuclear region. After agonist stimulation the membrane receptors are concentrated in small intracellular vesicles that resemble endosomes. The characteristic punctate pattern of receptor fluorescence (lower panel) was exclusively observed in the presence of agonist. For details, see Section 4.A. Bar = 10 μm.

antibody, M2 (Eastman Kodak Co., Cat. No. IB 13081), that specifically recognizes the Flag octapeptide epitope. M2 was used at a concentration of 10 μg/ml and detected with FL-conjugated affinity-purified donkey antimouse IgG (H+L) (Jackson ImmunoResearch Laboratories, Cat. No. 715-095-150) at a dilution of 10 μg/ml. Samples were analyzed on a Zeiss Axiovert 405 microscope with Zeiss Plan-NEOFLUAR 63× 1.25 NA oil immersion objective using fluorescein-sensitive optics. Images were recorded on 35-mm film (Kodak TMAX 400 ASA, 12 second exposure). After development, film negatives were scanned at 1012 pixels/inch. Digitized images were then processed using Adobe PhotoShop.

4.B. Dual Labeling

4.B.a. With Two Primary Antibodies. Nonvisual arrestins are important in the GPCR desensitization and sequestration events that follow agonist stimulation (Goodman et al., 1998). They also bind assembled clathrin *in vitro* (Krupnick et al., 1997; Goodman et al., 1996, 1997). Therefore, nonvisual arrestins may facilitate receptor endocytosis by acting as an adaptor between activated-phosphorylated receptors and coated pit clathrin.

To evaluate this hypothesis morphologically *in vivo,* we transiently co-transfected Flag-β_2AR and β-arrestin genes in COS-1 cells and studied the distribution of surface Flag-β_2AR and β-arrestin with respect to the distribution of clathrin-coated pits before and after agonist stimulation.

As shown in Figure 11.2, surface-localized Flag-β_2AR (left panels) is initially diffusely distributed on the cell surface (– ISO); however, on agonist treatment with isoproterenol (+ ISO), the β_2AR signal becomes punctate. This fine fluorescence coincides substantially with clathrin-coated pits (right upper two panels). Similar to β_2AR, β-arrestin (right lower two panels) also shifts from a diffuse (– ISO) to a punctate distribution following agonist treatment (+ ISO), with many of the fluorescence dots now coincident with the β_2AR ones.

This result provides evidence in support of a functionally important role *in vivo* for the interaction between clathrin and (nonvisual) arrestins observed *in vitro.*

Experimental Details

COS-1 cells were transfected as specified in Section 4.A. and were grown overnight on glass coverslips. Two sets of two coverslips were prepared. From each set, one coverslip was incubated with 10 μM (–)-isoproterenol in Hepes-DMEM for 10 minutes at 37°C. Cells from one set were then stained to detect β_2AR and clathrin while cells from the second set were stained to detect β_2AR and arrestin.

Flag-β_2ARs were detected with M2 (10 μg/ml) as mentioned above (4.A.); β-arrestin was detected with a rabbit polyclonal antibody (KEE) raised against a C-terminal 16 amino acid peptide specific for β-arrestin, 1:250 (Sterne-Marr et al., 1993); clathrin was detected using rabbit serum 27004 (Keen et al., 1981) (1:100). Primary antibodies were visualized using FL-conjugated affinity-purified donkey antimouse IgG (Jackson ImmunoResearch Laboratories, Cat. No. 715-095-150) and LR-conjugated affinity-purified donkey antirabbit IgG (Jackson ImmunoResearch Laboratories, Cat. No. 711-085-152), both used together at a dilution of 1:100.

Because we wanted to restrict our analysis to events occurring at the plasma membrane, we specifically visualized β_2AR present on the cell surface by staining fixed, unpermeabilized cells with anti-Flag antibody. This was accomplished by modifying protocol 2.H. with the insertion of new steps (I–V) between steps 3 and 4 as follows:

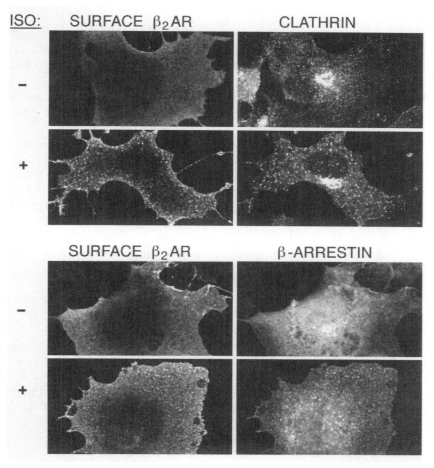

Figure 11.2. Redistribution of surface Flag-β_2AR and β-arrestin to clathrin-coated pits induced by agonist stimulation. COS-1 cells transiently transfected with Flag-β_2AR and ß-arrestin were incubated with serum-free media alone ("–" panels) or with media containing 10 μM isoproterenol (ISO) ("+" panels) for 10 minutes. The diffuse distribution of surface β_2AR in the absence of agonist becomes punctate on isoproterenol addition and largely coincident with clathrin. In concert with the change in β_2AR localization on agonist treatment, the previously diffuse β-arrestin distribution becomes largely punctate and co-localized with surface β_2AR in agonist-treated cells. Flag-β_2AR was detected by mouse M2 primary antibody followed by FL-tagged secondary antibody; β-arrestin was detected with rabbit KEE primary antibody and clathrin with rabbit 27004 primary antibody, both followed by LR-tagged secondary antibody. For details, see Section 4.B.a. Bar = 10 μm.

I. *No detergent block:* 4% NGS in PBS, 10 minutes at RT

II. *M2 Ab:* M2 are diluted in 4% NGS in PBS and incubated with the sample for 30 minutes at 37°C

III. *Wash:* 3× with PBS at RT then PBS, 20 minutes at 37°C

IV. *Fix:* 3.7% formaldehyde in PBS, 3 minutes at RT

V. *Wash:* 3× with PBS at RT

Steps 4–13 are as in the protocol using antiarrestin antibody for one set of coverslips and anticlathrin antibody for the other set.

Confocal analysis was performed on a Bio-Rad MRC-600 Ar/Kr laser scanning confocal microscope interfaced to a Zeiss Axiovert 100 microscope with Zeiss Plan-Apo 63× 1.40 NA oil immersion objective. Samples were analyzed with simultaneous excitation at 488 and 568 nm with proper filters to visualize FL and LR signals.

4.B.b. With One Primary Antibody and a Fluorescent Ligand.

The treatment of transfected COS-1 cells with agonist induces the appearance of β_2AR in small intracellular vesicles. To characterize the nature of the β_2AR-containing vesicles, we incubated Flag-β_2AR and β-arrestin expressing COS-1 cells with fluorescein-label transferrin (Fig. 11.3). Following stimulation with isoproterenol (Fig. 11.3B), we detected β_2AR appearance in the same intracellular vesicles with the fluorescein-labeled transferrin (FL-Tf). As shown, the two labels are highly coincident, indicative of the early endosomal nature of the compartment.

Experimental Details

Transferrin internalization was measured by following the endocytosis of fluorescein-labeled transferrin (available from Molecular Probes). Briefly, COS-1 cells co-transfected with β_2AR and β-arrestin (as described in Section 4.A.) were grown overnight on coverslips and then incubated at 37°C for 10 or 30 minutes in Hepes-DMEM medium containing 50 µg/ml fluorescein-labeled transferrin and 10 µM (−)-isoproterenol or only 20 µg/ml FL-Tf, respectively. Cells were then washed three times with PBS, fixed in 3.7% formaldehyde, and processed for immunostaining as specified in Section 4.A. following the protocol highlighted in Section 2.H. LR-conjugated affinity-purified donkey antimouse IgG (1:100) (Jackson ImmunoResearch Laboratories, Cat. No. 715-085-151) was used to visualize the M2 mouse anti-Flag antibody.

Confocal analysis was performed as specified above (Section 4.B.a.).

4.C. Triple Labeling With Primary Antibodies

Double-labeling experiments (Section 4.B.) can show that in agonist-treated cells β-arrestin becomes largely coincident with β_2AR and that β_2AR was mostly co-localized with the characteristic clathrin staining. To finally determine if all the components of this system are found in the same clathrin-coated pits, we detected the position of β_2AR, β-arrestin, and AP2 (the other structural component of clathrin-coated pits [Keen, 1990]) in the same cell following stimulation with agonist. Figure 11.4 shows that the receptor is indeed found in clathrin-coated pits with arrestin.

Experimental Details

COS-1 cells are transfected and stimulated as specified in Section 4.B.a. Flag-β_2AR is detected with M2 mouse monoclonal antibody, β-arrestin with rabbit polyclonal KEE affinity-purified serum, and AP.2 with biotinylated AP.6 mouse monoclonal antibody. Cy5-conjugated streptavidin (Cat. No. 016-170-084), FL-conjugated affinity-purified donkey antimouse (Cat. No. 715-095-

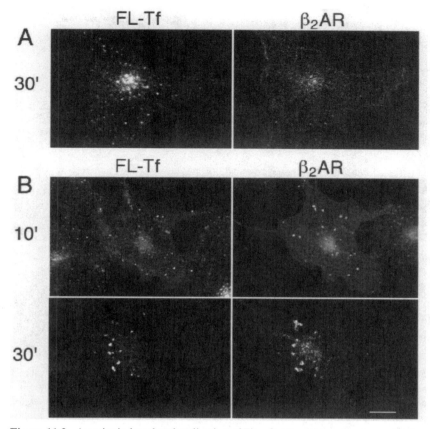

Figure 11.3. Agonist-induced co-localization of Flag-β_2AR with transferrin-containing endosomes. COS-1 cells transiently expressing Flag-β_2AR and ß-arrestin were incubated with fluorescein-tagged transferrin (FL-Tf) and stimulated with isoproterenol for the indicated time, as specified in the text. After fixation, the cells were immunostained with mouse M2 primary antibody and LR-tagged secondary antibody to localize the Flag-β_2AR. Direct observation with fluorescein filters reveals the localization of the endocytosed FL-Tf. The substantial coincidence of the two signals following agonist treatment demonstrates that β_2AR enters the endocytic pathway after stimulation. For details, see Section 4.B.b. Bar = 10 μm.

150), and LR-conjugated affinity-purified donkey antirabbit (Cat. No. 711-085-152) were purchased from Jackson ImmunoResearch Laboratory. All the solutions were spun 10 minutes at 13,500g at 4°C before use. Treated and untreated transfected cells were processed for the detection of surface Flag-β_2AR first as follows.

Steps 1, 2, and 3 are as in the protocol described in Section 2.H., followed by steps

   I. *No detergent block:* 4% NGS in PBS, 10 minutes at RT

   II. *M2 Ab:* M2 (10 μg/ml) in 4% NGS in PBS, 30 minutes at 37°C

   III. *Wash:* 3× with PBS at RT then PBS, 20 minutes at 37°C

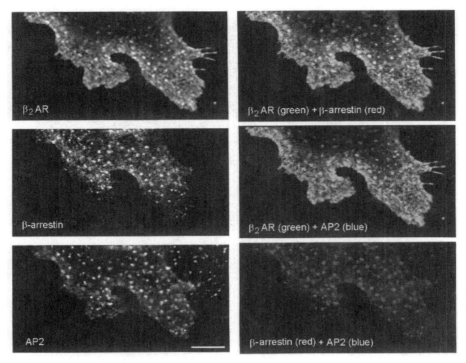

Figure 11.4. Flag-β₂AR and β-arrestin are present in the same clathrin-coated pits on agonist treatment. The punctate pattern of β₂AR and β-arrestin induced by isoproterenol coincides with coated pits, as indicated by the co-localization of AP.2, a marker for plasma membrane clathrin-coated pits. Flag-β₂AR was detected by mouse M2 mono-clonal antibody followed by FL-tagged secondary antibody; β-arrestin was detected with rabbit KEE primary antibody followed by LR-tagged secondary antibody; coated pit AP.2 were revealed by mouse biotinylated-AP.6 followed by Cy5-tagged streptavidin. For details, see section 4.C. Bar = 10 μm. (See color insert.)

   IV. *Fix:* 3.7% formaldehyde in PBS, 3 minutes at RT

   V. *Wash:* 3× with PBS at RT

Steps 4–10 are performed as in Section 2.H., using KEE (1:250) in Step 7 and FL-conjugated affinity-purified donkey antimouse and LR-conjugated affinity-purified donkey antirabbit (together, each at 1:100 dilution) in Step 9. Subsequently, the following steps are performed:

   VI. *Fix:* 3.7% formaldehyde in PBS, 3 minutes at RT

   VII. *Block:* 10 μg/ml normal mouse serum in PBS with 0.05% Triton X-100 15 minutes at RT

   VIII. *Bio-AP.6:* Biotinylated AP.6 (30 μg/ml) in 4% NGS in PBS with 0.05% Triton X-100 supplemented with 10 μg/ml mouse serum, 1 hour at 37°C

IX. *Wash:* 3× with PBS with 0.05% Triton X-100 then PBS with 0.05% Triton X-100 for 30 minutes at 37°C

X. *Cy5-strep:* Cy5-streptavidin (2 μg/ml) in 4% NGS in PBS with 0.05% Triton X-100 for 30 minutes at 37°C

XI. *Wash:* 3× with PBS with 0.05% Triton X-100 then PBS with 0.05% Triton X-100 for 30 minutes at 37°C

Steps 11–13 are as in Section 2.H.

Confocal analysis was performed on the system described in Section 4.B.a. Samples were analyzed using simultaneous excitation at 488 nm and 568 nm to visualize FL and LR signals, followed by excitation at 647 nm to visualize Cy5.

ACKNOWLEDGMENTS

Work in the laboratory was supported by NIH grant GM-28526.

REFERENCES

Cheng P-C (1994): Multidimentional Microscopy. New York: Springer-Verlag.

Giloh H, Sedat JW (1982): Fluorescence microscopy: Reduced photobleaching of rhodamine and fluorescein protein conjugates by *n*-propyl gallate. Science 217:1252–1255.

Goodman OB Jr, Krupnick JG, Gurevich VV, Benovic JL, Keen JH (1997): Arrestin/clathrin interaction. Localization of the arrestin binding locus to the clathrin terminal domain. J Biol Chem 272:15017–15022.

Goodman OB Jr, Krupnick JG, Santini F, Gurevich VV, Penn RB, Gagnon AW, Keen JH, Benovic JL (1996): Beta-arrestin acts as a clathrin adaptor in endocytosis of the beta2-adrenergic receptor. Nature 383:447–450.

Goodman OB Jr, Krupnick JG, Santini F, Gurevich VV, Penn RB, Gagnon AW, Keen JH, Benovic JL (1998): Role of arrestins in G-protein–coupled receptor endocytosis. In Goldstein DS, Eisenhofer G, McCarty R (eds): Advances in Pharmacology. New York: Academic Press, pp 429–433.

Gratton E, van de Ven MJ (1990): Handbook of Biological Confocal Microscopy, 2nd ed. New York: Plenum.

Harlow E, Lane D (1988): Antibodies. A Laboratory Manual. Cold Spring Harbor, NY: Cold Spring Harbor Laboratory.

Jacob MC, Favre M, Bensa JC (1991): Membrane cell permeabilization with saponin and multiparametric analysis by flow cytometry. Cytometry 12:550–558.

Johnson GD, Davidson RS, McNamee KC, Russell G, Goodwin D, Holborow EJ (1982): Fading of immunofluorescence during microscopy: A study of the phenomenon and its remedy. J Immunol Methods 55:231–242.

Johnson GD, Nogueira Araujo GM (1981): A simple method of reducing the fading of immunofluorescence during microscopy. J Immunol Methods 43:349–350.

Keen JH, Willingham MC, Pastan I (1981): Clathrin and coated vesicle proteins. J Cell Biol 256:2538–2544.

Keen JH (1990): Clathrin and associated assembly and disassembly proteins. Annu Rev Biochem 59:415–438.

Krupnick JG, Goodman OB Jr, Keen JH, Benovic JL (1997): Arrestin/clathrin interaction: Localization of the clathrin binding domain of non-visual arrestins to the carboxy terminus. J Biol Chem 272:15011–15016.

Menon AK, Holowka D, Baird B (1984): Small oligomers of immunoglobulin E (IgE) cause large-scale clustering of IgE receptors on the surface of rat basophilic leukemia cells. J Cell Biol 98:577–583.

Michl J, Pieczonka MM, Unkeless JC, Silverstein SC (1979): Effects of immobilized immune complexes on Fc- and complement-receptor function in resident and thioglycollate-elicited mouse peritoneal macrophages. J Exp Med 150:607–621.

Santini F, Keen JH (1996): Endocytosis of activated receptors and clathrin-coated pit formation—Deciphering the chicken or egg relationship. J Cell Biol 132:1025–1036.

Sterne-Marr R, Gurevich VV, Goldsmith P, Bodine RC, Sanders C, Donoso LA, Benovic JL (1993): Polypeptide variants of beta-arrestin and arrestin3. J Biol Chem 268:15640–15648.

Wang XF, Herman B (1996): Fluorescence Imaging Spectroscopy and Microscopy. New York: Wiley.

Willingham MC, Pastan I (1985): An Atlas of Immunofluorescence in Cultured Cells. London: Academic Press.

CHAPTER 12

CHARACTERIZATION OF RECEPTOR TRAFFICKING WITH GREEN FLUORESCENT PROTEIN

NADYA I. TARASOVA, ROLAND D. STAUBER, and
STEPHEN A. WANK

Regulation of G Protein–Coupled Receptor Function and Expression,
Edited by Jeffrey L. Benovic.
ISBN 0-471-25277-8   Copyright © 2000 Wiley-Liss, Inc.

I. INTRODUCTION

The green fluorescent protein (GFP) cloned from the bioluminescent jellyfish *Aequorea victoria* (Prasher et al., 1992) has been used extensively as a unique tool to monitor dynamic processes in living cells and translucent whole organisms noninvasively in real time as well as in formalin-fixed cells (for detailed reviews, see Chalfie and Kain, 1998; Cubitt et al., 1995; Gerdes and Kaether, 1996; Misteli and Spector, 1997). GFP has been used for a variety of applications, ranging from protein localization and trafficking, tracking of cell lineages during development, and assessing gene transfer and expression to monitoring of human immunodeficiency virus entry into cells. Due to the relatively low level of G protein–coupled receptor (GPCR) expression, methods used to study their function and regulation have been technically challenging as well as limiting.

Traditional antibody-based methods are laborious and nonquantitative, usually perturb the system, have limited access to cell compartments, and may lack the necessary sensitivity and specificity. Similarly, fluorescent-labeled ligands used to tag the receptor are problematic. They may be functionally altered by the attached fluorophores and dissociate from the receptor. Recently, several mutational improvements in the primary structure of GFP have significantly enhanced its fluorescence spectra, intensity, stability, and expression. As a result, GFP has recently also proved to be a powerful tool for real-time imaging of the trafficking of GPCRs in living cells (Barak et al., 1997a; Tarasova et al., 1997a, 1998; McClintock et al., 1997; Kallal et al., 1998).

This chapter provides the reader with the necessary information to form a rational approach toward the successful use of GFP in the study of GPCR trafficking. A background in the fluorescent properties of wild-type GFP as well as the advantages of the most recently developed enhanced mutant forms are discussed in the context of their application toward GFP-tagged GPCR imaging. The choice of cell systems to be used for GPCR expression as well as the appropriate vectors and optimal transfection methods are considered. However, because of the large number of receptors in the GPCR superfamily and their individual requirements for a variety of experimental systems available, no one vector or transfection method is optimal for all applications, and so the reader should refer to any one of a number of manuals (e.g., Ausubel et al., 1994) for the necessary protocols once the optimal system has been chosen. A detailed protocol for the fusion of GFP to the C terminus of any GPCR is also offered with the CCKAR as a specific example. The microscopic equipment and software along with protocols that we used in our studies for the detection and quantitation of GFP-tagged CCKA receptors are presented with reference to and full acknowledgement that other systems may be equally effective. Finally, cell compartment fluorescent markers are described along with specific examples for their use in studying the trafficking of GPCRs in a variety of living cells in real time.

2. PROPERTIES OF GFP AND GFP MUTANTS

GFP, in contrast to other bioluminescent molecules, fluoresces independently of cofactors, substrates, and other gene products. It is active in a variety of organisms and can be detected rapidly and easily upon ultraviolet (UV) illumination. The polypeptide of 238 amino acids contains a heterocyclic tripeptide (Ser[65]-dehydroTyr[66]-Gly[67]) minimal chromophore that forms post-translationally and spontaneously by oxidation (Cody et al., 1993). The chromophore is only fluorescent when embeded within the tightly packed β-can structure of the complete GFP protein (amino acids 7–229 required for full fluorescence; Li et al., 1997).

The weak fluorescent signal made wild-type GFP less attractive for practical applications. Thus, a variety of GFP mutants with enhanced fluorescence and different spectral properties have been developed. In addition, the introduction of enhanced GFP mutants emitting blue light (blue fluorescent protein [BFP]) allows the use of BFP and GFP as a dual color tagging system (Stauber et al., 1998). Table 12.1 and Figure 12.1 characterize some of the commonly used and commercially available GFP/BFP mutants.

The introduced amino acid changes not only increase the adsorption coefficients but also improve protein solubility, chromophore maturation, and stability. Codon usage has also been optimized for higher eukaryotic expression (Muldoon et al., 1997). Thus, the brighter fluorescence is the sum of several

TABLE 12.1. Properties of GFP Mutants[a]

Protein	Excitation Max (nm)	Emission Max (nm)	Increased Green Fluorescence[d]	Increased Blue Fluorescence[d]	Commercial Source
wtGFP	398	509			Clontech[b]/ Quantum[c]
S65Tb[e]	490	511	6×		Clontech[i]
P64L S65T[f]	488	507	35×		
(GFPsg25)[g] P64L/S65C/ I167T/ K238N	474	509	50–100×		Quantum (rsGFP)[j]
BFP (Y66H)[h]	383	447			Clontech/ Quantum
(BFPsg50)[g] P64L/Y66H/ V163A	386	450		63×	Quantum (BFP)[j]

[a] The numbering corresponds to amino acids of the wild-type GFP protein.
[b] Clontech Inc., Palo Alto, CA (http://www.clontech.com).
[c] Quantum Biotech. Inc., Quebec, Canada (http://www.qbi.com).
[d] Increase in green fluorescence (at maximum emission) is in comparison to wild-type GFP, and increase in blue fluorescence (at maximum emission) is in comparison to BFP(Y66H).
[e] Values as described by Heim et al. (1995).
[f] Values as described by Cormack et al. (1996).
[g] Values as described by Stauber et al. (1998).
[h] Values as described by Heim et al. (1996).
[i] Optimized for human codon-usage preferences.
[j] Improved for human codon-usage preferences.

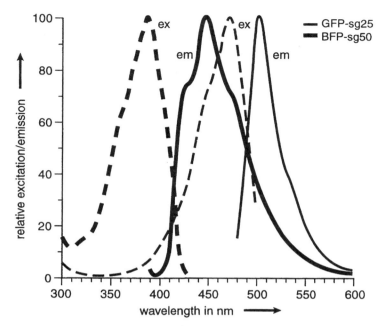

Figure 12.1. Excitation (ex) and emission (em) spectra of GFPsg25 and BFPsg50.

beneficial factors resulting in a higher percentage of active GFP molecules in a given population under physiological conditions.

3. DETECTION OF GFP AND GFP-TAGGED PROTEINS

The success of GFP tagging depends not only on the expression of a functional and highly fluorescent hybrid protein but also on the technical equipment optimized for signal detection. Ideally, for the highest signal-to-noise ratio, the wavelength used for excitation should match as close as possible the excitation maximum of the GFP mutants utilized. Likewise, detection of the emission maximum requires the appropriate filter set up (Fig. 12.1). This is essential when working with dim GFP signals or BFP.

The UV light needed for BFP excitation produces high background autofluorescence, and the signal of enhanced BFP is generally weaker than GFP. The major excitation peak of the enhanced GFP mutants encompasses the excitation wavelength commonly used for FACS analysis (argon ion laser at 488 nm), confocal laser scanning microscopy, and fluorescence microscopy (FITCS filter, 450–500 nm). Several companies offer specific filter sets for GFP/BFP application that, in addition, can be optimized for individual needs (GFP: Zeiss O9, excitation 450–490 nm, beam splitter 510 nm, emission filter >520 nm, Carl Zeiss, Thornwood, NY; High Q FITCS filter set, Chroma Technology Corporation, Brattleboro, VT; BFP: Zeiss 02, excitation maximum 365 nm, beam splitter 460 nm, emission filter >470 nm).

For high-resolution fluorescence microscopy, cells may be grown on glass coverslips, coated 50-mm glass-bottom microwell dishes (MatTek Cor., Ashland, MA), or Lab-Tek chamber slides (Nalgene Nunc International, Thornwood, NY). When possible, microscopy should be performed in medium lacking phenol red or in PBS, which reduces background autofluorescence.

With optimized technical equipment, the detection limits of enhanced GFP mutants were reported to be ~4,000 cytoplasmic molecules or ~700 molecules expressed on the cell surface per cell (Piston, 1997). One hundred thousand CCKAR receptor molecules tagged with GFP provided sufficient signal intensity to be detected and studied easily by fluorescence microscopy (Tarasova et al., 1997a).

The use of a digital CCD camera (e.g., Photometrix, AZ) together with the appropriate software (IPLab Spectrum software; Scanalytics, Vienna, VA) highly improves the sensitivity of signal detection and facilitates image analysis and presentation. Tagging of a variety of proteins has revealed that the resulting fluorescent signal intensity cannot be generalized but may vary from very bright fusions to constructs that give only dim signals. Most likely the weak fluorescence is not caused by a protein conformation unfavorable for GFP function but by factors influencing the expression level of the hybrid proteins (e.g., translation efficiency, protein maturation, degradation). Thus, the signal intensity has to be verified for the individual hybrid protein. However, the experimental procedures (e.g., transfection, filter, and microscope set up) can be verified and easily optimized using a vector expressing unfused GFP or a GFP hybrid protein previously reported.

4. EXPRESSION OF GFP-TAGGED GPCR

4.A. Expression Vectors

The choice of the most appropriate expression vector entails multiple considerations such as the species and cell type to be studied, the desired level of recombinant protein expression, the transfection/infection efficiency requirements, and the duration of expression. Most studies of GPCR function and trafficking are performed in eukaryotic cells at levels approximating wild-type receptor densities. If individual cells are to be examined for relatively short periods of time, then transient transfection by plasmid or viral vectors is satisfactory and convenient to achieve recombinant receptor expression in a fraction of cells for a transient period of time. While transiently transfected cells representing multiple individual clones overcomes the potentially misleading unique peculiarities of individual stably transfected clones, examination of multiple stably transfected clones can overcome this variability and offers selection of clones approximating wild-type receptor density as well as stability and convenience for prolonged and repeated studies.

The necessity for a particular cell type compatible with the receptor under study will also influence the vector selection according to its host range of transfection or infection. Viral vectors offer an increased species and cell-type range, while variation in the viral multiplicity of infection, promoter strength, as well as FACS sorting allows the appropriate level of expression in the suitable cell type as illustrated by the expression of CXCR4/GFP in U-937, Hela, CEM, and NIH/3T3 cells by an MLV-derived retroviral vector (Tarasova et al., 1998).

The important consequences of some of these considerations are exemplified in the results of our studies of CCKAR-GFP. Overexpression of CCKAR-GFP transiently in COS-1 cells resulted in aberrant targeting of the receptor to the cytoplasm, lysosomes, and mitochondria in addition to the appropriate expression in the endoplasmic reticulum, plasma membrane, and endosomes seen in stably transfected NIH/3T3, CHO-K1, and HeLa cell clones expressing fewer receptors. Even among the stably transfected cells, the cell type influenced the level of constitutive internalization with negligible internalization in CHO and HeLa cell clones compared with NIH/3T3 cell clones expressing similar numbers of receptors (Tarasova et al., 1997a).

4.B. Expression Systems

4.B.a. *Transient Expression.* To verify localization and transient expression of the hybrid protein, the use of a cell line that allows high transfection and expression efficiency (e.g., 293 cells, an adenovirus-transformed human embryonic kidney cell line) is straigthforward. In addition to electroporation, reagents for several transfection methods, including calcium phosphate, DEAE-dextran, and liposome transfection, have been developed and are commercially available from a variety of companies along with their recommended protocols (e.g., Clontech, Palo Alto, CA; Gibco BRL, Bethesda, MD). However, the efficiency of transfection may vary for different host cells. Thus, it is advisable to compare and optimize transfection protocols (Ausubel et al., 1994) for the efficiency and expression levels required for the individual application.

This can be accomplished rapidly (for GFP and BFP) using a fluorescence plate reader (e.g., Cytofluor II, PerSeptive Biosystems, Framingham, MA), which can quantitate GFP fluorescence even in living cells (Stauber et al., 1998). A highly effective and controllable approach for gene transfer, especially in cells refractory to other transfection methods, is the use of pseudotyped retroviral particles (Stauber et al., 1998). In general, care should be taken to avoid overexpression of the hybrid protein, which might titrate out cellular interaction partners important for function or result in aberrant protein localization. Thus, the generation of stable cell lines expressing low levels of the GFP-tagged protein is highly recomended for functional studies.

4.B.b. *Stable Expression.* Most of the commercial GFP-expression vectors already contain a marker for drug resistance (e.g., neomycin resistance). In addition, the presence of GFP allows for double selection to establish clonal cell lines. The selection can be accelerated by enrichment of positive GFP-expressing cells using flow cytometry bulk or single-cell sorting without the time-consuming need to select first for drug resistance (Tarasova et al.,

1997a; Stauber et al., 1998). Because the expression of the transgene correlates with the intensity of the GFP signal, the FACS approach allows one to focus on specific cell populations (i.e., high, medium, or low expressors). In addition, GFP helps to monitor continuously the expression of the transgene to detect changes in expression levels caused, for example, by promoter inactivation. In contrast to transient transfections, the fluorescent signal in stable cell lines is generally weak due to low levels of protein expression. Thus, the use of highly fluorescent GFP mutants is essential for fusion protein detection and study. Tagging the CCKAR with GFPsg25 allowed for the generation of stable cell lines displaying enough signal for studying receptor trafficking and localization, whereas tagging with the commonly used GFPS65T only allowed detection in transient transfections (Tarasova et al., 1997a; Stauber et al., 1998).

5. DEVELOPMENT OF A GFP-TAGGED GPCR

A variety of vectors for the expression of GFP in many kinds of hosts are commercially available, and the cloning strategy depends on the vector used. For mammalian cells, the expression of GFP is often under the control of the CMV immediate early promoter. GFP has been expressed as fusions to a variety of proteins (Chalfie et al., 1994; Misteli and Spector, 1997). In many cases, N- or C-terminal GFP fusions assume the localization and functional properties of the fusion partner. However, if the protein of interest contains functional domains at the N or the C terminus, GFP should be attached to the end of the protein less important for function.

The GFP can be fused with equal ease at either its N or C terminus without compromising its fluorescent properties. However, unlike smaller epitopes, fusion of a sizeable protein such as GFP to a GPCR may interfere with both GPCR targeting to the plasma membrane and functional properties, such as ligand affinity and signal tranduction. While the N terminus may be important for ligand interaction and membrane targeting and the C terminus for signal transduction, desensitization, and internalization for some GPCRs, it is not possible to predict the effect GFP fusion will have on these functions. Therefore, the choice and success of N- and C-terminal fusions must be empirically determined.

To date only eight GPCRs fused to GFP have been reported, *odr*-10, *Caenorhabditis elegans* odorant receptor (Sengupta et al., 1996), the β_2-adrenergic receptor (Barak et al., 1997a; Kallal et al., 1998), the cholecystokinin type A receptor (Tarasova et al., 1997a), the α_1-adrenergic receptor (Hirasawa et al., 1997), the chemokine receptors CXCR4 (Deng et al., 1996: Tarasova et al., 1998) and CCR5 (Amara et al., 1997), and a GPCR from *Saccharomyces cerevisiae* (Yun, 1997). All of these receptors were fused to GFP at their C terminus. While the investigators of the other seven receptors did not comment on their potential experience with N-terminal GPCR fusions to GFP, Tarasova et al. (1997a) reported that the CCKAR coupled at its N terminus to GFP did not target to the plasma membrane.

6. FUSION OF GFP TO THE C TERMINUS OF A GPCR

In all instances, the termination codon of the GPCR must be removed before fusion. This can be accomplished either by *in vitro* mutagenesis or polymerase chain reaction (PCR) with primers complementary to the C terminus and without a termination codon. Fusion of the GPCR with GFP must be performed while maintaining the correct reading frame for both proteins. Therefore, the restriction endonuclease sites must be chosen and placed correctly.

The restriction sites must be unique to the polycloning sites of the vector and not present in both the GPCR and GFP. The restriction sites can be placed either by *in vitro* mutagenesis or PCR by incorporating them at the ends of the primers. If PCR is chosen, the restriction endonulease site within the primers should be capped by the number of nucleotides required for efficient cleavage by the chosen restriction endonuclease. Whether *in vitro* mutagenesis or PCR is employed, all resulting constructs should be verified by nucleotide sequencing due to the small but finite infidelity inherent in the polymerases. Alternatively, vectors containing different mutant forms of GFP prepared for either N-terminal or C-terminal fusion (in all three reading frames in the case of EGFP) are commercially available (Clontech Laboratories Inc, Palo Alto, CA; Quantum Biotechnologies Inc., Quebec, Canada). A protocol for fusion of the C terminus of the CCKAR to GFP is given below, illustrating the use of PCR to prepare both the CCKAR and GFP for fusion in a proprietary neomycin-containing, eukaryotic-expressing, plasmid vector, pCDL-SRα/Neo (Takebe et al., 1988) (Fig. 12.2).

1. PCR amplification of the rat CCKAR coding sequence (cds)
   - 5′-primer: (5′-GACTAGCCG-GAATTC-ATGGACGTGGTCGACA-GCCTT-3′) containing a 9-bp cap, *Eco*RI restriction site, and the first 21 bp of the CCKAR-specific 5′cds (12.5 μM), 2 μl
   - 3′-primer: (5′-ACTGACTAG-TCTAGA-GGGGGGTGGAGCGAG-GTGCT-3′) containing a 9-bp cap, *Xba*I restriction site, and the last 21 bp of the CCKAR-specific 3′cds without the TGA stop codon (12.5 μM), 2 μl

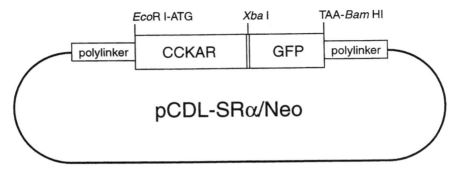

Figure 12.2. Schematic of the CCKAR-GFP construct in the plasmid vector pCDL-SRα/Neo.

- 0.2 pm of RCCKAR, 1 μl
- dNTPs (25 μM each), 5 μl
- 10× standard PCR buffer (Perkin-Elmer/Cetus), 5 μl
- H₂O, 33 μl
- *Taq* polymerase (Perkin Elmer/Cetus, Foster City, CA), (2.5 U)/pfu polymerase (Stratagene, La Jolla, CA) (0.25 U), 2 μl
- 20 thermocycles: 94°C, 30 seconds; 60°C, 15 seconds; 72°C, 2 minutes with final cycle for 15 minutes

2. PCR amplification of phGFP-(F64L, T65C, I167T) coding sequence (cds)

   - 5′-primer: (5′-ACTGACTAG-TCTAGA-ATGGTGAGCAAGGGC-GAGGAGC-3′) containing a 9-bp cap, *Xba*I restriction site, and the first 22 bp of the GFP-specific 5′cds (12.5 μM), 2 μl
   - 3′-primer: (5′-ACTGACCGC-GGATCC-TTACTTGTACAGCTCGT-CCATGC-3′) containing a 9-bp cap, *Bam*HI restriction site, and the last 23 bp of the GFP-specific 3′cds including a TAA stop codon (12.5 μM), 2 μl
   - 0.2 pm of GFP, 1 μl
   - dNTPs (25 μM each), 5 μl
   - 10× standard PCR buffer (Perkin-Elmer/Cetus), 5 μl
   - H₂O, 33 μl
   - *Taq* polymerase (2.5 U)/pfu polymerase, (0.25 U), 2 μl
   - 20 thermocycles: 94°C, 30 seconds; 60°C, 15 seconds; 72°C, 2 minutes with final cycle for 15 minutes.

3. Restriction digestion of the CCKAR and GFP PCR products and pCDL-SRα/Neo

   - The CCKAR and GFP PCR products were each purified using Qiaquick columns (Qiagen Corp, Santa Clarita, CA) and eluted in 50 μl H₂O
   - Double digests were performed using either 10× *Eco*RI buffer for the CCAKR and pCDL-SRα/Neo or 10× *Bam*HI buffer for GFP, 7 μl
   - *Eco*RI 100 U plus *Xba*I 100 U; or *Xba*I 100 U plus *Bam*HI 50 U were added to the CCKAR and pCDL-SRα/Neo or GFP, respectively
   - Samples were digested at 37°C for 2 hours
   - Digested CCKAR, pCDL-SRα/Neo, and GFP were each purified again on a Qiaquick column

4. Ligation of CCKAR into pCDL-SRα/Neo

   - The *Eco*RI/*Xba*I–digested CCKAR is ligated to *Eco*RI/*Xba*I–digested pCDL-SRα/Neo in a 3:1 ratio using 20 U T4 DNA ligase in 10 μl at 20°C for 2 hours

5. Digestion of pCDL-SRα/Neo/CCKAR following bacterial transformation, amplification, and plasmid purification

   - pCDL-SRα/Neo/CCKAR is digested with *Xba*I 100 U plus *Bam*HI 50 U and processed as described in Step 3 above for GFP

6. Ligation of GFP to pCDL-SRα/Neo/CCKAR C terminus
   - *Xba*I/*Bam*HI–digested GFP is ligated to *Xba*I/*Bam*HI–digested pCDL-SRα/Neo/CCKAR in a 3:1 ratio as described in Step 4
7. The ligated product pCDL-SRα/Neo/CCKAR-GFP is transformed and amplified in *Escherichia coli,* plasmid purified, and sequenced

Due to the relatively large size of GFP (238 amino acids), it is remarkable that a GPCR-GFP fusion protein expresses and functions normally. In fact, unlike smaller epitopes, GFP fused to the N terminus of the CCKAR results in cytoplasmic rather than plasma membrane receptor targeting. Nonetheless, C-terminal fusions of GPCR to GFP have been shown to target and function similar to native receptors for the β_2-adrenergic receptor (Barak et al., 1997a; Kallal et al., 1998), the CCKA receptor (Tarasova et al., 1997a), and the α_1-adrenergic receptor (Hirasawa et al., 1997). Unfortunately, it is not possible to predict the effect that GFP fusion will have on receptor function and therefore it is necessary to demonstrate that the GPCR-GFP fusion protein functions normally.

Typically, this requires pharmacological assessment of agonist and antagonist interaction and appropriate assays for assessing G protein coupling and ligand-induced internalization. For example, CHO cells stably expressing CCKAR-GFP (Tarasova et al., 1997a) behaved similar to CHO cells expressing

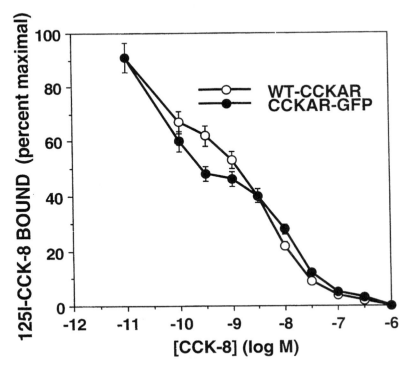

Figure 12.3. Displacement of ^{125}I-CCK-8 binding to CHO cells stably expressing wild-type CCKAR and CCKAR-GFP. Data are presented as the percent saturable binding in the absence of unlabeled ligand.

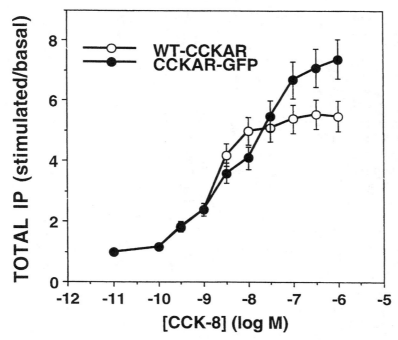

Figure 12.4. Ability of CCK-8 to increase total inositolphosphates (IP) in CHO cells stably expressing wild-type CCKAR and CCKAR-GFP. Data are shown as the CCK-8–stimulated increase in total inositolphosphates divided by the unstimulated basal total inositolphosphates.

wild-type receptor on the basis of receptor density (1.75 ± 0.63 versus $1.70 \pm 0.29 \times 10^5$ receptors/cell, respectively), ligand affinity (5.93 ± 2.19 versus $3.78 \pm 0.71 \times 10^9$ M, respectively) (Fig. 12.3), CCK-8 stimulated–increase in total inositol phosphates (5.4 ± 1.0 versus 7.36 ± 3.3-fold, respectively) (Fig. 12.4), and ligand-induced internalization (data not shown). In the case of the β_2-adrenergic receptor fused to GFP, agonist and antagonist binding, adenylyl cyclase stimulation, receptor phosphorylation, and ligand internalization studies were indistinguishable from wild-type receptor when expressed in HEK 293 cells (Barak et al., 1997a; Kallal et al., 1998).

7. APPLICATIONS OF GFP-TAGGED GPCR

Tagging of GPCRs represents a major technical advance in the study of receptor localization, trafficking, and sorting from the ligand in living cells in real time. Confocal laser scanning microscopy allows for the identification of intracellular compartments bearing tagged receptor molecules when used in combination with a variety of fluorescent organelle markers. Many organelle markers have been developed that can be applied directly to living cells and make it no longer necessary to fix, permeablize, and stain with antibodies to visualize the organelles.

A variety of microscopes can be used for GPCR-GFP trafficking studies. We have utilized a Zeiss inverted confocal laser scanning microscope (CLSM) LSM 410, although other brands of inverted CLSMs can serve the purpose well. If an inverted microscope is not available, the observation of the cells can be performed with a conventional CLSM, although it is less convenient and limits possible manipulations with live cells.

If an inverted microscope is used, the cells are grown on either Lab-Tek chamber slides (two or four chamber) (Nalgene Nunc International, Thornwood, NY) or coated 50-mm cover glass-bottom microwell dishes (MatTek Corp., Ashland, MA). In case of the use of conventional CLSM, the cells can be grown on any glass support. We have utilized Nunc chamber slides with removable walls. The pregrown cells treated with necessary markers are rinsed, the chamber walls and the sealing gaskets are removed, and the cells adherent to the glass bottom are covered with a drop of PBS and a coverslip. The excess of fluid between the slide and the coverslip is removed with a pipette tip. When observing the cells, precautions should be taken to avoid contact of lens oil with the medium. Avoid using nail polish to seal the coverslip because it can interfere with GFP fluorescence.

Growing the cells in medium without phenol red is important because phenol red produces significant background fluorescence in a broad range of wavelengths.

8. LOCALIZATION OF GFP-TAGGED GPCR

The following procedures describe labeling with the variety of fluorescent markers that can be used for co-localization with the GFPsg25 mutant used in our study (Fig. 12.5). Within the limitations required by the GFP mutants and fluorescent markers chosen, the exact choice of microscopy conditions depends to some degree on the lasers and filters installed on the microscope to be used. GFP fluorescence was excited with a 488-nm argon/krypton laser and emission detected with a 515–540 nm band pass filter. For all markers described below, we have found a 568-nm helium/neon laser (for excitation) and a 590-nm cutoff filter (for fluorescence detection) to be optimal. The pinhole that defines the width of the layer under observation should be kept as small as possible for co-localization experiments. Close positioning of organelles (one on top of the other) frequently leads to false co-localization of fluorescent labels residing in different compartments. The wider the layer of observation, the higher the probability of such overlays. The lower limit for the pinhole is defined by the intensity of fluorescence and the sensitivity of detection. When using a 63× oil immersion lens and working with stably transfected cell lines, we were unable to reduce the pinhole below 20 without compromising the intensity of the fluorescent signal. The range between 25 and 40 appeared to be optimal.

8.A. Cell Surface Membranes

Concanavalin A (ConA) labels only cell surface glycoproteins when the cells are applied at temperatures below 10°C. For the labeling, the cells were washed

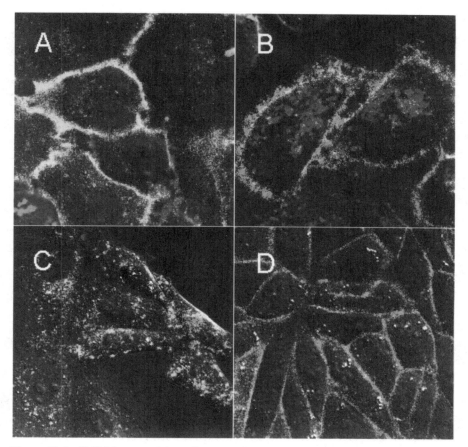

Figure 12.5. CLSM images of the cells expressing GFP-tagged GPCRs. **A:** Hela cells expressing CXC chemokine receptor 4 (in green) after labeling the cell surface with tetramethyl rhodamine ConA (red). **B:** Hela cells expressing CC-chemokine receptor 5 (green) with endosomes labeled with tetramethyl rhodamine transferrin (red). **C,D:** Co-localization of CCKAR-GFP with the ligand, Cy3.29–CCK-8 (red) in CHO cells 15 minutes after addition of the peptide (C) or 60 minutes after removal of the ligand (D). (See color insert.)

with cold PBS and exposed for 2 minutes to tetramethyl rhodamine ConA solution in PBS (10 μg/ml). After rinsing with cold PBS, the cells are immediately observed under CLSM. If observation cannot be done within minutes after labeling, the cells should be fixed briefly (for 5 minutes) with cold 4% paraformaldehyde to prevent internalization of ConA.

8.B. Early Endosomes

The majority of cells in cell cultures express transferrin receptors known to undergo reversible spontaneous endocytosis. Thus, transferrin derivatives can be used for labeling of early endosomes. Tetramethyl rhodamine–labeled transferrin represents a convenient marker for cells expressing GFP-tagged

GPCRs. Transferrin, bound to its receptor, recycles together with the GFP-tagged receptor, and therefore labeled transferrin should be present in the medium during observation of the cells so that endosomes remain labeled at the highest possible level. It should be noted that endosomes exist in constant movement and can travel a distance equal to their diameter in less than 5 seconds. Therefore, when observing the cells with CLSM, it is preferable to scan with two detectors simultaneously. If red and green fluorescence are not detected at the same time, the co-localization with the marker can be underestimated. Alternatively, the cells may be briefly fixed with cold 4% formaldehyde for 5 minutes to prevent movement of the vesicles during observation.

8.C. Late Endosomes and Lysosomes

LysoTracker Red DND-99 (Molecular Probes Inc., Eugene, OR) is a cationic dye concentrated in acidic vesicles. In the cells pretreated with the dye, it accumulates in late endosomes and lysosomes after a prolonged incubation. For labeling of late endosomes and lysosomes, the cells are incubated with the dye (supplied as a solution in DMSO) diluted 1:10,000 in PBS for 15 minutes at 37°C. The dye is rinsed away and the cells left in PBS in the incubator for an additional 30–60 minutes. It should be noted that GFP fluorescent intensity decreases at low pH (pH 5.5–7.0); however, if GFP is tagged to the C terminus of a GPCR, it should be exposed to the physiologic pH of the cytoplasm.

8.D. Endoplasmic Reticulum

For labeling of endoplasmic reticulum, the cells are incubated with 1 mM solution of the hexyl ester of rhodamine B (Molecular Probes Inc.) in PBS for 20 minutes at 37°C, rinsed with PBS, and used for observation. We have noticed that this dye, when illuminated with the laser beam, undergoes photochemical changes accompanied by broadening of emission spectra. To avoid this side effect, which may result in spectral overlap with GFP, it is important to avoid scanning the same area with the lasers more than two times.

8.E. Mitochondria

We have found both MitoTracker Red CMSRos and dihydrorhodamine 6G (Molecular Probes Inc.) to work efficiently in labeling mitochondria. Dihydrorhodamine 6G is not fluorescent until it is oxidized in living cells. Working with fresh solutions of the dye is important because it can undergo spontaneous oxidation with molecular oxygen, which leads to increased background. Fresh dihydrorhodamine 6G, 100 nM solution in PBS, is prepared from a stock in DMSO and used for the treatment of the cells for 30 minutes in a CO_2 incubator. After rinsing, the cells are ready for observation. Alternatively, 50 nM MitoTracker CMSRos in PBS is incubated with the cells for 40 minutes, also in a CO_2 incubator. The cells are then rinsed and analyzed with CLSM.

8.F. Translocation of GFP-Tagged GPCR

The translocation of GFP-tagged proteins can be characterized either by observation of changes in the localization of the tagged receptor with time in a single cell or a group of cells, or, alternatively, by quantitative estimation of changes in the degree of tagged receptor co-localization with an organelle marker determined for a large number of cells for each time point. The first method, the analysis of the same group of cells without moving the sample from a microscope stage, is more suited for fast processes, occurring within 10 minutes or less (e.g., receptor internalization). The second approach, using quantitative microscopy, is more suitable for the study of receptor trafficking processes that require longer cell incubations (e.g., receptor recycling, receptor translocation from endoplasmic reticulum, or receptor-ligand sorting).

8.G. Single-Cell Microscopy

Although single-cell experiments are visually appealing, they are difficult to perform due to multiple technical limitations. Bleaching of the fluorescent markers during multiple scans with lasers are difficult to avoid. We were unable to perform more than seven scans of stably transfected cells without significant loss of GFP fluorescence. Organelle markers and synthetic fluorescent tags are usually even less light stable than GFP. Nevertheless, five to seven scans still can allow for determination of a time span of receptor translocation, especially if preliminary experiments are performed to estimate the optimal time intervals between the scans. The development of two-photon microscopy should overcome the problem of rapid bleaching and make single-cell microscopy much easier to perform.

We have found that the time needed for receptor internalization and recycling in different cells can differ even within the same colony of cells. For estimation of a mean time, analysis of many cells is necessary, which makes the whole procedure very laborious.

Minor vibrations of the microscope table (e.g., due to nearby centrifuges, pumps, or other vibrating equipment) and changes in the temperature of the sample during observation result in the changes in the focal plane that can interfere with the experiment and should be avoided.

Conducting the studies under conditions that are optimal for cell function and growth is always desirable. This is best achieved through the use of a heated microscope stage equipped with a miniature CO_2 chamber to control and vary experimental conditions during observations. Such stages are commercially available now; however, their application is limited by their high cost.

9. QUANTITATIVE MICROSCOPY

Determination of the degree of co-localization of a tagged protein with an organelle marker constitutes one of the major applications of quantitative microscopy. We have used co-localization of GPCR-GFP with tetramethyl rhodamine ConA for monitoring recycling of GPCRs to the cell surface after ligand-induced internalization. This approach allowed the observation of significant differences in recycling patterns between different GPCRs (Fig. 12.6). Variations were also observed for the same receptor expressed in different cell types, although they were not as dramatic. The changes in the rate of endocytosis can be estimated with the help of the marker for early endosomes, tetramethyl rhodamine transferrin. This technique conveniently permits the distinction between spontaneous and ligand-induced endocytosis (Tarasova and Michejda, unpublished results).

Cell labeling for studies of the dynamics of receptor translocation are done essentially in the same way as described for localization studies except that the cells are pretreated with a the protein synthesis inhibitor, cycloheximide (25 μg/ml), for at least 15 minutes before the experiment as well as during the experiment to prevent *de novo* synthesis of receptor molecules. Several dishes (one for each time point) are set up, pretreated with cycloheximide, and simultaneously exposed to the agent thought to induce receptor translocation. At various times the dishes are removed from the incubator and used for observation.

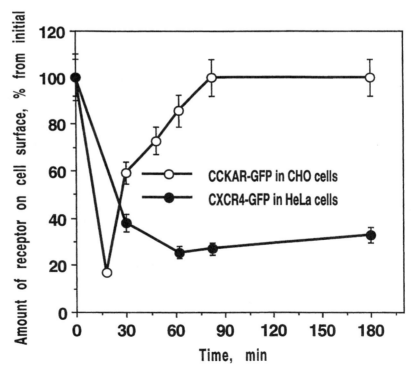

Figure 12.6. Recycling of CCKAR and CXCR4. The cells were exposed to 100 nM CCK-8 and SDF-1a, respectively, for 10 minutes at 37°C, rinsed, and left in the incubator in medium containing cycloheximide for varying time intervals. The relative amount of receptor molecules on the cell surface was determined by co-localization with the cell surface marker tetramethyl rhodamine ConA. Note significantly different efficiency of CCKAR and CXCR4 recycling.

Special precautions should be taken to perform the labeling with organelle markers in the same conditions for all time points. Images of at least 80–100 cells should be taken for each point (usually 10-12 images of the cells grown to near confluence).

Image processing software that allows the calculation of the degree of co-localization has been introduced by several companies (e.g., Optimas Corp. Bothell, WA; Universal Imaging Corp., West Chester, PA). We have used the Zeiss LSM program mainly because it does not require the retrieval and refor-matting of the images that were obtained and stored using the same program. The procedure is performed in the following way:

1. Load GFP image in the green (G) channel, an organelle marker image in the red (R) channel, and switch to the RGB channel to obtain an overlay.

2. Choose "Colocalization" in the "Function" section of the main menu.

3. Press the "Mark area" button on the screen and then the "Draw" button. Image display will now show the diagram on which Y and X axes represent the intensity of green and red fluorescence, respectively. The areas with equal red and green fluorescence will be represented by the dots on the diagonal. Define, with the help of the left mouse button, the dots on the diagram that will correspond to the areas having comparable GFP and marker fluorescence, and press the middle mouse button inside the defined area. Three overlaid images will now appear on the image display: GFP image in green, the marker in red, and co-localization mask in blue (it can be viewed separately by going into the blue [B] channel).

4. Choose the green (G) channel on the image control panel.

5. Select "Area Measure" in the "Function" section of the main menu. Three numbers will appear on the screen. The first (M) is the one to be used. It is a mean intensity of the fluorescence on the image that is currently loaded on the image display. "S" is deviation, and "A" is the area of the image.

6. Choose the blue (B) channel on the image control panel. It corresponds to a co-localization mask.

7. Repeat operations from Step 5. The mean intensity obtained now will correspond only to the co-localization areas. By dividing it by the number obtained for the green image in Step 5, one estimates the ratio of GFP-GPCR fluorescence intensity in the co-localization areas to the total GFP-GPCR fluorescence on the image.

To obtain consistent and reproducible results, it is important to make identical selections of the cut-off values for the green and red fluorescence on Step 3 for all measurements within the same experiment. Unfortunately, the current version of LSM software does not allow storage of the selection and subsequent application toward all analyzed images. Marking of the area on the intensity diagram needs to be done by hand for each new image examined. However, simple marks on the screen help to make selections fast and in the same manner.

10. RECEPTOR AND LIGAND SORTING

Labeling receptors with GFP is presently the only technique available for real-time simultaneous observation of receptor and ligand trafficking. Commercially available reactive fluorescent dyes allow for relatively simple preparation of a fluorescent ligand. The fluorescent tag for the modification of the ligand should have the highest possible fluorescence intensity; chemical, metabolic, and light stability; and a spectrum that does not overlap with GFP. Among currently available labels that have good spectral separation from GFP, Cy3.29 has the highest fluorescence intensity, but the dye fluorescence fades in lysosomes. New Alexa dyes that became available recently from Molecular Probes Inc. are reported to have higher quantum yields than Cy3.29 and better chemical stability. Depending on the chemistry and structure–function relationship of the ligand, the label can be introduced either in solution or during the solid-phase synthesis of the peptide. We have labeled the N termini of gastrin and CCK-8 with Cy3.29 succinimidyl ester using a mixture of acetonitrile and 0.2 M sodium bicarbonate buffer, pH 9.0, as a solvent (Czerwinski et al., 1995) and have added rhodamine B to a number of peptides during the synthesis on the resin (N. Tarasova, unpublished results).

Observation and trafficking studies with fluorescent ligand and GFP-tagged receptors can be performed using the same conditions and image processing techniques as described for receptor translocation. The ligand will substitute for the organelle marker. Initial incubation with the ligand is usually performed for a period sufficient to visualize receptor and ligand internalization (5–15 minutes). After that, the ligand is rinsed off and the incubation is continued. We were able to observe rather efficient sorting of CCKAR-GFP and Cy3.29–CCK-8 with a half-time of 25 minutes (Tarasova et al., 1997a).

11. SUMMARY

GFP has already proven to be an extremely useful tool that is applicable to virtually all fields of biology. Fortunately, the improved properties of the enhanced mutants of GFP have extended its usefullness for tagging proteins expressed at low levels, such as GPCRs. Remarkably, tagging of GPCRs has, for at least the small group of GPCRs reported thus far, not altered their pharmacological and biochemical behavior and therefore indicates that GFP should be a powerful tool for understanding the biology of GPCRs in living cells in real time.

Already, GFP tagging of GPCRs has begun to reveal its potential for localization and trafficking studies within cells at a level that have not been practical or even possible in the past. The development of different spectral properties of other GFP mutants such as BFP already permits observations of co-localization of different molecules. However, the spectral properties of GFP and BFP, as well as the likelihood of even better mutants in the future now that the crystal structure of GFP has been determined, raise the probabilty of real-time observations of actual interactions between GPCRs themselves with known molecular interactants through the application of fluorescent resonance energy transfer experiments. The recent report of GFP-tagged arrestin retaining functional translocation and interaction with the C terminus of the β-adrenergic receptor (Barak et al., 1997b) suggests a tempting opportunity for just such an experiment utilizing fluorescence energy transfer between the BFP-tagged C terminus of the receptor and a GFP-tagged arrestin. Although the field of GPCR is already quite advanced, exciting experiments such as these as well as countless others suggest that GFP will bring new light to the end of the tunnel.

ACKNOWLEDGMENTS

We thank Eric H. Hudson for assistance with CLSM and Dr. Margherita Rosati for preparation of cells expressing CCR5-GFP and CCR4-GFP. Research was sponsored in part by National Cancer Institute, DHHS, under contract with ABL.

REFERENCES

Amara A, Le Gall S, Schwarz O, Salamero J, Montes M, Loetscher P, Baggiolini, M, Virelizier J-L, Arezana-Seisdedos F (1997): HIV-1 coreceptor downregulation as antiviral principle. J Exp Med 186:139–146.

Ausubel FM, Brent R, Kingston RE, Moore DD, Seidman JG, Smith JA, Struhl K (1994): Current Protocols in Molecular Biology. New York: John Wiley and Sons, Inc.

Barak LS, Ferguson SS, Zhang J, Martenson C, Meyer T, Caron MG (1997a): Internal trafficking and surface mobility of a functionally intact β₂-adrenergic receptor-green fluorescent protein conjugate. Mol Pharmacol 51:177–184.

Barak LS, Ferguson SS, Zhang J, Caron MG (1997b): A beta-arrestin/green fluorescent protein biosensor for detecting G protein–coupled receptor activation. J Biol Chem 272: 27497–27500.

Chalfie M, Kain S R (1998): GFP: Green Fluorescent Protein Strategies, Applications, and Protocols. New York: John Wiley and Sons, Inc..

Chalfie M, Tu Y, Euskirchen G, Ward WW, Prasher DC (1994): Green fluorescent protein as a marker for gene expression. Science 263:802–803.

Cody CW, Prasher DC, Westler WM, Prendergast FG, Ward WW (1993): Chemical structure of the hexapeptide chromophore of *Aequoria* green-fluorescent protein. Biochemistry 32:1212–1218.

Cormack BP, Valdivia RH, Falkow S (1996): FACS-optimized mutants of the green fluorescent protein (GFP). Gene 173:33–38.

Cubitt AB, Heim R, Adams SR, Boyd AE, Gross LA, Tsien RY (1995): Understanding, improving and using green fluorescent proteins. Trends Biochem Sci 20:448–455.

Czerwinski G, Wank SA, Tarasova NI, Hudson EA, Resau JH, Michejda CJ (1995): Synthesis and properties of three fluorescent derivatives of gastrin. Lett Peptide Sci 1:235–242.

Deng H, Liu R, Ellmeier W, Choe S, Unutmaz D, Burkhart M, Marzio PD, Marmon S, Sutton RE, Hill CM, Davis CB, Peiper SC, Schall TJ, Littman DR, Landau NR (1996): Identification of a major co-receptor for primary isolates of HIV-1. Nature 381:661–666.

Gerdes HH, Kaether C (1996): Green fluorescent protein: Applications in cell biology. FEBS Lett 389:44–47.

Heim R, Cubitt AB, Tsien RY (1995): Improved green fluorescence. Nature 373:663–664.

Heim R, Tsien RY (1996): Engineering green fluorescent protein for improved brightness, longer wavelengths and fluorescence resonance energy transfer. Curr Biol 6:178–182.

Hirasawa A, Sugawara T, Awaji T, Tsumaya K, Ito H, Tsujimoto G (1997): Subtype-specific differences in subcellular localization of alpha1-adrenoceptors: Chlorethyl-clonidine preferentially alkylates the accessible cell surface alpha1-adrenoceptors irrespective of the subtype. Mol Pharmacol 52:764–770.

Kallal L, Gagnon AW, Penn RB, Benovic JL (1998): Visualization of agonist-induced sequestration and down-regulation of a green fluorescent protein-tagged beta2-adrenergic receptor. J Biol Chem 273:322–328.

Li X, Zhang G, Ngo N, Zhao X, Kain SR, Huang CC (1997): Deletions of the *Aequorea victoria* green fluorescent protein define the minimal domain required for fluorescence. J Biol Chem 272:28545–28549.

McClintock TS, Landers TM, Gimelbrant AA, Fuller LZ, Jackson BA, Jayawickreme CK, Lerner MR (1997): Functional expression of olfactory-adrenergic receptor chimeras and intracellular retention of heterologously expressed olfactory receptors. Brain Res 48:270–278.

Misteli T, Spector DL (1997): Application of the green fluorescent protein in cell biology and biotechnology. Nature Biotech 15:961–964.

Muldoon RR, Levy JP, Kain SR, Kitts PA, Link CJ Jr (1997): Tracking and quantitation of retroviral-mediated transfer using a completely humanized, red-shifted green fluorescent protein gene. Biotechniques 22:162–167.

Piston DW (1997): Personal communication. Clontechniques 12:22.

Prasher DC, Eckenrode VK, Ward WW, Prendergast FG, Cormier MJ (1992): Primary structure of the *Aequorea victoria* green-fluorescent protein. Gene 15:229–233.

Sengupta P, Chou JH, Bargmann CI (1996): Odr-10 encodes a seven transmembrane domain olfactory receptor required for responses to the odorant diacetyl. Cell 84:899–909.

Stauber RH, Horie K, Carney P, Hudson EA, Tarasova NI, Gaitanaris GA, Pavlakis GN (1998): Development and applications of enhanced Green Fluorescent Protein mutants. Biotechniques 24:462–471.

Takebe Y, Seiki M, Fujisawa JI, Hoy P, Yokota K, Arai KI, Yoshida M, Arai N (1988): SR alpha promoter: An efficient and versatile mammalian cDNA expression system composed of the simian virus 40 early promoter and the R-U5 segment of human T-cell leukemia virus type 1 long terminal repeat. Mol Cell Biol 8:466–472.

Tarasova NI, Stauber RH, Michejda CJ (1998): Spontaneous and ligand-dependent trafficking of CXC-chemokine receptor 4. J Biol Chem 273:15883–15886.

Tarasova NI, Stauber RH, Choi JK, Hudson EA, Czerwinski G, Miller JL, Pavlakis GN, Michejda CJ, Wank SA (1997a): Visualization of G-protein receptor trafficking with the aid of green fluorescent protein: Endocytosis and recycling of cholecystokinin receptor type A. J Biol Chem 272:14817–14824.

Tarasova NI, Wank SA, Hudson EA, Romanov VI, Czerwinski G, Resau JH, Michejda CJ (1997b): Endocytosis of gastrin in cancer cells expressing gastrin/CCK-B receptor. Cell Tissue Res 287:325–333.

Yun CW, Tamaki H, Nakayama R, Yamamoto K, Kumagai H (1997): G-protein coupled receptor from yeast *Saccharomyces cerevisiae*. Biochem Biophys Res Commun 240:287–292.

CHAPTER 13

THE ANALYSIS OF POST-TRANSCRIPTIONAL REGULATION OF THE EXPRESSION OF G PROTEIN–COUPLED RECEPTORS

BABY G. THOLANIKUNNEL and CRAIG C. MALBON

Regulation of G Protein–Coupled Receptor Function and Expression,
Edited by Jeffrey L. Benovic.
ISBN 0-471-25277-8   Copyright © 2000 Wiley-Liss, Inc.

I. INTRODUCTION

G protein–coupled receptors (GPCRs) transduce vital physiology, are regulated via hormones, neurotransmitters, and therapeutic drugs, and transduce sensory perception, such as visual excitation and olfaction (Birnbaumer et al., 1990; Dohlman et al., 1991; Hadcock and Malbon, 1993). Molecular cloning has revealed the primary sequence of more than 300 members of this superfamily of GPCR (GeneBank, 10/96). Two parameters are paramount in determining the responsiveness of these biological processes in normal and pathophysiological states; receptor abundance and function. Many laboratories devoted to the question of how receptor expression is regulated in cell signaling have employed β-adrenergic receptors (βARs) as prototypic of GPCR, and research to date supports the value of these molecules as targets for analysis of the properties of GPCR superfamily members (Collins et al., 1991; Malbon and Hadcock, 1993; Bahouth and Malbon, 1994).

I.A. Regulation of Expression of GPCRs

The expression of GPCR can be modulated either in a positive (up-regulation) or negative (down-regulation) manner. βARs provide excellent examples, their expression being up-regulated by glucocorticoids (β_2AR, β_3AR), retinoids (β_2AR), thyroid hormones (β_1AR), β-agonists (β_2AR via cAMP response element), androgens (β_2AR), and cytokines (β_2AR) while down-regulated by chronic β-agonists (β_1AR, β_2AR, β_3AR), glucocorticoids (β_1AR), thyroid hormones (β_2AR), and various disease states such as congestive heart disease (β_1AR, β_2AR). In most instances, the physiological response reflects the changes in receptor expression (e.g., the cAMP response to β-agonists reflects increased expression of β_2AR four- to fivefold by excess glucocorticoids (Hadcock and Malbon, 1991) or hypothyroidism (β_2AR; Malbon, 1980). At the level of mRNA, steady-state levels of GPCR mRNA are typically low (~1 attomol/μg total cellular RNA) and parallel levels of receptor expression (Hadcock et al., 1989c), suggesting that *in vivo* receptor expression is tightly coupled with mRNA levels. Receptor mRNA levels are dictated by transcriptional regulation and mRNA degradation. In some instances, alterations in receptor expression can be described in some detail. The glucocorticoid-induced elevation of β_2AR expression, for instance, represents an increased synthesis of receptor protein resulting from steroid-induced elevation (four- to eightfold) of β_2AR mRNA levels (Hadcock and Malbon, 1988a) via activation of receptor gene transcription (Collins et al., 1988; Malbon and Hadcock, 1989). Deletion of a consensus GRE in the most proximal, 5′ untranslated region (UTR) domain of the hamster β_2AR abolishes the transcriptional activation (Malbon and Hadcock, 1988). Glucocorticoids suppress, rather than activate, the β_1AR gene, decreasing receptor synthesis, mRNA levels, and rates of gene transcription (Guest et al., 1990; Zhong and Minneman, 1993; Kiely et al., 1994; Bahouth et al., 1994). A consensus GRE in the 5′ UTR of the human β_1AR gene provides glucocorticoid-induced suppression of reporter gene (luciferase) expression, and its deletion abolishes the regulation by steroid. This complex regulation of receptor expression is not confined to βARs, as similar observations have been reported for α_1ARs. Glucocorticoids increase αARs in DDT_1MF-2 cells via gene activation (Sakaue and Hoffman, 1991), and increased expression in response to elevated cAMP implicates a CRE in the promoter region of the $\alpha_{2A}AR$ gene (Shilo et al., 1994).

Much less is known about the degradation of rare mRNA species, like those of βARs (Hadcock et al., 1989b). In DDT_1MF-2 smooth muscle cells, C6 glioma, and hepatocytes, agonist treatment leads to a decline (50%–80%) in receptor mRNA levels (i.e., agonist-induced down-regulation of mRNA) (Hadcock and Malbon, 1988b; Collins et al., 1989). This decline occurs at the post-transcriptional level, reflecting a change in the stability of pre-existing mRNA (Hadcock et al., 1989b; Hadcock and Malbon, 1993). How does agonist-induced down-regulation occur at the molecular level? A recent lead has come from the analysis of RNA-binding proteins recognizing receptor mRNAs (Port et al.,1992) and the identification of a 35,000 molecular weight species by ultraviolet (UV) cross-linked label transfer from uniformly radiolabeled, capped, and polyadenylated mRNAs (Port et al., 1992; Huang et al., 1993). Analysis of the destabilization, the nature of the 35,000 molecular weight species, and the determinants for protein–nucleic acid interactions are important goals toward the understanding of agonist-induced down-regulation of receptor expression.

Efforts were extended from the observation that chronic stimulation of β_2AR in cells results in a reduction in the expression of receptor, as measured by either radioligand binding or quantitative immunoblotting. DNA-excess solution hybridization (and later a riboprobe approach) has been adapted to assay β_2AR mRNA levels in cells displaying agonist-induced down-regulation (Hadcock et al., 1989c; Bahouth et al., 1994). GPCR (e.g., β_2AR and $\alpha_{1b}AR$) mRNA levels have been quantified and found to be "rare" (i.e., steady-state levels at 1–2 attomol/μg total cellular RNA) (Hadcock and Malbon, 1988a,b). In response to agonist treatment, β_2AR mRNA levels decline by 50%–60% in DDT_1MF-2 smooth muscle cells and by 80% in C6 glioma cells (Hough and Chuang, 1990). The response (reversible with antagonist) is nearly maximal within 12 hours in the former (Hadcock and Malbon, 1988b) and less than 2 hours in the latter (Hough and Chuang, 1990).

I.B. Transcriptional Versus Post-Transcriptional Regulation of Receptor Expression

The reduction in β_2AR mRNA induced by agonist and cAMP could reflect either transcriptional repression or destabilization or both. Analysis of mRNA half-lives revealed a marked decline in the stability of the β_2AR message in cells persistently stimulated by agonist (Hadcock et al., 1989a,b; Bouvier et al., 1989). Two distinct approaches have been attempted historically, with differing levels of success. Reconstitution of cell extracts with full-length, uniformly labeled, capped, and polyadenylated mRNAs has been achieved with mRNAs from a variety of receptors, including those that display agonist-induced destabilization of receptor mRNA and those that did not. We have examined the ability of S100, polysome-enriched, and nuclear fractions prepared from cells in culture to degrade labeled mRNA in an *in vitro* system described earlier (Bernstein et al., 1992). Ross and colleagues have reported the use of a similar system that could discriminate among rapidly (c-fos) as compared with slowly (β-globin) turning-over transcripts (see Brewer, 1991; Bernstein et al., 1992). In our laboratory we devoted considerable effort to this approach, but were disappointed with the performance, even when conditions had been optimized. Although discriminating among rapidly turning-over transcripts (c-fos and c-myc) and more stable ones (β-globin), between polyadenylated or capped compared with noncapped and nonpolyadenylated, as well as between β_2AR and α_{1b}AR transcripts, the *in vitro* reconstitution with fractions from agonist down-regulated compared with naive cells did not reproducibly display accelerated degradation of target mRNAs. The degradation rates *in vitro,* not unexpectedly, were much greater than *in vivo*. Attempts to reduce the apparent rates of degradation have been successful (increasing salt concentration and addition of carrier RNA), but reproducibility remains a formidable obstacle to routine use of this assay to explore degradation of GPCR mRNAs.

I.C. Identification of a β-Adrenergic Receptor mRNA-Binding Protein

The second approach is aimed at identifying RNA-binding proteins regulating or participating in destabilization of β_2AR mRNA (βARB protein). The operating premise was that, having constructed templates allowing preparation of uniformly labeled, capped, and polyadenylated transcripts for β_2AR, β_1AR, and α_{1b}AR, as well as β-globin, c-fos, and c-myc, we could initiate a search for RNA-binding proteins "specific" to the destabilization response. A candidate protein would be expected to discriminate between the transcripts that undergo destabilization and those that do not. The approach to identification was based on UV-catalyzed cross-linking and label transfer (for references, see Port et al., 1992). Labeled transcripts are incubated with S100, polysome-enriched, and nuclear fractions and then cross-linked by UV irradiation (Port et al., 1992; Huang et al., 1993). The samples are digested with RNAase and the proteins subjected to SDS-PAGE and made visible by autoradiography. Several proteins (M_r 52,000-doublet, ~70,000, and 85,000) are observed in UV-irradiated preparations of S100 cytosolic fractions incubated with α_{1b}AR, β-globin, and β_2AR labeled transcripts. In contrast to these common species is a prominent 35,000 M_r RNA-binding protein(s) that selectively bound β_2AR and β_1AR mRNA, but neither α_{1b}AR nor globin messages.

The binding of the β_2AR transcript to the 35,000 M_r protein is sensitive to homopolymers of U, but not poly-A, -C, or -G. The specificity of this 35,000 M_r RNA-binding protein recognition of the β_2AR mRNA (operationally termed *βARB protein*), but not of those failing to display agonist-induced destabilization generated considerable excitement. This was the first identification of an RNA-binding protein implicated in signaling via GPCRs and a member of a rather small group of newly discovered RNA-binding proteins involved in reg-ulatable mRNA biology.

The relative levels of the 35,000 M_r βARB protein have been examined in de-tail for DDT_1MF-2 smooth muscle cells in culture challenged with β-agonist (10 μM isoproterenol). Within 24 hours of challenge, the level of βARB protein in-creased two- to threefold, peaking at 48 hours at four- to fivefold over basal and then declining to 72 hours. The $β_2AR$ mRNA, in contrast, declines by 50%–60% within 24 hours and by >60% at 48 hours postchallenge in these cells. Gluco-corticoids, on the other hand, increase $β_2AR$ mRNA levels and stimulate a frank decline in βARB protein. βARB protein levels declined by 50%–60% at 24 hours in response to dexamethasone (500 nM). Thus, the apparent abundance of this 35,000 M_r RNA- binding protein selective for the $β_2AR$ mRNA varies in-versely with the level of receptor mRNA, being increased by agonists that down-regulate receptor mRNA and receptor expression (Port et al., 1992).

Delineating the molecular basis for agonist-induced destablization of mRNA of GPCR is fundamental to understanding agonist-induced down-regu-lation of receptor. Investigation of the molecular determinants of the $β_2AR$ mRNA contributing to the recognition by βARB has been a high priority (Huang et al., 1993). One approach is to assess the ability of known mRNA motifs to compete with the labeled $β_2AR$ for binding to βARB using the UV cross-linking strategy. The focus of an initial analysis should be investigations using the well-characterized AU-rich domains of 3′ UTR of c-fos, c-myc, and human granulocyte-macrophage colony-stimulating factor (GM-CSF) as well as adenovirus IVa2 RNA. An AUUUA-motif, found in many regulatable mRNA species, and an AUUUUA hexamer, have been identified as molecular determinants for βARB protein using this strategy. A 10-fold molar excess of the 3′ UTR of GM-CSF effectively abolished binding of $β_2AR$ mRNA to βARB protein (Fig. 13.1), whereas the ΔGM-CSF RNA deletion mutant devoid of AUUUA pentamer competed only weakly (Huang et al., 1993; Tholanikun-nel et al., 1995). To more fully assess the interaction between the $β_2AR$ mRNA and competing RNA for βARB-protein one can prepare uniformly labeled probes of the 3′ UTRs of c-myc, c-fos, GM-CSF, and other regulatable mRNAs and then evaluate the ability of the labeled probes to bind RNA-binding pro-teins as well as the ability of the unlabeled $β_2AR$ mRNA to compete for bind-ing, using the UV-catalyzed cross-linking approach. The 3′ UTRs of c-myc, c-fos, and GM-CSF can be covalently cross-linked to the 35,000 M_r species βARB-protein as well as to several other prominent RNA-binding proteins. This observation contrasts with similar experiments performed with labeled $β_2AR$ mRNA in which βARB protein is the prominent labeled species. In ad-dition, the 35,000 M_r βARB displays marked sensitivity to competition by un-labeled $β_2AR$ mRNA, whereas the other RNA-binding proteins are found to be relatively insensitive to competition by the $β_2AR$ message. The roles of AU-rich flanking sequences and AUUUA pentamer in recognition by ARB-protein agree well with an analysis of the 3′ UTR of the $β_2AR$ mRNA that was found to possess both an AUUUA pentamer and flanking U-rich regions.

RNA-binding proteins (or activities) have been described over a wide range of molecular size (15,000–77,000 M_r). Some of these proteins in the 32,000–44,000 M_r range (Bonnieu et al., 1990) recognize AU-rich sequences.

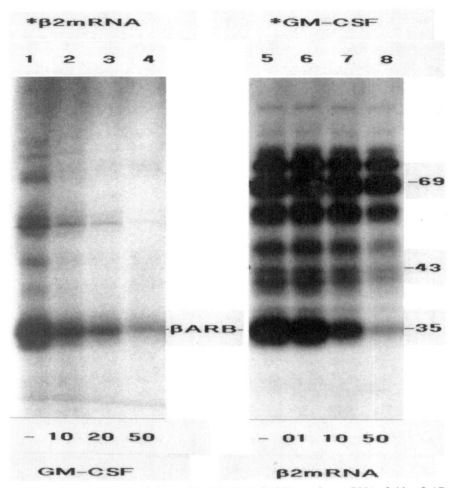

Figure 13.1. UV cross-linking of the 35,000 M_r βARB protein to mRNA of either β_2AR or GM-CSF: competition studies with unlabeled, excess RNA. Autoradiogram of UV cross-linking between S100 cytosolic fractions from DDT_1-MF2 cells and full-length, capped, uniformly labeled *in vitro* transcribed mRNAs corresponding to hamster β_2AR (lanes 1–4) and to GM-CSF (lanes 5–8). Equal amounts of S100 cytosolic protein and equimolar concentrations for each radiolabeled mRNA were added into the appropriate lane. The radiolabeled β_2AR mRNA (*β_2mRNA) was incubated with S100 in the absence (lane 1) or presence of increasing molar excess of unlabeled, full-length, capped mRNA of GM-CSF as a competitor (lanes 2–4). The same approach was used with full-length, capped, uniformly labeled *in vitro* transcribed mRNAs to GM-CSF (*GM-CSF, lanes 5–8) and unlableled mRNA from β_2mRNA at 1–50 molar excess. Note that the cross-linking of *β_2mRNA identifies the predominant species as the βARB protein. Cross-linking studies show that *GM-CSF, in contrast, labels a variety of RNA-binding proteins, including βARB protein, which is uniquely sensitive to competition with unlabeled receptor mRNA. (Reproduced with permission Huang et al., 1993.)

βARB protein differs from these known proteins (activities) by the following criteria: βARB protein levels are induced by agonist or cAMP, and the others are not; βARB protein binding requires both an AUUUA pentamer/AUUUUA hexamer *and* poly-U flanking sequences (e.g., the β_3AR mRNA has AU-rich domains, but no pentamer and does not bind to βARB protein); βARB protein binding of mRNA does not increase based on increased numbers of pentamers (Huang et al., 1993; Tholanikunnel et al., 1995), as observed with other RNA-binding proteins (Bonnieu et al., 1990).

To further probe the AU-rich domains implicated in post-transcriptional regulation (destabilization) of RNA, one can study the ability of a series of point mutations in adenovirus (Ad) IVa2 mRNA to compete for recognition by RNA-binding proteins of interest, such as βARB protein (Huang et al., 1993). U to G substitutions of AUUUA abolish the ability of the unlabeled AdIVa2 RNA to compete with β_2AR mRNA binding to βARB protein. UV cross-linking experiments performed with labeled wild-type versus the U to G AdIVa2 mutants confirm the loss of βARB protein recognition. U to G substitutions of the U-rich flanking regions of the AUUUA pentamer/AUUUUA hexamer suggest a critical role of these flanking sequences in recognition by βARB. U to G substitutions at 4–5 bases flanking either 3′ or 5′ of the AUUUA pentamer affect binding to βARB. It has been shown that for c- myc mRNA a 3′ UTR AUUUA motif is not sufficient to permit regulatable degradation and that a 39-base U-rich domain adjacent to, but distinct from, the AUUUA is part of the recognition domain. For c-myc degradation, a cytoplasmic, protease-sensitive "activity" in Balb/c3T3 cells is a prominent binding element. This type of study demonstrates a correlation between mRNA half-life and U content in the immediate vicinity of AUUUA pentamers.

The β_3AR displays agonist-induced down-regulation. Labeled β_3AR mRNA has been employed to probe binding to βARB protein. Virtually no cross-linking of the β_3AR message to βARB is observed (Fig. 13.2), and unlabeled β_3AR mRNA does not compete with β_2AR mRNA-binding to βARB protein (Tholanikunnel et al., 1995). Further analysis revealed that the agonist-induced decline in β_3AR mRNA did not invovle destabilization, but rather transcriptional repression. Definition of the genomic sequence of the rat β_3AR revealed no AUUUA pentamer, only AU-rich motifs. Thus the β_3AR, unlike the β_1AR and β_2AR, does not display agonist-induced destabilization of receptor mRNA, nor those molecular determinants identified to date for βARB recognition (Tholanikunnel et al., 1995).

Scanning the GPCR subgroup of GenBank against the destabilization sequence data obtained from the study of β_2AR mRNA revealed candidate GPCRs, including the thrombin receptor, muscarinic M_2 and M_3 receptors, angiotensin II receptors, to name several. One of these, the thrombin receptor, has been evaluated addressing the predictive value of the destabilization motif to define regulatability of the receptor mRNA (Tholanikunnel et al., 1995; Tholanikunnel and Malbon, 1997). The thrombin receptor is an exciting target because its mode of activation (via proteolytic cleavage revealing a tethered ligand) differentiates it from β_2AR and most members of the GPCR known. Analysis of the regulation of the thrombin recently revealed both agonist– and cyclic AMP–inducible destabilization of the thrombin receptor mRNA (Fig. 13.3).

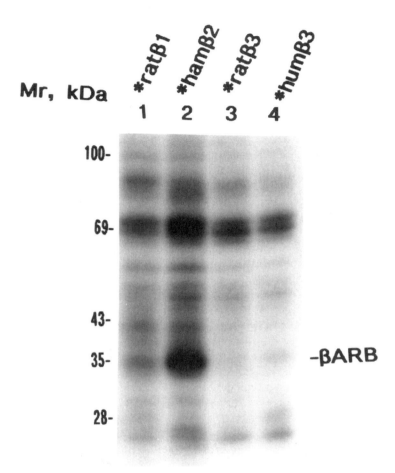

Figure 13.2. UV cross-linking of the 35,000 M_r βARB protein(s) to mRNA of β-adrenergic subtypes. Autoradiogram of UV cross-linking between S100 cytosolic fractions from DDT$_1$-MF2 cells and full-length, capped, uniformly labeled *in vitro*–transcribed mRNAs corresponding to rat β$_1$ (lane 1), hamster β$_2$ (lane 2), rat b$_3$ (lane 3), and human b$_3$ (lane 4). Equal amounts of S100 cytosolic protein and equimolar concentrations for each radiolabeled mRNA were added into the appropriate lane. Note the prominent labeling of βARB protein by labeled mRNA from the β$_1$AR and β$_2$AR. The β$_3$AR regulation, in contrast, is transcriptional, and the mRNA for the β$_3$AR does not bind to βARB protein. (Reproduced with permission Tholanikunnel et al., 1995.)

I.D. Identification of βARB Protein-Binding Domain of β$_2$AR mRNA

Previous studies using β$_2$AR mRNA show the importance of both AUUUA pentamers and poly-U regions in the binding of this protein. Two such regions are present in the 3′ UTR of β$_2$AR mRNA. There is one AUUUA pentamer that is not flanked by a poly-U region and another AUUUUA hexamer that is flanked on either side by a poly-U sequence. These candidate sequences were

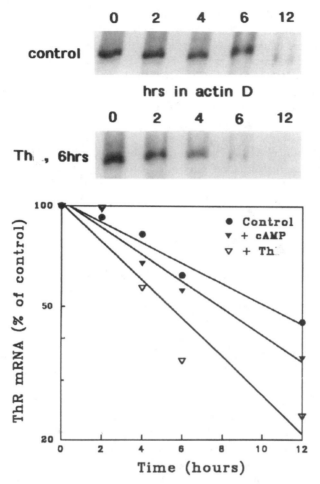

Figure 13.3. Cells expressing thrombin receptor mRNAs display agonist-induced destablization of receptor mRNA. Autoradiograms obtained from RNAse protection analysis of thrombin receptor (ThR) mRNA isolated from HEL cells. Cells were challenged with either no agent (Control), CPT-cAMP (cAMP), or thrombin (Th) for 6 hours and the half-life of ThR mRNA determined as described in this chapter. The RNAse-resistant bands were quantified by phosphorimaging analysis of each band. The autoradiograms are representative of at least three replicate experiments. Quatitative data are the mean values ± SD of three replicate experiments for each. Note that both agonist and cAMP were able to provoke a destabilization and accelerated loss of the thrombin receptor mRNA. These data demonstrate the predictive value of identification of the 20-nucleotide A+U-rich motif in the 3' UTR of mRNAs and the ability of both agonist and cAMP to destablize the mRNA. The thrombin receptor mRNA harbors the cognate binding motif and avidly binds to the βARB protein.

mutated by replacing the middle U by G and tested the ability of *in vitro* transcribed RNA made from these mutants to bind βARB protein. Four different mutations were made, which identified the βARB protein-binding domain as an AUUUUA hexamer that is flanked on either side by a poly-U sequence. Mutation of the AUUUA pentamer that is not flanked by a poly-U region has no effect on the binding of βARB protein. Mutating the two U residues in the hexamer region to G abolishes the binding of βARB protein more than 90%. Replacing two more U residues by G in the poly-U region almost abolished the binding of this protein (Tholanikunnel et al., 1995; Tholannikunnel and Malbon, 1997). Mutation of the hexamer with consequent expression *in vivo* yield a loss of the ability of the receptor mRNA to undergo either agonist-induced or cyclic AMP–induced destabilization, demonstrating the role of a 3' UTR motif as functional in the regulatable nature of a GPCR.

2. ISOLATION OF CELLULAR RNA

Routinely, total cellular RNA is isolated by a single-step guanidine isothiocyanate/phenol-chloroform extraction method (Chomczynski and Sacchi, 1987) from cells in culture using the commercially available RNA STAT-60 (TEL-TEST "B," Inc., Tyler, TX) reagent, as described below. DDT_1-MF2 hamster smooth muscle cells are grown in monolayers, lysed directly in the culture dish (P-100) by adding the RNA STAT-60, and passing the cell lysate several times through a Pasteur pipette. Cells grown in suspension such as HEL cells must first be collected by centrifugation, washed with phosphate-buffered saline (PBS), and lysed in the RNA STAT-60 by repetitive pipetting. These crude homogenates are allowed to stand at room temperature for 5 minutes.

Chloroform is added to the homogenates (0.2 ml of chloroform/1 ml of RNA STAT-60), the samples vortexed for 15 seconds, and the mixtures allowed to stand at room temperature for 2–3 minutes. The samples are centrifuged at 12,000g for 15 minutes at 4°C and the aqueous phase removed into fresh tubes and extracted again with 1 ml of chloroform:isoamyl alcohol (24:1). Samples are subjected to centrifugation as in the previous step, the aqueous phase removed, and the RNA precipitated by addition of isopropanol (0.5 ml of isopropanol/ml of the RNA STAT-60 used for the homogenization). Samples are stored at room temperature for 5–10 minutes and centrifuged at 12,000g for 10 minutes. The well-formed white pellet is washed using 75% ethanol by vortexing the mixture vigorously and by subsequent centrifugation at 7,500g for 5 minutes. The pellet containing the cellular RNA is allowed to air-dry for 20–30 minutes and the pellet then dissolved in RNase-free water. This final preparation of RNA is free of protein and DNA contamination. The ratio of 260/280 should be greater than 1.8. The integrity of RNA isolated can be tested directly on agarose gels by gel electrophoresis.

3. NORTHERN BLOT ANALYSIS OF RECEPTOR RNA

The method employed is essentially as reported by Ausubel et al. (1994) and modified as described previously (Tholanikunnel et al., 1995).

3.A. Reagents

1. 10× MOPS running buffer (to 400 ml of RNAse-free water, add 42 g MOPS and adjust pH to 7.0 with NaOH [0.4 M MOPS])

   Add 16.6 ml of 3 M sodium acetate (0.1 M sodium acetate)

   Add 10 ml of 0.5 M of EDTA, pH 8 (RNAase free)

   Bring the final volume to 500 ml with RNAase-free water

2. 37% formaldehyde

3. Deionized formamide (deionize 100 ml formamide on approximately 10 g Amberlite ion-exchange resin. Filter the eluate and store in aliquots at −20°C)

4. Denaturing solution

   500 μl formamide

   100 μl 10× MOPS

   150 μl formaldehyde

   Bring volume to 1.0 ml

5. Formaldehyde loading buffer

   1 mM EDTA, pH 8.0

   0.25% (w/v) bromphenol blue

   0.25% (w/v) xylene cyanol

   50% (v/v) glycerol

6. Hybridization buffer: 50% formamide in

   100 mM NaPO4, pH 7.0

   0.02% PVP-ficoll

   0.02% bovine serum albumin (BSA)

   0.75 M NaCl in 0.075 M sodium citrate (pH 7.0)

   1% sodium dodecylsulfate (SDS)

   100 μg/ml salmon sperm DNA

3.B. Gel Electrophoresis

1. To a 500-ml flask, add 2.0 g agarose in 144 ml of water. Heat in the microwave until the agarose is in solution, and then cool the vessel and contents to ~60°C. Add 20 ml of 10× MOPS running buffer and 36 ml of 12.3 M formaldehyde. Be sure to use caution and perform these steps in a certified-fume hood.

2. Pour the agarose gel into the proper gel bed holder and allow the gel to set in an RNAase-free gel casting unit with the comb in place. Remove the comb, place the gel in the gel tank, and add sufficient 1× MOPS running buffer to cover to a depth of ~1 mm.

3. Pipette 40–50 μg of total RNA into a sterile Eppendorf tube and dry by use of a Speed-Vac. Resuspend the pellet in 50 μl of denaturing solution, mix by vortexing, spin briefly in a microcentrifuge to collect the liquid, and incubate 15 minutes at 55°C.

4. Add 10 µl of formaldehyde loading buffer, vortex, and spin to collect liquid. Load the samples onto the gel in the preformed combs. A duplicate gel is prepared for ethidium bromide staining to check the integrity of RNA in the final preparation.

5. Run the gel at 5 V/cm until the bromphenol dye has migrated from one-half to two-thirds of the length of the gel (running time is approximately 3 hours).

6. Place the gel in an RNAase-free glass dish and rinse with several changes of deionized water. Allow the water to cover the gel; then rinse and let sit in 20× SSC buffer and soak for 45 minutes.

3.C. Transfer of RNA From Gel to Membrane by Capillary Action

1. *Handle gel and membranes using gloves and blunt-ended forceps.* Transfer of RNA from gel to nitrocellulose is best achieved by capillary transfer. This can accomplished using a glass jar and a solid support with wicks made out of Whatman 3MM paper. Cut and place three pieces of Whatman 3MM paper to the same size as the gel. Place them on the solid support placed in the glass jar and wet them with 20× SSC buffer. Place the gel on the filter and squeeze out air bubbles by rolling a sterile plastic pipette over the surface.

2. Cut and place a piece of nitrocellulose membrane just the size of the gel and wet it with water and replace water by 20× SSC buffer and leave another 10 minutes. Place the wetted membrane on the surface of the gel. By carefully rolling a sterile plastic pipette over the membrane, remove any air bubbles trapped between the gel and the membrane.

3. Add 20× SSC buffer on top of the membrane, and put five sheets of Whatman 3MM paper of the same size as the membrane on top of the membrane. Cut paper towels of the same size as that of the gel (or sightly smaller) and stack on top of the Whatman 3MM paper to a height of 5 to 7 cm. Put a glass plate on top of the structure and add a weight to hold everything in place and leave overnight.

3.D. Preparation of Membrane for Hybridization With Radioprobe

1. Remove the paper towels, membrane, and flattened gel. Mark in pencil the position of the wells on the membrane and the orientations. The transfer efficiency is checked routinely by staining the gel with ethidium bromide.

2. Rinse the membrane in 2× SSC buffer, and then place it on a sheet of Whatman 3MM paper and allow to dry.

3. Bake the membrane in a vacuum oven for 2 hours at 80°C.

3.E. Hybridization

1. The membrane hybridization is carried out in a polyethylene bag using heat-sealing. Add 10 ml hybridization buffer to the bag with the membrane and incubate at 42°C for 1-2 hours

2. Remove the buffer and add hybridization buffer containing the radioactive probe. The blots are hybridized under conditions of high stringency with a random, primer-generated, ^{32}P-labeled probe. Incubate at 42°C for 12–16 hours with shaking immersed in the water bath. For example, the thrombin receptor 700 bp-coding region from the full-length receptor cDNA was obtained by restriction digestion followed by recovery from agarose gel. The DNA is labeled by random oligonucleotide priming to a specific activity of $>10^8$ dpm/µg and the unincorporated nucleotides removed. The double-stranded probe is denatured by heating in a water bath for 10 minutes at 93°–95°C and immediately transferred to an ice slurry.

3. Two washes, each for 5 minutes, are performed in 2× SSC buffer and 0.03 M sodium citrate containing 0.5% SDS for 5 minutes each at room temperature

4. Two more washes using prewarmed (60°C) 0.1× SSC buffer containing 0.1% SDS are performed for 30 minutes each at 60°C. These steps were used for thrombin receptor RNA and may vary depending on the nature of the probe.

5. Remove the membrane, rinse with 2× SSC buffer at room temperature, blot excess liquid, and cover in plastic wrap and set up for autoradiography.

4. QUANTITATION OF RECEPTOR mRNA BY RIBONUCLEASE PROTECTION ASSAY

This basic protocol was adapted from the earlier work by Hod (1992).

4.A. Determination of β$_2$AR mRNA Half-Life

Because the levels of β$_2$AR and thrombin receptor mRNA are low, RNAse protection assay is employed often for quantification of these messages. Chinese hamster ovary cells in culture are pretreated with isoproterenol (10 µM), chlorophenylthio (CPT)–cyclic AMP (10 µM), or vehicle. After 24 hours of treatment with isoproterenol or after 12 hours of treatment with CPT–cyclic AMP, actinomycin D (5 µg/ml) is added to arrest transcription at specific times (Fig. 13.3). Total RNA is extracted from individual cell culture dishes (P-100) at time periods ranging from 30 minutes to 24 hours and the amount of receptor mRNA established by use of an RNAse protection assay. Two different regions from β$_2$AR cDNA (Dixon, et al., 1986) corresponding to 600 (+740–1,338) and 285 (+1,201–1,486) nucleotides from the coding region) were polymerase chain reaction (PCR) amplified and cloned into pSP70 plasmid vector. Radiolabeled antisense riboprobes corresponding to the above regions were made and employed in RNAse protection assay. For studies of the thrombin receptor (Bahou et al., 1993), antisense riboprobes corresponding to 550 (+222–772) nucleotides were amplified by PCR and then cloned into the pSP70 plasmid vector and employed in RNAse protection assay of thrombin receptor mRNA. Analysis of other receptor mRNAs will require the same empirical analysis to determine the proper probe for studies of RNA.

4.B. Ribonuclease Protection Assay

1. Total RNA (20–40 μg) is extracted, re-dissolved in water, and then combined with approximately 2.5×10^5 (1–2 ng) cpm of labeled RNA probe and dried down using a Speed Vac.

2. The dried RNA is suspended in 20 μl of hybridization buffer (80% deionized formamide, 40 mM 1,4-piperazinediethane sulfonic acid [PIPES], pH 6.4, 0.4 M NaCl, 1mM EDTA).

3. The samples are heated at 90°–93°C for 2–3 minutes to denature secondary structures. Hybridization is performed at 48°C for 16 to 20 hours.

4. At the end of the hybridization period, the samples are supplemented with 200 μl of RNAse digestion buffer (Ambion reagent, Cat. No. 8533G) and with RNase A + T1 mix (Ambion reagent, Cat. No. 2286) and mixed by inverting the tubes. The incubation is performed at 30°–34°C for 45 minutes.

5. Add 300 μl of RNAse inactivation precipitation solution (Ambion reagent, Cat. No. 8539G), add 2 μl yeast tRNA (RNAse-free yeast tRNA, 5 μg /ml), mix by inverting the tube a couple of times, and then incubate 10 minutes at –60°C.

6. Precipitate/collect the RNA by centrifugation in a microcentrifuge a maximal speed for 15 minutes.

7. The supernatant must be removed carefully using a very thin pipette tip (such as a gel-loading tip) and centrifuge the tube briefly again to bring all the residual solution to the bottom of the tube. Remove the last traces of the precipitation buffer and air dry the precipitate.

8. The protected hybrid RNA is resuspended in 10 μl of loading buffer (80% formamide, 10 mM EDTA, 1 mg/ml bromophenol blue, 1 mg/ml xylene cyanol) and denatured by heating for 4 minutes at 90°C, and an aliquot is loaded on a 5% polyacrylamide, 8 M urea gel used for manual sequencing.

5. PREPARATION OF PROBES FOR IDENTIFICATION OF RNA-BINDING PROTEINS BY *IN VITRO* TRANSCRIPTION

The wild-type (Zhang et al., 1993) and mutant cDNAs for the hamster β_2AR routinely are inserted in pSP70 plasmid vector, which is linearized. Transcription is performed *in vitro* using SP6 DNA-directed RNA polymerase to produce full-length, 5′-capped, uniformly labeled, poly-A$^+$ mRNAs, based on the technique of Melton et al. (1984). Briefly, mRNAs are transcribed in the presence of RNasin (Promega), radiolabeled α[^{32}P]-UTP (800 Ci/mmol, New England Nuclear), nucleotide, and buffer conditions as detailed by Promega. Co-transcriptional capping was performed by using the cap analogue m^7(5′)Gppp(5′)G (New England Biolabs) at a concentration that was 10-fold in excess to the concentration of GTP.

Prepare the mixture in RNAse-free sterile tubes as follows:

5× Transcription buffer	5 µl
100 mM DTT	2 µl
Acetylated BSA (1mg/ml)	1 µl
rRNasin (ribonuclease inhibitor)	1 µl
Nucleotides (add 1.25 µl each of 10 mM ATP and CTP and 1 mM GTP plus 10 mM m^{7}G[5′]ppp[5′]G)	5 µl
100 uM UTP	2.5 ul
Linearized template (0.5–1.0 mg/ml)	1 µl
α[^{32}P]UTP (800 Ci/mmol, Dupont NEN)	5 ul
SP6 RNA polymerase	1 µl
RNAse-free water	1.5 µl
Total	25.0 µl

All the reagents are to be added in the order specified, and the enzyme is added last. The mRNA is transcribed at temperatures ranging from 30° to 37°C depending on the length of the transcript to be made. mRNAs longer than 2 kb are made by carrying out transcription at 30°C in order to get full-length transcript. After the mRNA was transcribed, RNAse-free DNAse was added to the mixture to remove template DNA. The labeled transcript is extracted with phenol and then with chloroform and precipitated finally with 2.5 volume ice-cold ethanol and 0.1 volume 3 M Na-acetate. The labeled transcripts are then washed with 70% ethanol and reconstituted in RNAse-free water, maintained at −80°C, and used within 24 hours of synthesis. The size and integrity of the transcripts were verified immediately before use by agarose/formaldehyde gel electrophoresis.

6. CROSS-LINKING AND LABEL TRANSFER TO RNA-BINDING PROTEINS USING ULTRAVIOLET IRRADIATION

6.A. Reagents and Solutions

The following additions are made to a 96-well cell culture plate:

Yeast tRNA (10 mg/ml)	5 µl
Heparin (10 mg/ml)	5 µl
DTT (100 mM)	2 µl
RNasin (40 U/µl)	1 µl
Unlabeled RNA (as required)	0–20 µl
2–4 × 10^6 cpm α[^{32}P]labeled mRNA or probe	2–4 µl
RNAse-free water	To make the volume to 50 µl
S100 cytosolic fraction (30–100 µg of protein)	10–15 µl

The above additions are made to a 96-well plate in the order specified (except the addition of the S100 fraction) and mixed, and the S100 cytosolic fraction is then added and mixed. In competition experiments unlabeled (cold) RNA is added in molar excesses and mixed with labeled RNA before addition of the S100 fraction. The mixture is allowed to incubate for 10 minutes at 22°C before UV cross-linking.

The mixture is placed in an ice slurry and then exposed to short-wave (254-nm) UV irradiation at a distance of 7 cm from the source for 30 minutes. The mRNA not cross-linked to protein is digested with RNAse A (0.5 mg/ml) and RNAse T1 (10 U/ml) at 37°C for 30 minutes.

6.B. SDS-PAGE of RNA Protein Adducts

Samples are solubilized in 50 µl (1:1) of Laemmli loading buffer (Chung et al., 1996) for 10 minutes at 68°–70°C. The samples are then loaded onto an SDS-PAGE (10% acrylamide separating gel with 5% acrylamide stack) and subjected to gel electrophoresis for 110 milliamp hours. Resolved proteins were stained with Coomaissie R-Blue and the gels destained, dried, and subjected to autoradiography for 12–24 hours. The relative intensity of radiolabeled species on the gel was quantified by direct analysis of radioactivity using a beta-phosphorimager. (See Figs. 13.1 and 13.2 for examples of autoradiograms of cross-linking of receptor mRNAs to RNA-binding proteins, especially to the βARB protein.)

6.C. Transfection of Chinese Hamster Ovary Cells

To perform analysis of RNA elements that regulate GPCR mRNAs *in vivo*, expression vectors must be utilized. Chinese hamster ovary (CHO) cells offer an excellent target for most studies. CHO wild-type cells were co-transfected using lipofectin (Life Technologies, Inc) and vectors harboring mutant or wild-type receptor cDNAs or empty vector plasmids, each in combination with a plasmid such as pCW1 containing the neomycin resistance gene. Stable transfectant clones are selected for neomycin resistance in Dulbecco's modified Eagle's medium (DMEM) containing 10% fetal bovine serum and the neomycin analogue G418 (400 µg/ml). Expression of GPCR in the stable transfectants is accomplished by radioligand binding. For the βARs, steady-state expression is determined for wild-type receptors and receptors with mutated 3′ UTRs using ^{125}I-iodocyanopindolol (ICYP) binding (Collins et al., 1992).

1. *Cell culture:* DDT$_1$-MF2 vas deferens smooth muscle cells can be best cultured in DMEM supplemented with 5%, heat-inactivated fetal bovine serum (Hyclone), penicillin (60 µg/ml), and streptomycin (100 µg/ml) as described by Scarpace et al. (see Brewer et al., 1993). CHO cells are propagated in DMEM supplemented with 10%, heat-inactivated fetal bovine serum (HyClone). Cells are treated with either drugs prepared in a vehicle or with the vehicle alone, as described in each protocol.

2. *Preparation of cytosolic (S100) extracts:* Following drug treatment, cells are washed twice with PBS and removed from the plate with 1.0 mM EDTA in PBS. Approximately 5×10^7 cells are collected gently by low-speed (1,000g)

centrifugation, resuspended in PBS, transferred to a sterile polypropylene ul-tracentrifuge tube, and collected again gently by centrifugation. The PBS is as-pirated from the cell pellet and 5 μl aliquots of each of the protease inhibitors (10 mg/ml) aprotinin and leupeptin are added to the cell pellet. The cells are then subjected to ultracentrifugation (100,000g) for 90 minutes at 4°C. The re-sulting supernatant fraction was transferred to Eppendorf tubes and maintained in an ice-bath for immediate use. This cytosolic fraction is referred to as the *S100 fraction*. Protein concentration of the fractions is conveniently determined by method of Lowry et al. (see Bohjanen et al., 1994).

3. *Mutagenesis and plasmid construction:* Mutagenesis of GPCRs, such as the β_2AR cDNA, is performed in pSP70 by overlap extension PCR. Briefly, mutagenic primers are constructed containing complementary sequences to β_2AR cDNA immediately 5′ or 3′ to the flanking 3′ UTR AUUUA pentamer (nucleotides 1520–1524), designated mutant 1, or to the AUUUUA hexamer nucleotide (1,598–1,603), designated mutant 2. PCR is performed on a β_2AR cDNA template using an SP6 or T7 promoter primer and one of the mutagenic primers. Products can be separated by agarose gel electrophoresis, made visi-ble by staining with ethidium bromide and UV irradiation, excised, and puri-fied with a GeneClean-11 kit, as described by the manufacturer (Bio 101 Inc., La Jolla, CA). Amplified fragments (5–10 ng) from the forward and reversed PCR are mixed, subjected to a second round of amplification by PCR, and the products separated and identified as above. The fragment corresponding to full-length, mutant β_2AR cDNA is excised and gel purified. The fragment is sub-jected to digestion with *Eco*RI and the *Eco*RI-digested fragment and then cloned into *Eco*RI sites of both pSP70 and pCMV5. After identification of the appropriate recombinants, orientation is determined by restriction digestion mapping. The mutated cDNAs are sequenced by a dideoxy method to verify the sequence for the appropriate base substitution. Plasmid vector pSP70, into which wild-type and various mutants of β_2AR cDNA are inserted, is used for *in vitro* transcription after linearization with a restriction enzyme that cleaves the plasmid immediately 3′ to the receptor cDNA insert.

Expression vector pCMV5 was used for stable transfection in CHO cells. Mutant 2 was used as the template for engineering both mutants 3 and 4 (Tholanikunnel et al., 1995; Tholanikunnel and Malbon, 1997). Plasmids con-taining 20-nucleotide AUUUUA hexamers flanked by poly-U regions as well as those containing a 20-nucleotide pentamer flanked by poly-U regions can be constructed readily by use of complementary synthetic oligonucleotides flanked by restriction sequences for *Hin*dIII at the 5′ end and Cla1 at the 3′ end. Complementary oligonucleotides are annealed and cloned into pSP70. The re-sultant plasmids are linearized immediately 3′ to the AU-rich region and em-ployed as templates for *in vitro* transcription.

7. CLOSING REMARKS

The post-transcriptional regulation of GPCRs offers new insights into the mech-anisms of receptor expression. 3′ UTRs of mRNA harbor motifs that target these mRNAs to RNA-binding proteins that regulate the stability of the messenger RNA. Analysis of steady-state expression of rare mRNAs is a formidable task, as is unlocking the cognate sequences that constitute the basis for establishing the half-life of a receptor messenger RNA in both the basal, unstimulated and the agonist-stimulated states. Agonist-promoted down-regulation of the βAR, thrombin receptor, and many other members of the superfamily of GPCRs reflects post-transcriptional mechanisms that alter the stability of pre-existing mRNAs.

Ultimately, the ability to mutate the sequence of the 3′ UTRs of the messenger RNAs that confer regulatability to the mRNAs and to study mRNA stability *in vivo* will allow a more detailed understanding of receptor regulation for the entire superfamily of GPCRs. Coupling the *in vivo* studies to *in vitro* analyses of the RNA-binding proteins, such as the βARB protein, will enable structure/function studies at this exciting level of regulation, mRNA stability.

REFERENCES

Bahou WF, Coller BS, Potter CL, Norton KJ, Kutok JL, Goligorsky MS (1993): The thrombin receptor extracellular domain contains sites crucial for peptide ligand–ligand activation. J Clin Invest 91:1405–1413.

Bahouth SW, Malbon CC (1994): Genetic (transcriptional and post-transcriptional) regulation of G-protein-linked receptor expression. In Sibley DR, Houslay MD (eds): Regulation of Cellular Signal Transduction Pathways by Desensitization and Amplification. West Sussex, England: John Wiley & Sons, Ltd., pp 99–112.

Bahouth SW, Parks EA, Beauchamp M, Cui X, Malbon CC (1994): Identification of a glucocorticoid repressor domain in the β_1-adrenergic receptor gene. Receptors Signal Transduction 6:141–149.

Bernstein PL, Herrick DJ, Prokipcak RD, Ross J (1992): Control of c-myc mRNA $t_{1/2}$ *in vitro* by a protein capable of binding to a coding region stability determinant. Genes Dev 6:642–654.

Birnbaumer L, Abramowitz J, Brown AM (1990): Receptor-effector coupling by G-proteins. Biochim Biophys Acta 1031:163–224.

Bohjanen PR, Petryniak B, June CH, Thompson CB, Lindsten T (1994): An inducible cytoplasmic factor (AU-B) binds selectively to AUUUA multimers in the 3′UTR of mRNA. Mol Cell Biol 11:3288–3295.

Bonnieu A, Roux P, Marty L, Jeanteur Ph, Piechaczyk M (1990): AUUUA motifs are indespensable for rapid degradation of the mouse c-myc RNA. Oncogene 5:1585–1588.

Bouvier M, Caron MG, Lefkowitz RJ (1989): Two distinct pathways for cAMP-mediated down-regulation of the β_2-adrenergic receptor: Phosphorylation and regulation of mRNA level. J Biol Chem 264:16786–16792.

Brewer G (1991): An A+U-rich element RNA-binding factor regulates c-myc m RNA stability *in vitro.* Mol Cell Biol 11:2460–2466.

Chomczynski P, Sacchi N (1987): Single-step method of RNA isolation by acid guanidinium thiocyanate-phenol-chloroform extraction. Anal Biochem 162:156–159.

Chung S, Jiang L, Chang S, Furneaux H (1996): Purification and properties of HuD, a neuronal RNA-binding protein. J Biol Chem 271:11518–11524.

Collins S, Bouvier M, Bolanowski MA, Caron MG, Lefkowitz RJ (1989): cAMP stimulates transcription of the β-adrenergic receptor gene in response to short-term agonist exposure. Proc Natl Acad Sci USA 86:4853–4857.

Collins S, Caron MG, Lefkowitz RJ (1988): Glucocorticoids regulation the transcription of the beta-2 adrenergic receptor gene. J Biol Chem 263:9067–9070.

Collins S, Caron MG, Lefkowitz RJ (1992): From ligand binding to gene expression: New insights into the regulation of G-protein–coupled receptors. TIBS 17:37–39.

Collins S, Lohse MJ, Caron MG, Lefkowitz RJ (1991): Structure and regulation of G-Protein–coupled receptors: The beta 2-adrenergic receptor as a model. Vitam Horm 46:1–39.

Dixon RA, Kobilka BK, Strader DJ, Benovic JL, Dohlman HG, Frielle T, Bolanowski MA, Bennett CD, Rands E, Diehl RE (1986): Cloning of the gene and cDNA for mammalian beta-adrenergic receptor and homology with rhodopsin. Nature 321:75–79.

Dohlman HG, Thorner J, Caron MG, Lefkowitz RJ (1991): Model systems for the study of seven-transmembrane segment receptors. Annu Rev Biochem 60:653–688.

Guest SJ, Hadcock JR, Watkins DC, Malbon CC (1990): β1- and β2-adrenergic receptor expression in differentiating 3T3-L1 cells: Independent regulation at the level of mRNA. J Biol Chem 265:5370–5375.

Hadcock JR, Malbon CC (1988a): Regulation of β-adrenergic receptors By "permissive" hormones: Glucocorticoids increase steady-state levels of receptor mRNA. Proc Natl Acad Sci USA 85:8415–8419.

Hadcock JR, Malbon CC (1988b): Down-regulation of β-adrenergic receptors: Agonist-induced reduction in receptor mRNA levels. Proc Natl Acad Sci USA 85:5021–5025.

Hadcock JR, Malbon CC (1991): Regulation of receptor expression by agonist: Transcriptional and post-transcriptional controls. Trends Neurosci 14:242–247.

Hadcock JR, Malbon CC (1993): Agonist regulation of gene expression of adrenergic receptors and G-proteins. J Neurochem 60:1–9.

Hadcock JR, Ros M, Malbon CC (1989a): Agonist regulation of β-adrenergic receptor mRNA: Analysis in S49 mouse lymphoma mutants. J Biol Chem 264:13956–13961.

Hadcock JR, Wang H-Y, Malbon CC (1989b): Agonist-induced destabilization of β-adrenergic receptor mRNA. Attenuation of glucocorticoid-induced up-regulation of β-adrenergic receptors. J Biol Chem 264:19928–19934.

Hadcock JR, Williams DL, Malbon CC (1989c): Physiological regulation at the level of mRNA: Analysis of the steady-state levels of specific mRNAs by DNA-excess solution hybridization. Amer J Physiol (Cell Physiol) 256:457–465.

Hod Y (1992): A simplified ribonuclease protection assay. Biotechniques 13:462–464.

Hough C, Chuang DM (1990): Agonist down-regulates β1- and β2-adrenergic receptor mRNA in C6 glioma cells. Biochem Biophys Res Commun 170:46–52.

Huang L-Y, Tholanikunnel BG, Vakalapoulou E, Malbon CC (1993): The 35,000-M_r β-adrenergic receptor mRNA-binding (ARB) protein induced by agonists requires both an AUUUA pentamer and U-rich domains for RNA recognition. J Biol Chem 268:25769–25775.

Kiely J, Hadcock JR, Bahouth SW, Malbon CC (1994): Glucocorticoids down-regulate β1-adrenergic receptor expression by suppressing transcription of the receptor gene. Biochem J 302:397–403.

Malbon CC (1980): Liver cell adenylate cyclase and β-adrenergic receptors: Increased β-adrenergic receptor number and responsiveness in the hypothyroid rat. J Biol Chem 255:8692–8699.

Malbon CC, Hadcock JR (1988): Evidence that glucocorticoid response elements in the 5′-noncoding region of the hamster β2-adrenergic receptor gene are obligate for glucocorticoid regulation of receptor in RNA levels. Biochem Biophys Res Commun 154:676–681.

Malbon CC, Hadcock JR (1993): Regulation of G-protein–linked receptors: Analysis from the genome to the cell membrane. In Brody JS, Center DM, Tkashuck VA (eds): Signal Transduction in Cells. New York: Marcel Dekker, pp 23–47.

Melton DA, Krieg PA, Rebagliati MR, Maniatis T, Zinn K, Green MR (1984): Efficient *in vitro* synthesis of biologically active RNA and RNA hybridization probes from plasmids containing a bacteriophage SP6 promoter. Nucl Acids Res 12:7035–7056.

Port JD, Huang L-Y, Malbon CC (1992): β-Adrenergic agonists that down-regulate receptor mRNA up-regulate a 35,000-M_r protein(s) that selectively binds to β-adrenergic receptor mRNAs. J Biol Chem 267:24103–24108.

Sakaue M, Hoffman BB (1991): Glucocorticoids induce transcription and expression of the alpha 1b-adrenergic receptor gene in DDT1 MF-2 smooth muscle cells. J Clin Invest 88:385–389.

Scarpace PJ, Baresi LA, Sanford DA, Abrass IB (1985): Desensitization and resensitization of beta-adrenergic receptors in a smooth muscle cell line. Mol Pharmacol 28:495–501.

Shilo L, Sakaue M, Thomas JM, Philip M, Hoffman BB (1994): Enhanced transcription of the human alpha 2A-adrenergic receptor gene by cAMP. Cell Signal 6:73–82.

Tholanikunnel BG, Malbon CC (1997): A 20-nucleotide rich (A+U)-rich element of beta-adrenergic receptor mRNA medicates binding to bARB protein and is obligate for agonist-induced destablization of receptor mRNA. J Biol Chem 272:11471–11478.

Tholanikunnel BG, Granneman J, Malbon CC (1995): The 35,000-M_r β-adrenergic receptor mRNA-binding (βARB) protein differentiates among subtype mRNAs. J Biol Chem 270:12787–12794.

Zhang W, Wagner BJ, Ehremen K, Schefer AW, DeMaria C, Crater D, DeHaven K, Long L, Brewer G (1993): Purification, characterization, and cDNA cloning of an AU-rich element RNA-binding protein, AUF1. Mol Cell Biol 13:7652–7665.

Zhong H, Minneman KP (1993): Close reciprocal regulation of β_1- and β_2-adrenergic receptors by dexamethasone in C6 glioma cells: Effects on catecholamine responsiveness. Mol Pharmacol 44:1085–1093.

INDEX